THE ALKALOIDS

Chemistry and Pharmacology

VOLUME 45

THE ALKALOIDS

Chemistry and Pharmacology

Edited by

Geoffrey A. Cordell

College of Pharmacy
University of Illinois at Chicago
Chicago, Illinois

Arnold Brossi

Department of Chemistry
Georgetown University
Washington, D.C.

VOLUME 45

ACADEMIC PRESS

A Division of Harcourt Brace & Company
San Diego   New York   Boston
London   Sydney   Tokyo   Toronto

This book is printed on acid-free paper. ∞

Academic Press, Inc.
525 B Street, Suite 1900, San Diego, California 92101-4495

United Kingdom Edition published by
ACADEMIC PRESS LIMITED
24-28 Oval Road, London NW1 7DX

International Standard Serial Number: 0099-9598

International Standard Book Number: 0-12-469545-0

PRINTED IN THE UNITED STATES OF AMERICA
94 95 96 97 98   QW   9 8 7 6 5 4 3 2 1

CONTENTS

CONTRIBUTORS

Numbers in parentheses indicate the pages on which the authors' contributions begin.

WILLIAM A. AYER (233), Department of Chemistry, University of Alberta, Edmonton, Alberta, Canada T6G 2G2

HERBERT BENZ (1), Organisch-chemisches Institut der Universität Zürich, 8057 Zürich, Switzerland

STEFAN BIENZ (1), Organisch-chemisches Institut der Universität Zürich, 8057 Zürich, Switzerland

GÁBOR BLASKÓ (127), EGIS Pharmaceutical Ltd., Budapest, Hungary

GÁBOR DÖRNYEI (127), Central Research Institute for Chemistry, Hungarian Academy of Sciences, H-1525 Budapest, Hungary

WOLFGANG FIEDLER (1), Organisch-chemisches Institut der Universität Zürich, 8057 Zürich, Switzerland

ARMIN GUGGISBERG (1), Organisch-chemisches Institut der Universität Zürich, 8057 Zürich, Switzerland

MANFRED HESSE (1), Organisch-chemisches Institut der Universität Zürich, 8057 Zürich, Switzerland

ANDREA SCHÄFER (1), Organisch-chemisches Institut der Universität Zürich, 8057 Zürich, Switzerland

CSABA SZÁNTAY (127), Central Research Institute for Chemistry, Hungarian Academy of Sciences, H-1525 Budapest, Hungary

LATCHEZAR S. TRIFONOV (233), Department of Chemistry, University of Alberta, Edmonton, Alberta, Canada T6G 2G2

PREFACE

This volume of *The Alkaloids: Chemistry and Pharmacology* brings together three quite dissimilar groups of alkaloids that in each case are reviewed by outstanding researchers in their respective fields. While on the surface these alkaloid groups, from spiders, higher plants, and club mosses, may appear disparate, they are bound by the common thread of medical significance.

In recent years, several groups around the world, both academic and industrial, have recognized the potential of sources other than plants for the discovery of biologically active alkaloids. Previously it had been thought that the biologically active constituents of the venoms of spiders and wasps were high-molecular-weight polypeptides. While in some cases this is certainly true, there are now numerous instances where the active principles have been identified as series of polyamine derivatives. Hesse and co-workers describe the isolation and synthesis of these derivatives and provide evidence of the significant medical potential for this group of compounds. Details of the biological aspects of these compounds will be described in a forthcoming volume.

Few groups of alkaloids have had the commercial success of the morphine alkaloids. In a detailed presentation, Szántay, Dörnyei, and Blaskó review the recent developments in the isolation, synthesis, and chemistry of the morphine alkaloids and describe the extensive activity still ongoing in this vibrant area of alkaloid chemistry.

The *Lycopodium* alkaloids have not been reviewed in this series since 1985. Given the recent substantial synthetic and biological interest in one member of this series, huperzine A, particularly for the treatment of Alzheimer's disease, the review by Ayer and Trifonov is both timely and appropriate.

Geoffrey A. Cordell
University of Illinois at Chicago

Arnold Brossi
Georgetown University

CHAPTER 1

POLYAMINE TOXINS FROM SPIDERS AND WASPS

ANDREA SCHÄFER, HERBERT BENZ, WOLFGANG FIEDLER,
ARMIN GUGGISBERG, STEFAN BIENZ, AND MANFRED HESSE

Organisch-chemisches Institut der Universität Zürich
8057 Zürich, Switzerland

I. Introduction

The venoms of spiders and wasps are complex composites of free amino acids, large proteinaceous toxins (>3000 Da), and relatively small polyamine toxins (<1000 Da). For a long time, the interest in such venoms primarily focused on the high molecular weight components, particularly those extracted from tropical spiders. Only relatively recently, with the development of improved analytical methodologies and because of the interesting neurotoxic activities of these compounds, have the polyamine venom components of common "garden variety" spiders, as well as wasps, attracted more attention. Since the mid-1980s, several laboratories have concentrated their research on the isolation, structure elucidation, synthesis, and pharmacology of such natural products, as well as the study of synthetic analogs.

In this chapter, we summarize all of the naturally occurring polyamine toxins from spider and wasp sources published up to the end of October 1992, and we discuss the basic problems of their isolation, structure elucidation, and synthesis. In addition, polyamine toxin analogs are described. Plausible structural variations of the original natural products are outlined and synthetic approaches to the respective artificial target molecules briefly shown. We do not focus in this review on the pharmacology of the toxins and their analogs. This will be the topic of a forthcoming chapter in this serial.

II. Structure and Nomenclature

A. GENERAL STRUCTURE AND NOMENCLATURE

Polyamine toxins of spiders and wasps·share two structural peculiarities: all possess a linear α,ω-diamino polyazaalkane backbone (part C of the schematic structure in Fig. 1) modified at one end with a more or less lipophilic unit, in most cases an aromatic acyl group (Fig. 1, part A). For some classes of toxins, the head portion A is separated from the polyamine

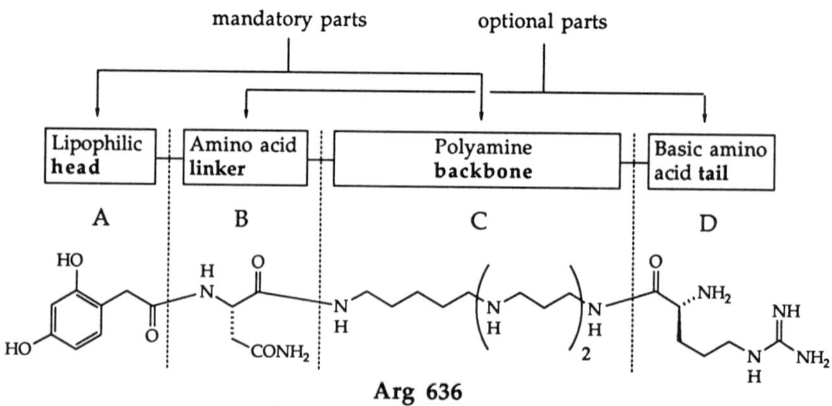

FIG. 1. Structures of polyamine spider and wasp toxins exemplified by Arg 636.

backbone by one or more α-amino acid moieties (Fig. 1, part B), and the most complex toxins are further modified at the tail of the polyamine backbone with an additional basic amino acid fragment (Fig. 1, part D). The structure of Arg 636, the most profoundly investigated toxin isolated from several *Argiope* species, consisting of all four parts A–D, exemplifies the general construction of this class of compounds.

It is reasonable to classify the spider and wasp toxins according to their source and the animal family, because the toxins found in different species of the same family are structurally closely related. Two families of spiders, Agelenidae and Araneidae, have been extensively investigated for their polyamine toxins. Several species from the two families have yielded 17 and 43 toxins, respectively, of the overall 68 toxins reviewed in this chapter. The structures or partial structures of these compounds, as far as proposed by the original authors, are summarized in order of increasing molecular weight in the Tables I and II, respectively. The few remaining polyamine toxins from several spiders and the wasp *Philanthus triangulum* are outlined in Table III. Table IV contains the names of all the natural toxins in alphabetical order with the corresponding reference to Tables I–III. To visualize the relations of the respective arthropods of interest, a portion of the genealogical tree of the animal kingdom is delineated in Fig. 2.

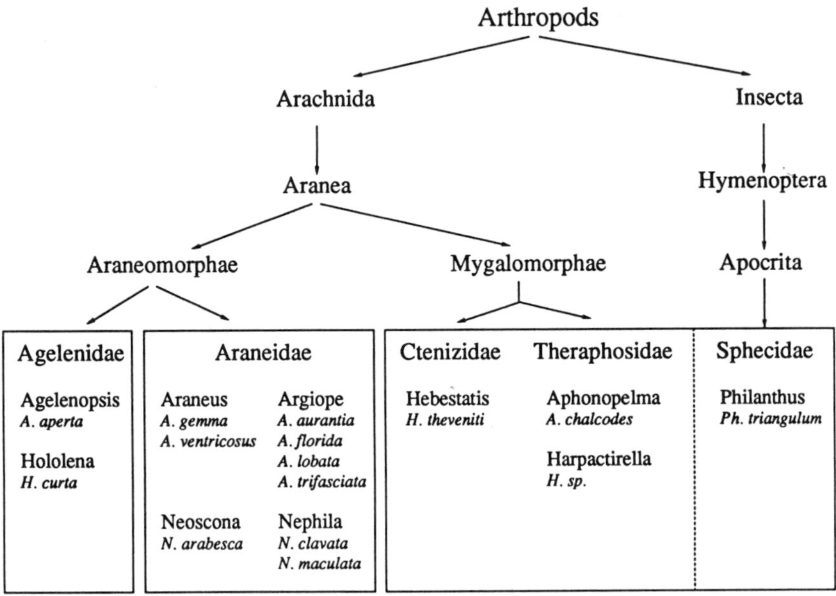

FIG. 2. Portion of simplified genealogical tree of the animal kingdom.

TABLE I
AGELENIDAE POLYAMINE TOXINS

Molecular weight	Structure	Trivial name and synonym(s)	Species isolation and structure	Synthesis (references)
359	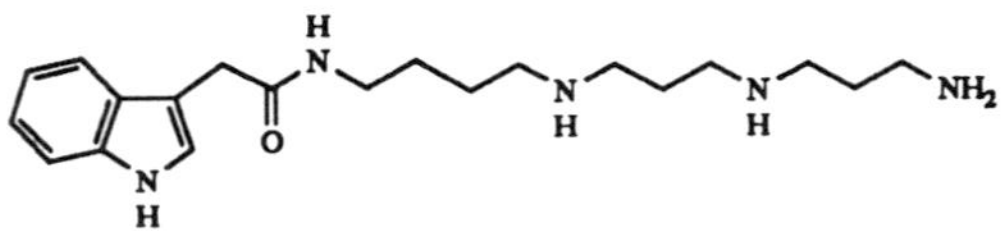	HO 359	*Hololena curta* (*113*)	
395	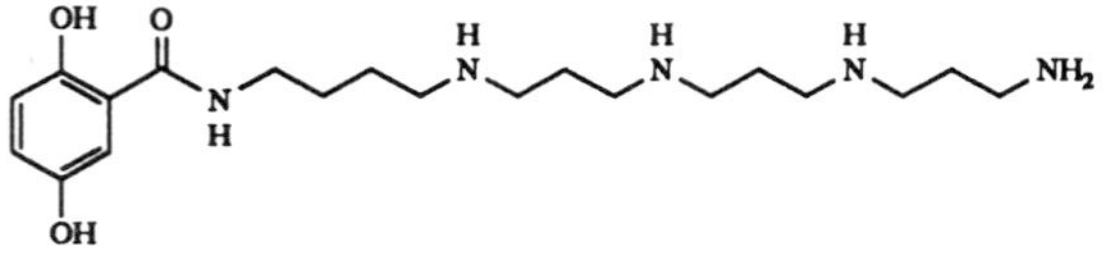	HO 395	*Hololena curta* (*113*)	
416	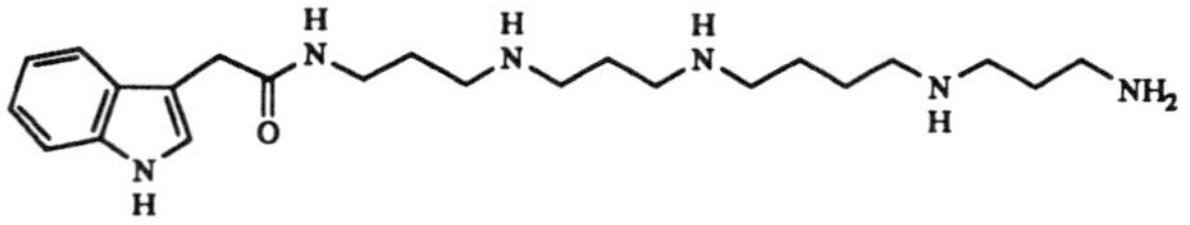	Agel 416 (AG 416, HO 416a)	*Agelenopsis aperta* (*3*) *Hololena curta* (*113*)	(*3*)
416	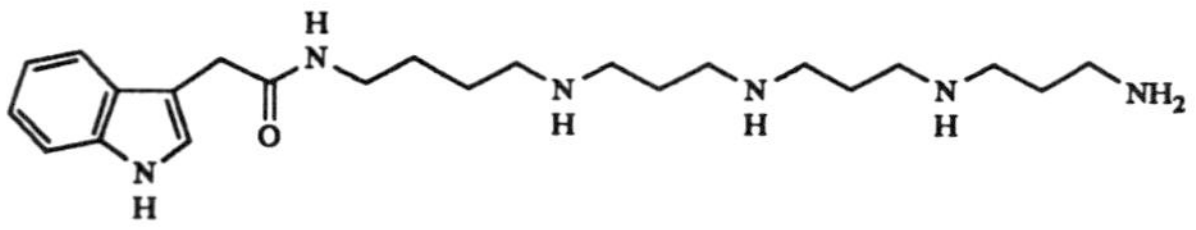	HO 416b	*Hololena curta* (*113*)	

MW	Structure	Name	Source	Ref.
448	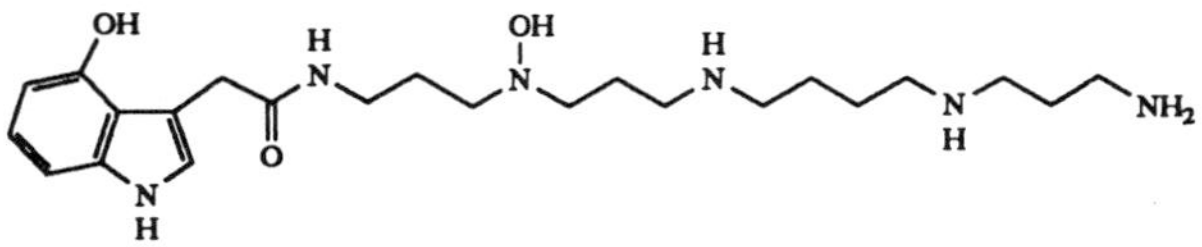	Agel 448 (AG 448)	*Agelenopsis aperta* (1,2)	(1,2)
448	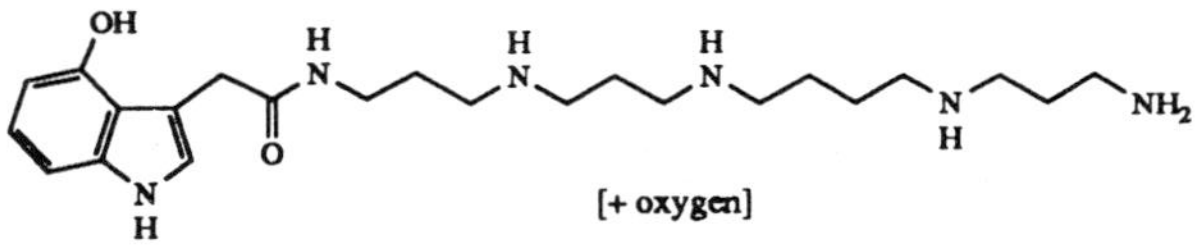 [+ oxygen]	HO 448	*Hololena curta* (113)	
452	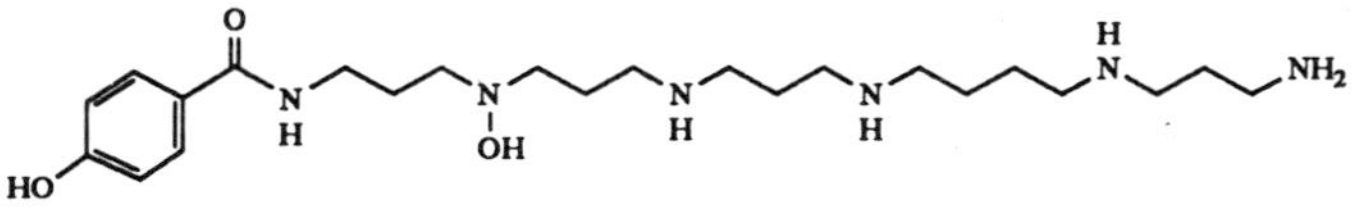	Agel 452 (AG 452)	*Agelenopsis aperta* (1,2,31,97)	(1,2)
452	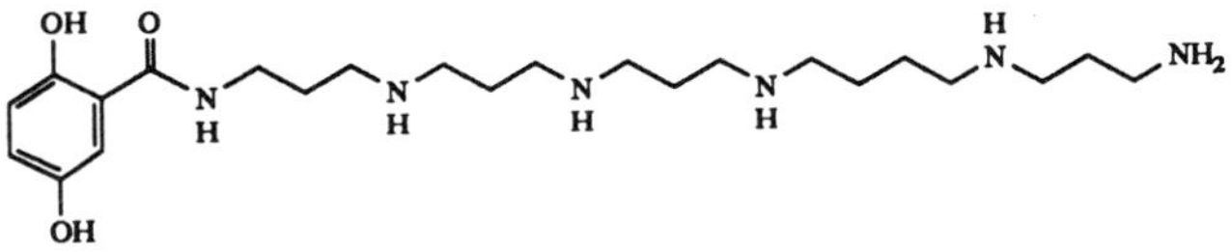	HO 452	*Hololena curta* (113)	
464	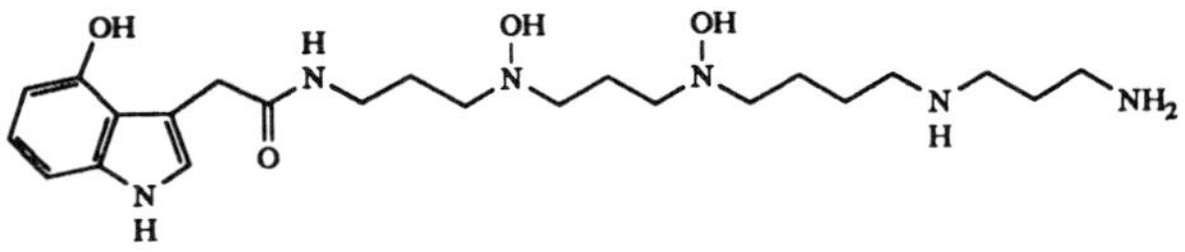	Agel 464	*Agelenopsis aperta* (3)	

(continues)

TABLE I *(Continued)*

Molecular weight	Structure	Trivial name and synonym(s)	Species isolation and structure	Synthesis (references)
468	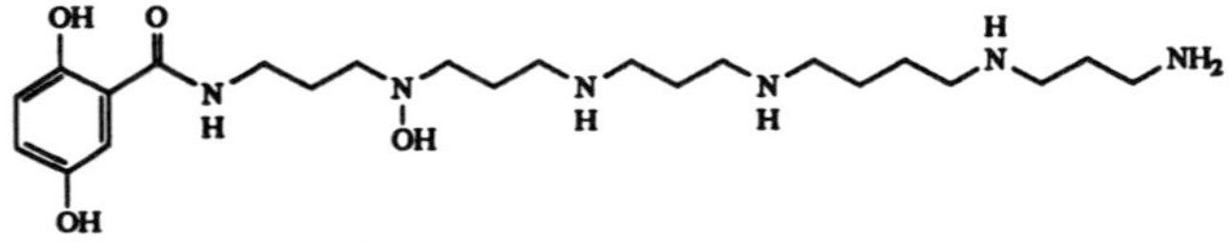	Agel 468 (AG 468, HO 468)	*Agelenopsis aperta* *(1,2)* *Hololena curta* *(113)*	*(1,2)*
473	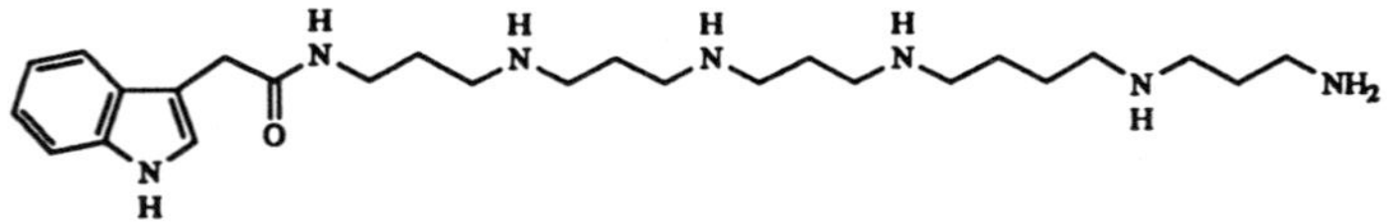	HO 473	*Hololena curta* *(113)*	
489	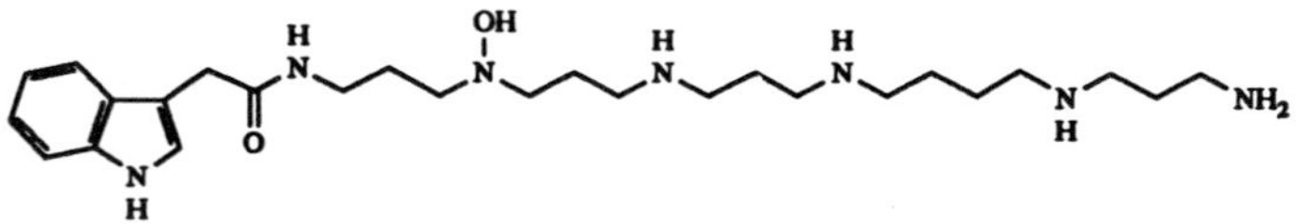	Agel 489 (AG 489, HO 489)	*Agelenopsis aperta* *(1,2,31,96,97)* *Hololena curta* *(113)*	*(1,2)*
489	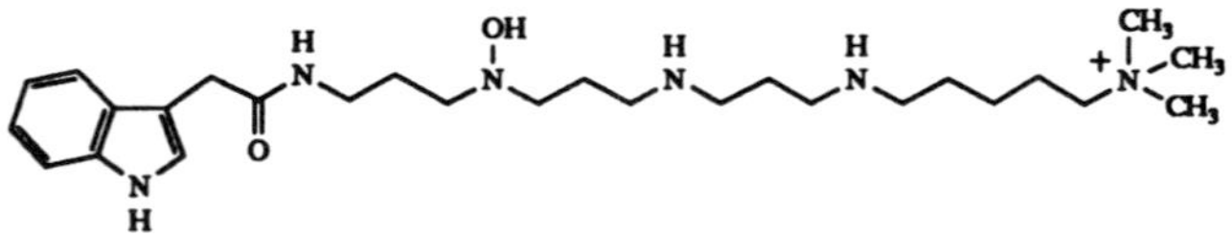	Agel 489a (AG 488)	*Agelenopsis aperta* *(22)* Wrong structure *(31,96,97)*	*(22)*

489	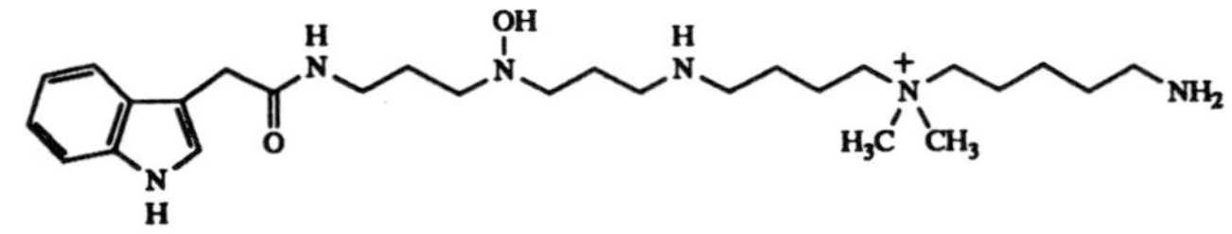 Agel 489(A)	*Agelenopsis aperta* *(3)*	
505	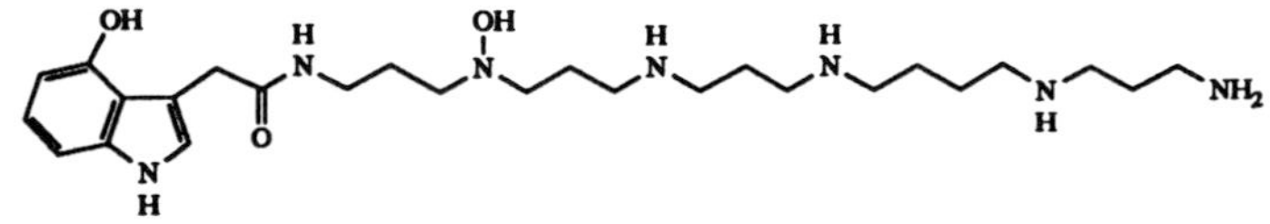 Agel 505 (AG 505, HO 505)	*Agelenopsis aperta* *(1,2,31,96,97)* *Hololena curta* *(113)*	*(1,2)*
505	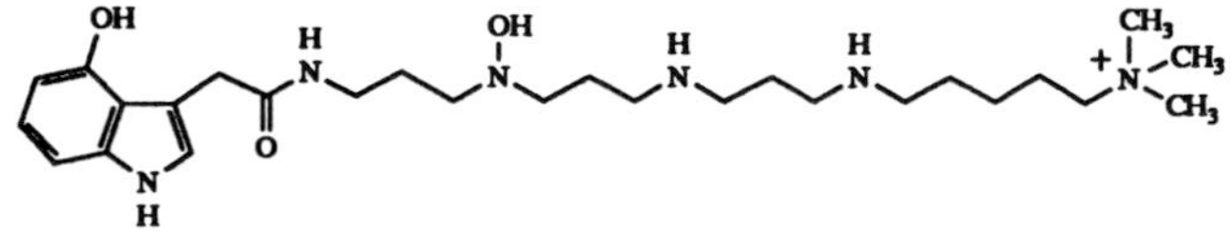 Agel 505a (AG 504, Agel 504)	*Agelenopsis aperta* *(2,22)* Wrong structure *(31,96,97)*	
521	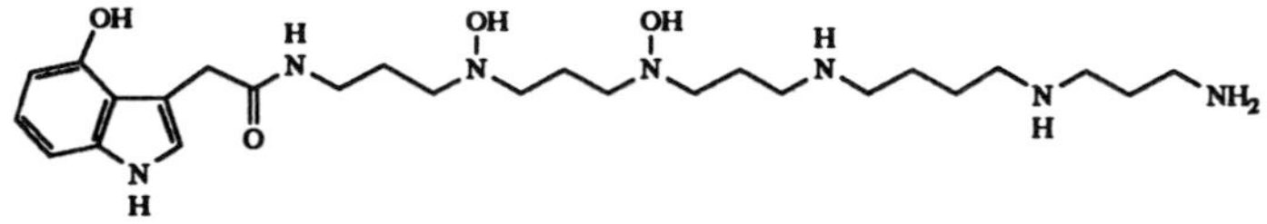 Agel 521	*Agelenopsis aperta* *(3)*	

TABLE II
Araneidae Polyamine Toxins

Molecular weight	Structure	Trivial name and synonym(s)	Species isolation and structure	Synthesis (references)
Unknown	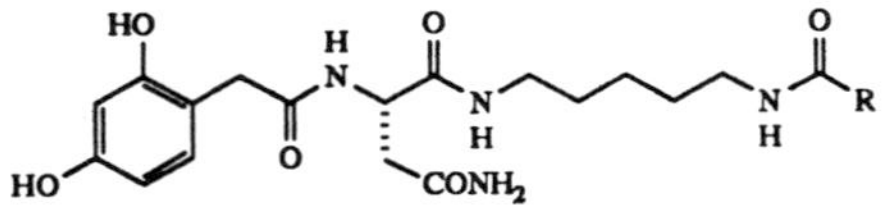	NSTX-1	*Nephila maculata (105,112)*	
373	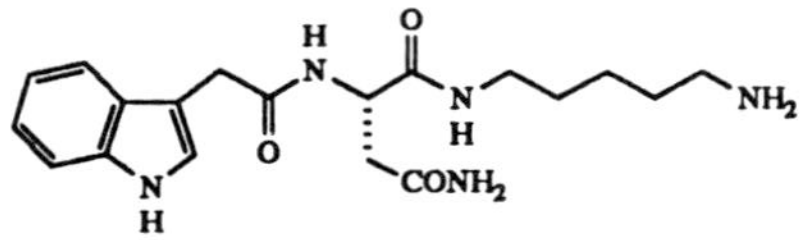	Arg 373 (pseudoar-giopinine III)	*Argiope lobata (94)*	
466	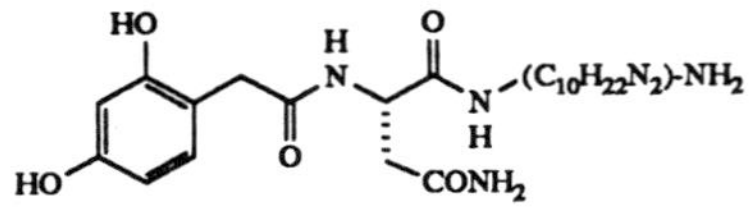	Arg 466	*Argiope florida (78) Argiope trifasciata (78)*	
466	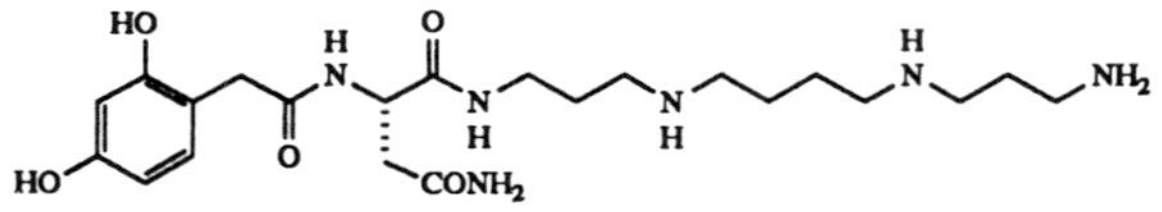	JSTX-1	*Nephila clavata (104,106,112)*	

480	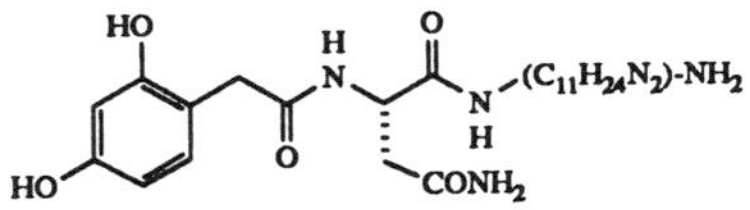	Arg 480	*Argiope florida* (78) *Argiope trifasciata* (78)	
487	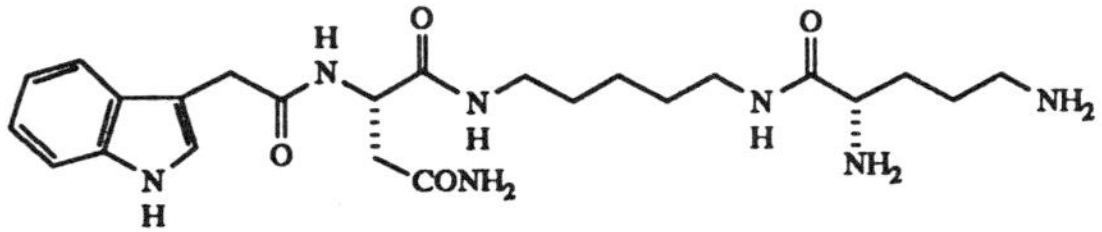	NPTX-11	*Nephila clavata* (107)	(52)
494	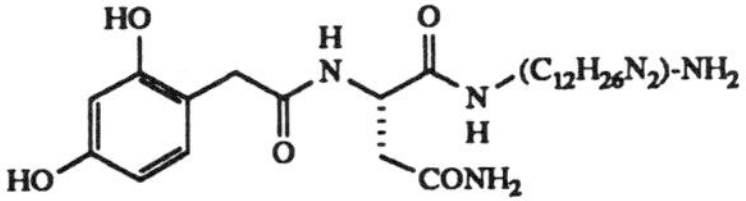	Arg 494	*Argiope florida* (78) *Argiope trifasciata* (78)	
503	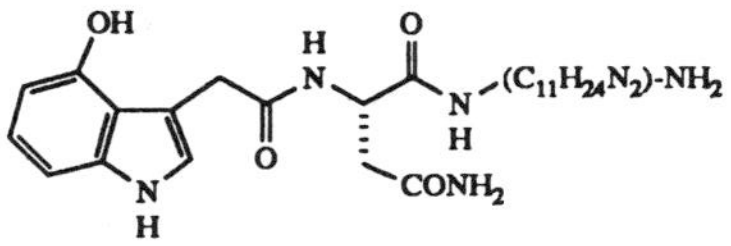	Arg 503	*Argiope florida* (78) *Argiope trifasciata* (78)	
515	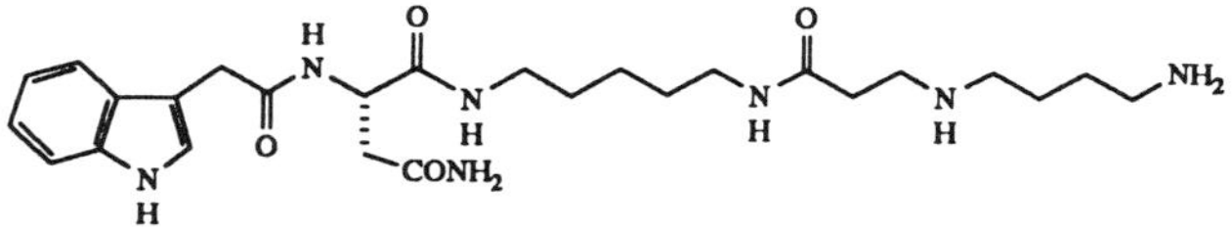	NPTX-12	*Nephila clavata* (107)	(53)

(*continues*)

TABLE II (*Continued*)

Molecular weight	Structure	Trivial name and synonym(s)	Species isolation and structure	Synthesis (references)
517	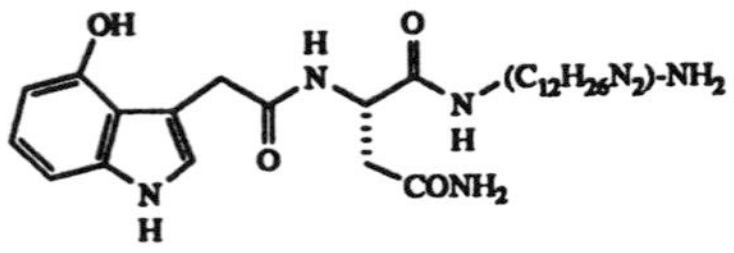	Arg 517	*Argiope florida (78) Argiope trifasciata (78)*	
522	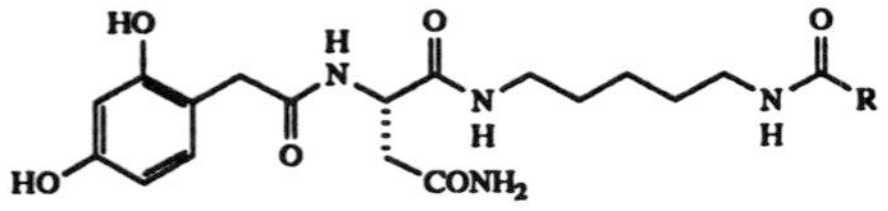	NSTX-2	*Nephila maculata (105,112)*	
565	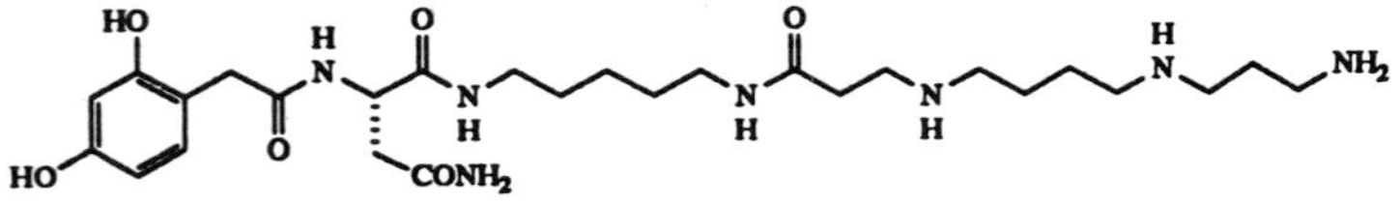	JSTX-3	*Nephila clavata (106,108,112)*	*(29,54)*
572	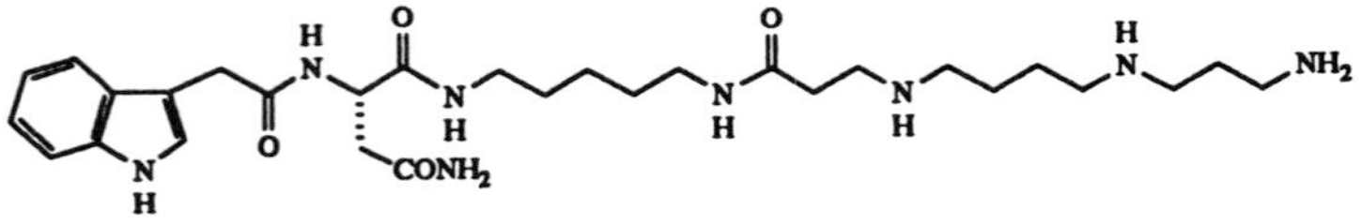	NPTX-8	*Nephila clavata (101,107)*	

10

588	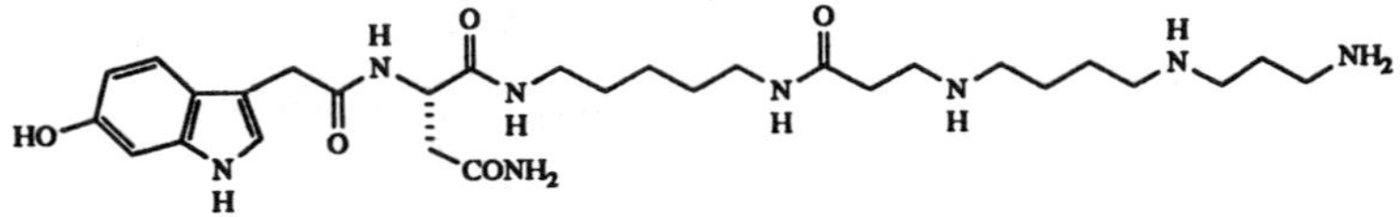	NPTX-1	*Nephila clavata (107)*	
622	Unknown structure	AN 622	*Araneus gemma (95,96)*	
622	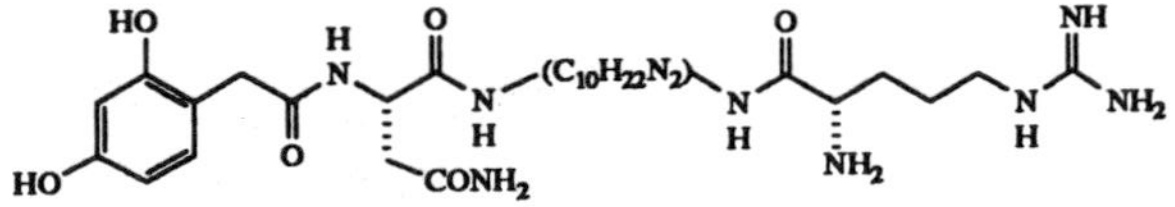	Arg 622	*Argiope florida (78,95)* *Argiope trifasciata (78,95)*	
630	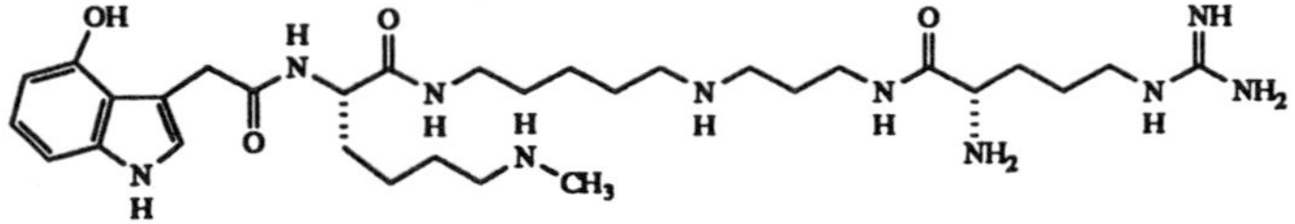	Arg 630 (argiopinine IV)	*Argiope lobata (94)*	
636	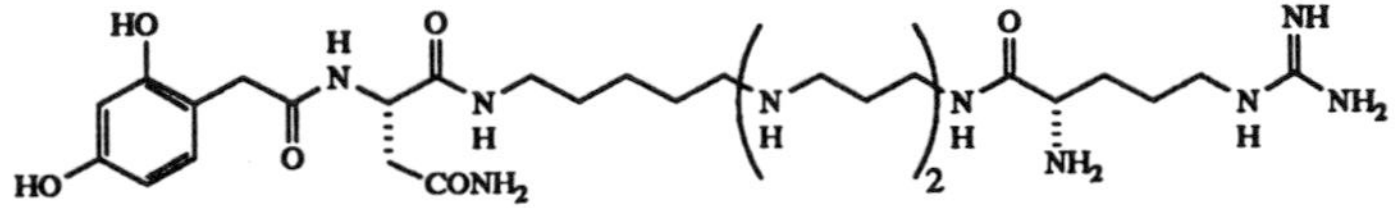	Arg 636 (argiopine, argiotoxin, argiotoxin 636, AR 636, ArgTX-636)	*Argiope aurantia (65,93,95,99)* *Argiope lobata (65,93,95,99)* *Argiope florida (78,85)*	*(28,63–65)*

(continues)

11

TABLE II (*Continued*)

Molecular weight	Structure	Trivial name and synonym(s)	Species isolation and structure	Synthesis (references)
			Argiope trifasciata (*78,85,95*) *Araneus gemma* (*78,85,95*)	
636	Unknown structure; compound differing from Arg 363 by its neurotoxic action mechanism	Arg 636b (A-Y[B]2[1])	*Argiope lobata* and other Araneidae spiders (*134*)	
636	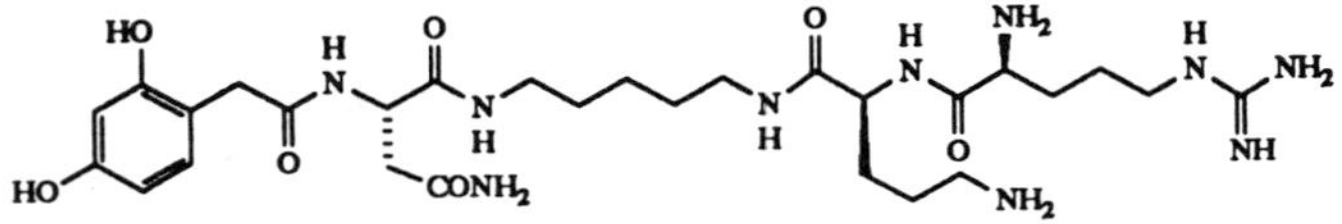	JSTX-4	*Nephila clavata* (*104,106,112*)	
643	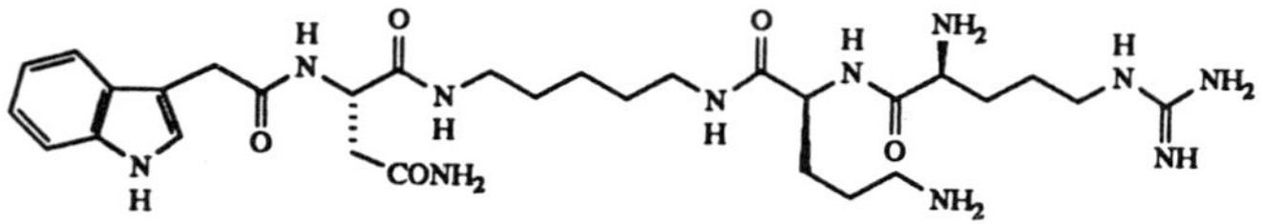	NPTX-9	*Nephila clavata* (*101,107*)	(*52*)
645	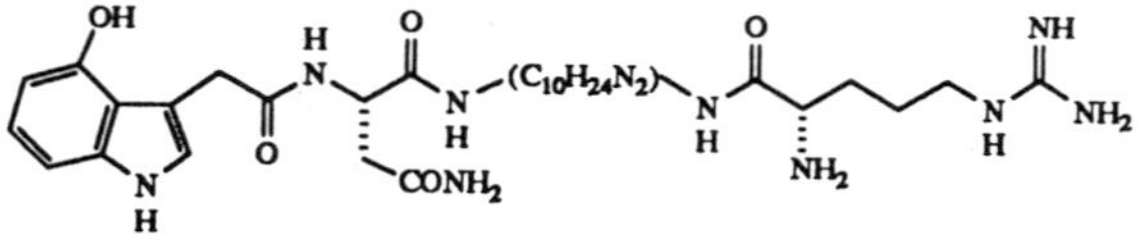	Arg 645	*Argiope florida* (*78*) *Argiope trifasciata* (*78*)	

650	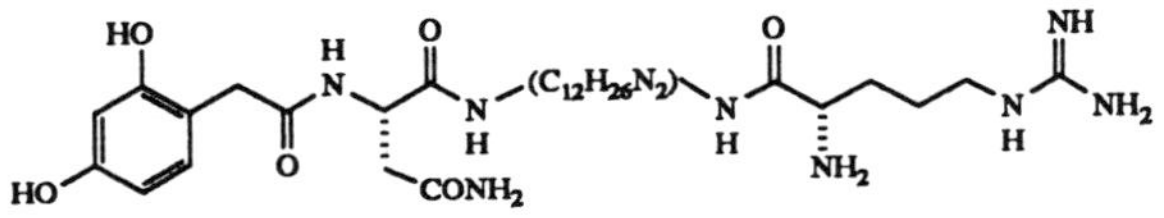 (chemical structure)	Arg 650	*Argiope florida (78)* *Argiope trifasciata (78)*
650	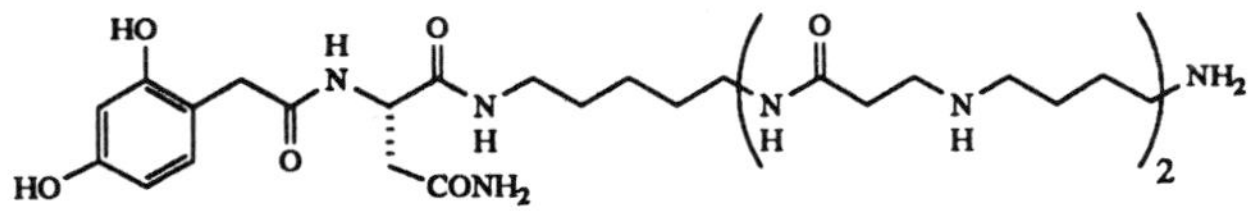 (chemical structure)	JSTX-2	*Nephila clavata (104,106,112)*
657	Unknown structure	Arg 657 (argiolobatine)	*Argiope lobata* and other Araneidae spiders (*134*)
658	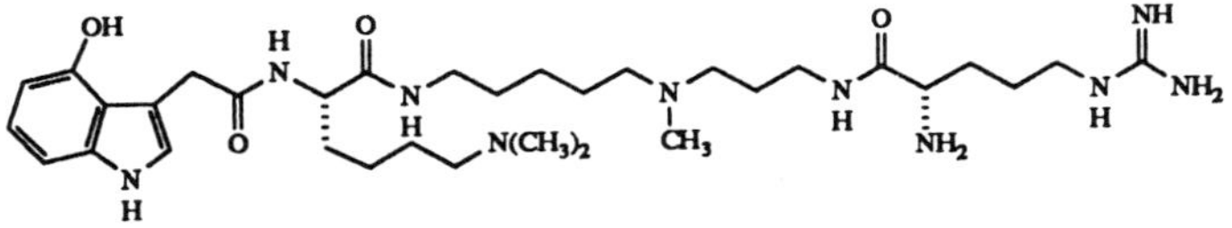 (chemical structure)	Arg 658 (argiopinine V)	*Argiope lobata (94)*
659	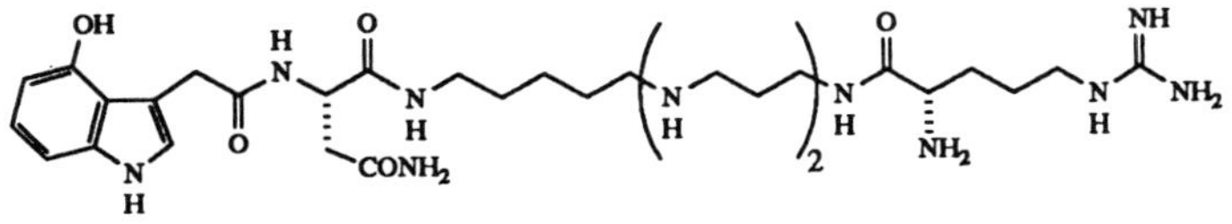 (chemical structure)	Arg 659 (argiopinine III, argiotoxin 659, AR 659)	*Argiope aurantia (65,94,133)* *Argiope lobata (65,94)* *Argiope florida (78)* *Argiope trifasciata (78)* (28,65)

(continues)

TABLE II (*Continued*)

Molecular weight	Structure	Trivial name and synonym(s)	Species isolation and structure	Synthesis (references)
659	Unknown structure	NA 659	*Neoscona arabesca* *(96)*	
664	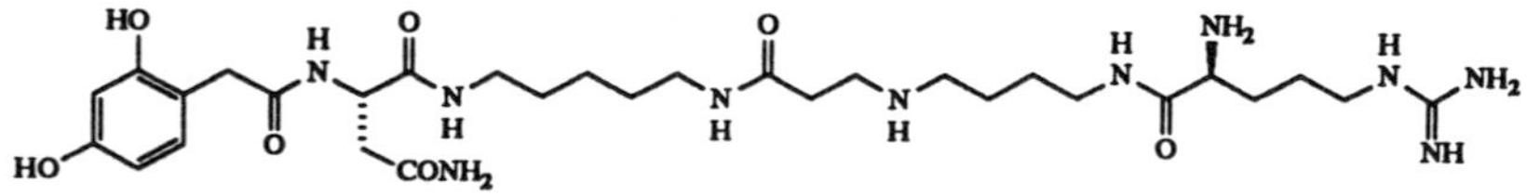	NSTX-3	*Nephila maculata (105,108,112)*	*(29,56–58)*
671	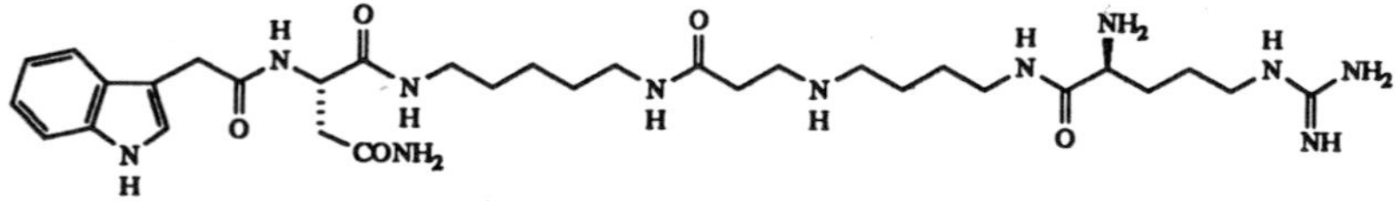	NPTX-10	*Nephila clavata (107)*	*(53)*
673	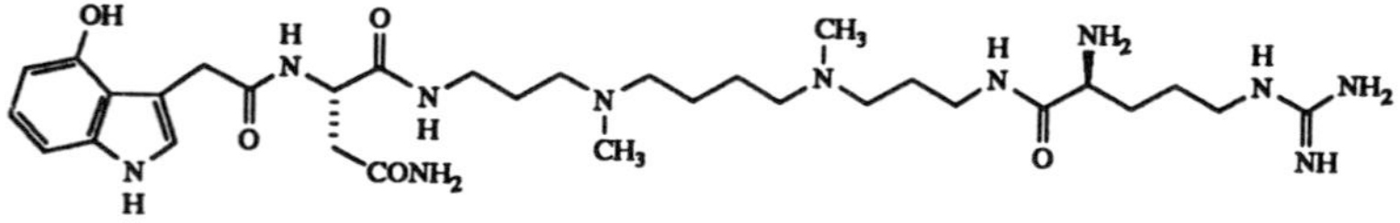	Arg 673 (argiotoxin 673 AR 673)	*Argiope aurantia (65) Argiope florida (78) Argiope trifasciata (78)*	*(28)*
673	Unknown structure	NA 673	*Neoscona arabesca* *(96)*	

687	Unknown structure	NA 687	*Neoscona arabesca (96)*
728	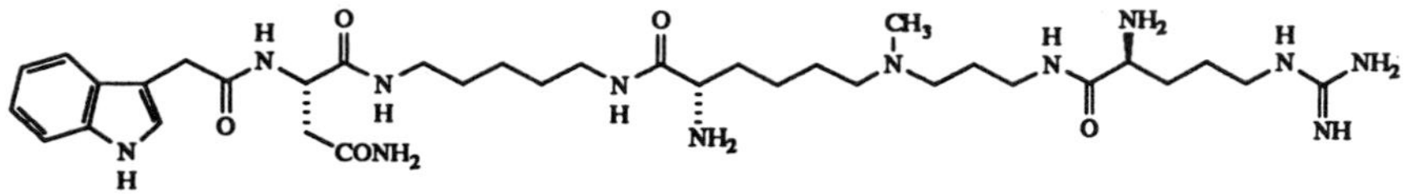	Arg 728 (pseudoar-giopinine II)	*Argiope lobata (94)*
743	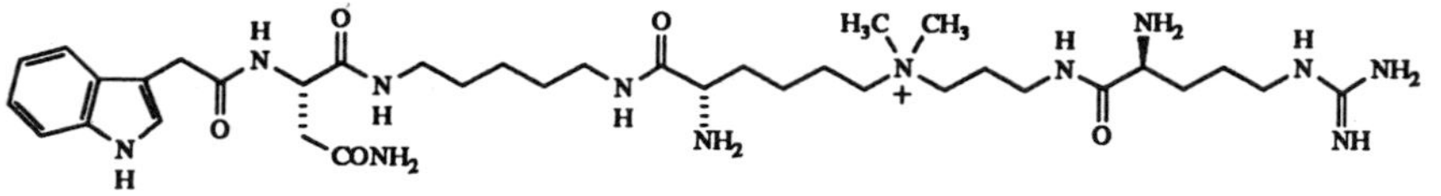	Arg 743 (pseudoar-giopinine I)	*Argiope lobata (94)*
744	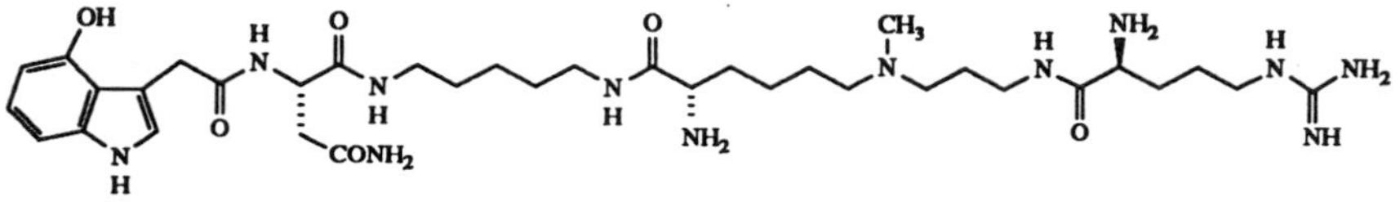	Arg 744 (argiopinine II)	*Argiope lobata (94)*
758	Unknown structure	AN 758	*Araneus gemma (96)*
759	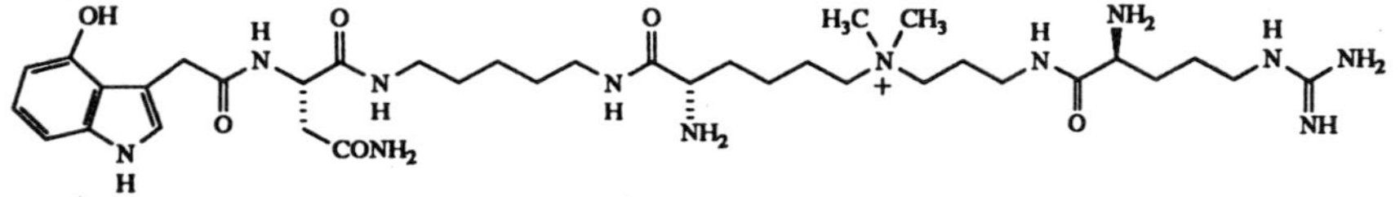	Arg 759 (argiopinine I)	*Argiope lobata (94)*

(continues)

15

TABLE II (*Continued*)

Molecular weight	Structure	Trivial name and synonym(s)	Species isolation and structure	Synthesis (references)
772	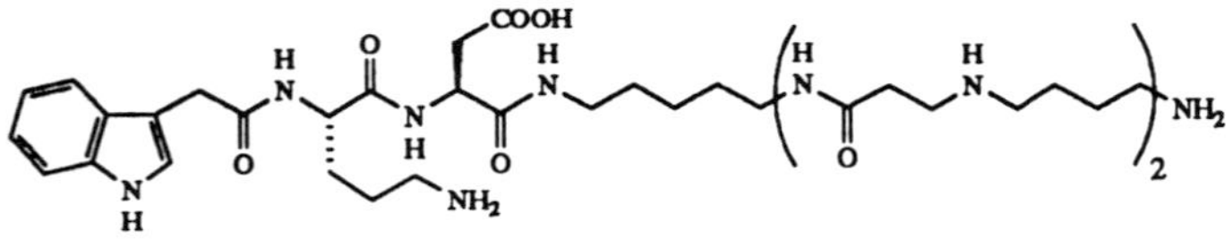	NPTX-7	*Nephila clavata* (*101,107*)	
792	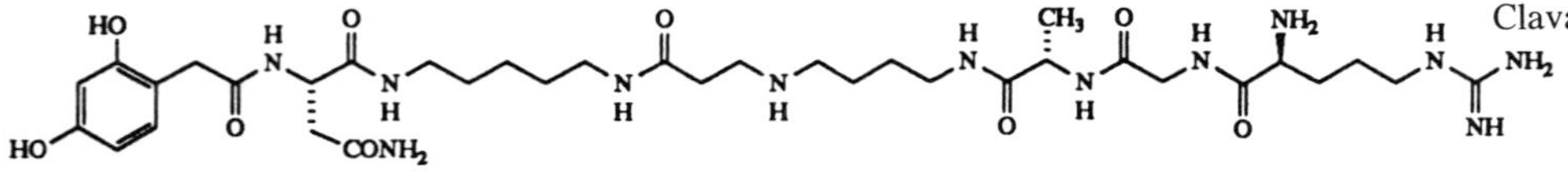	Clavamine	*Nephila clavata* (*115*)	(*59*)
801	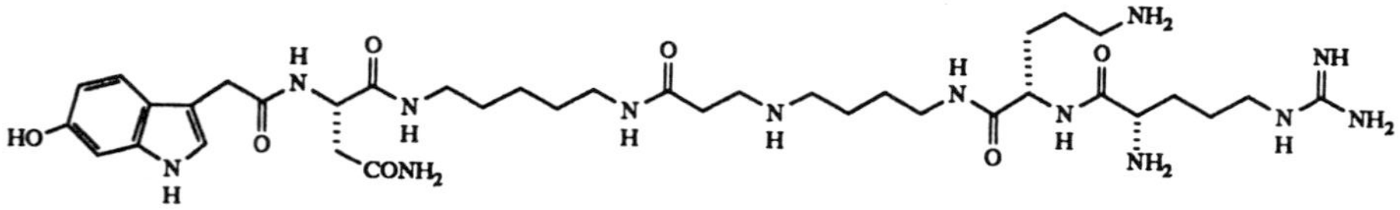	NPTX-2	*Nephila clavata* (*107*)	
815	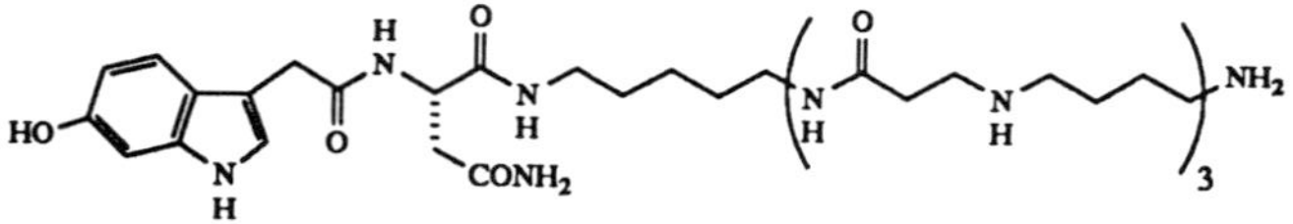	NPTX-5	*Nephila clavata* (*107*)	

819	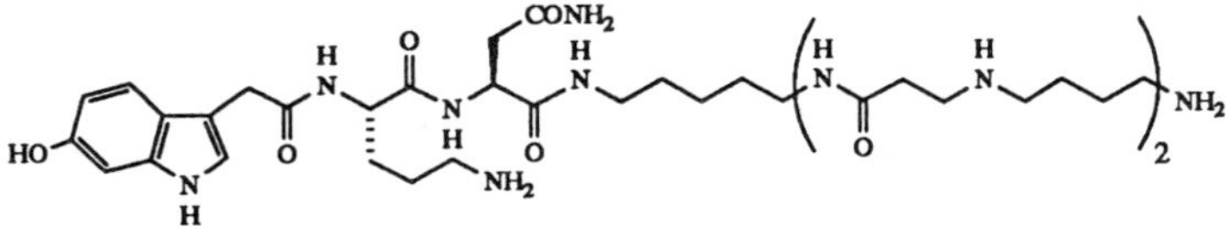	NPTX-3	*Nephila clavata* (*107*)
929	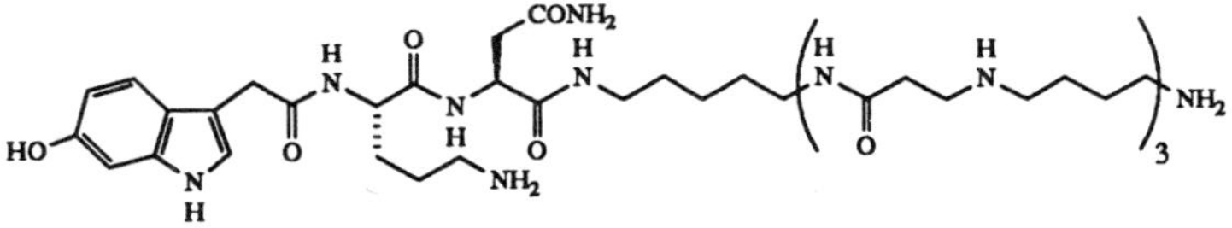	NPTX-4	*Nephila clavata* (*107*)
957	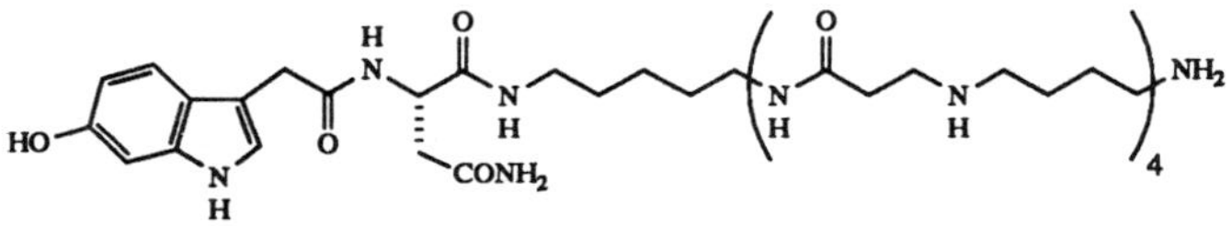	NPTX-6	*Nephila clavata* (*107*)

TABLE III
MISCELLANEOUS SPIDER AND WASP POLYAMINE TOXINS

Molecular weight	Structure	Trivial name and synonym(s)	Species isolation and structure	Synthesis (references)
243	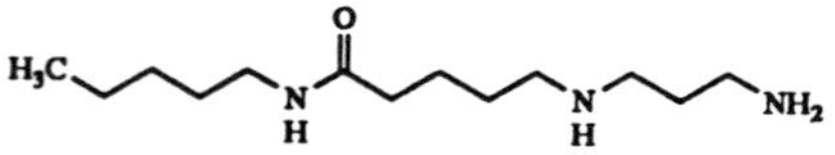	β-PhTX (β-philanthotoxin)	*Philanthus triangulum* (68,86)	(68)
389	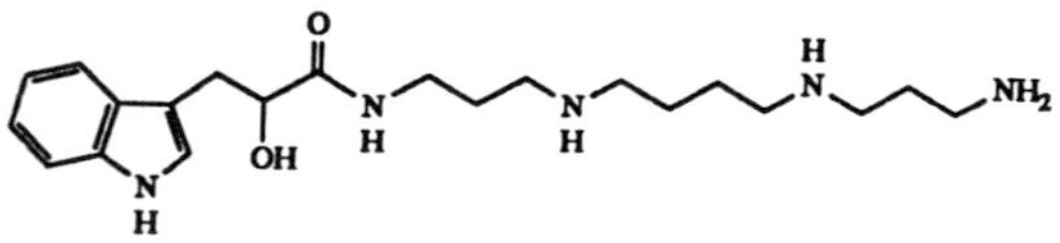	Het 389	*Hebestatis theveniti* (4) *Harpactirella* sp. (4)	(4)
403	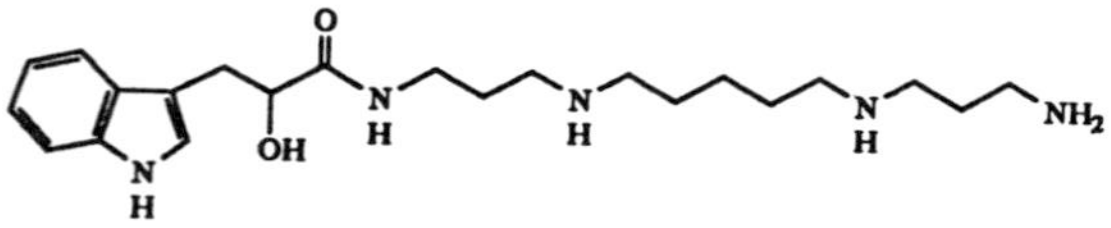	Het 403	*Hebestatis theveniti* (4)	

18

435	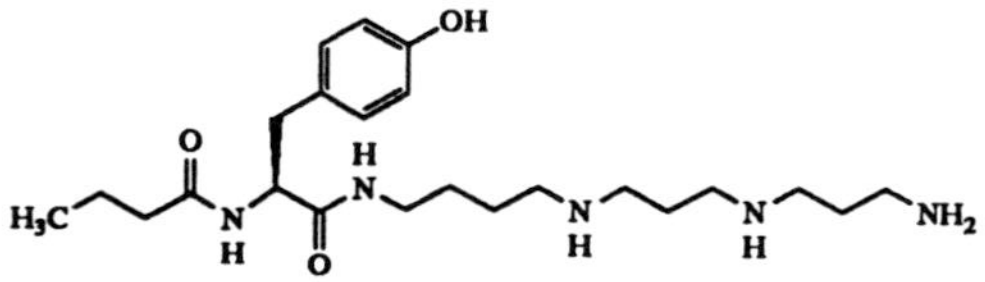 (chemical structure)	δ-PhTX (δ-philanthotoxin, philanthotoxin-433, PhTX-433, PTX-433)	*Philanthus triangulum* *(37,40)*	*(37,38,40, 41,45)*
600	Unknown structure (polyamine analysis: PA3, PA343; no amino acids; UV: similar to tyramine)	Apc 600	*Aphonopelma chalcodes*	*(4)*
728	Unknown structure (polyamine analysis: PA3, PA343; no amino acids; UV: similar to tyramine)	Apc 728	*Aphonopelma chalcodes*	*(4)*

TABLE IV
Alphabetical List of All Spider and Wasp Toxins[a]

Toxin	Molecular weight	Table	Toxin	Molecular weight	Table	Toxin	Molecular weight	Table	Toxin	Molecular weight	Table
Agel 416	416	I	Arg 622	622	II	A-Y^{B}-2^1	657	II	NPTX-1	588	II
Agel 448	448	I	Arg 630	630	II	Clavamine	792	II	NPTX-2	801	II
Agel 452	452	I	Arg 636	636	II	Het 389	389	III	NPTX-3	819	II
Agel 464	464	I	Arg 636b	636	II	Het 403	403	III	NPTX-4	929	II
Agel 468	468	I	Arg 645	645	II	HO 359	359	I	NPTX-5	815	II
Agel 489	489	I	Arg 650	650	II	HO 395	395	I	NPTX-6	957	II
Agel 489(A)	489	I	Arg 657	657	II	HO 416a	416	I	NPTX-7	772	II
Agel 489a	489	I	Arg 658	658	II	HO 416b	416	I	NPTX-8	572	II
Agel 504	505	I	Arg 659	659	II	HO 448	448	I	NPTX-9	643	II
Agel 505	505	I	Arg 673	673	II	HO 452	452	I	NPTX-10	671	II
Agel 521	521	I	Arg 728	728	II	HO 468	468	I	NPTX-11	487	II
AN 622	622	II	Arg 743	743	II	HO 473	473	I	NPTX-12	515	II
AN 758	758	II	Arg 744	744	II	HO 489	489	I	NSTX-1	Unknown	II
Apc 600	600	III	Arg 759	759	II	HO 505	505	I	NSTX-2	522	II
Apc 728	728	III	Argiolobatine	657	II	JSTX-1	466	II	NSTX-3	664	II
Arg 373	373	II	Argiopine	636	II	JSTX-2	650	II	PhTX-433	435	III
Arg 466	466	II	Argiopinine I	759	II	JSTX-3	565	II	β-PhTX	243	III
Arg 480	480	II	Argiopinine II	744	II	JSTX-4	636	II	δ-PhTX	435	III
Arg 494	494	II	Argiopinine III	659	II	NA 659	659	II	Pseudoargiopinine I	743	II
Arg 503	503	II	Argiopinine IV	630	II	NA 673	673	II	Pseudoargiopinine II	728	II
Arg 517	517	II	Argiopinine V	658	II	NA 687	687	II	Pseudoargiopinine III	373	II

[a] AG is a synonym for Agel; argiotoxin, ArgTX, and AR are synonyms for Arg; and philanthotoxin and PTX are synonyms for PhTX.

The nomenclature of the polyamine toxins is not homogeneous throughout the literature. Systematic names, according to IUPAC rules, are generally not used for these compounds. Frequently, the exact structures, and therefore the corresponding systematic names, could not be given directly after isolation and purification. As a result, the structures of some of the toxins listed in the tables are still only partially known or completely unknown, and it is therefore appropriate to use trivial names for these natural products. Most names in Tables I–III are abbreviations based on the genus of the arthropod from which they are derived and the molecular weight of the respective compounds* or the order of increasing retention time in high-performance liquid chromatography (HPLC). Arg 636 thus means the *Argiope* toxin with the molecular mass 636 and JSTX-3 the Joro spider toxin (*Nephila clavata*) eluting as the third component. Within the *Nephila* toxins, the abbreviation NSTX stands for the toxins from *Nephila maculata* (all possessing the same chromophore in the head portion), whereas both JSTX and NPTX are used for toxins isolated from *Nephila clavata*, the divergent names indicating different chromophores (a dihydroxyphenyl group with JSTX and an indolyl or hydroxyindolyl group with NPTX). The names in Tables I–III were introduced by the researchers who first isolated the natural products or were assigned in a similar fashion by us (names based on the molecular weight only). For certain compounds, more than one name has been given by different authors. All of these names, except for differences in capitalization, are listed in the tables. In the following text, however, the first listed abbreviations are used exclusively.

It is worth mentioning at this point that the choice of an abbreviation system based on the source genus and the molecular weight of the toxin is rather unfortunate. There exist already several pairs of different toxins from the same genus with equal molecular mass, which had to be differentiated by an appendix. The choice of appendix is rather heuristic, however, and it is to be foreseen that with the increasing number of researchers working in the field of polyamine toxins, and with the increasing number of toxins with identical molecular weights likely to be found in the future, confusion in nomenclature will arise.

---

* The digits 433 in the abbreviated name PhTX-433 used by some authors for a *Philanthus* toxin designate, in accordance with accepted polyamine naming conventions, the number of methylene units between the respective nitrogen atoms of the polyamine backbone. To prevent confusion with the nomenclature based on molecular masses, we prefer the original abbreviation δ-PhTX for this compound.

B. Agelenidae Toxins

The Agelenidae polyamine toxins isolated from *Agelenopsis aperta* and *Hololena curta* are among the hitherto known spider and wasp toxins of the simplest structures. They consist solely of the two parts A and C of the schematic structure mentioned above, that is, they embrace simply a lipophilic aromatic acyl head bound directly to a polyamine backbone (Fig. 1).

The lipophilic head involves one of the four aromatic carboxylic acids **1–4** (Fig. 3), which is connected to the polyamine portion by an amide bond to a terminal amino group. Whereas head derivatization with the carboxylic acids **1** and **2** is also found in polyamine toxins from spiders of other families, the head modification with **3** and **4** (Agel 452 and HO 395, Agel 468) seems to be typical for Agelenidae toxins.

The polyamine portion for a given toxin contains four to six nitrogen atoms separated generally by three or four, rarely five, methylene units. In some compounds one or more nitrogen atoms of the chain are hydroxylated [Agel 448, 452, 464, 468, 489, 489a, 489(A), 505, 505a, 521, and HO 448], and in three cases one nitrogen atom of the backbone (Fig. 1, part C) is permethylated to form a quaternary ammonium ion [Agel 489, 489(A), and 505a]. Hydroxylamino functionalities are unique to Agelenidae toxins, but methylation or permethylation of the nitrogen atoms of the polyazaalkane moiety is also found within other classes of polyamine toxins.

The structures of the toxins from *Agelenopsis aperta* and *Hololena curta* are not substantially different from one another. Except for the polyamine portion PA334, which has so far been found only in HO 359 from *H. curta*, all structural units are shared by toxins of both species. In

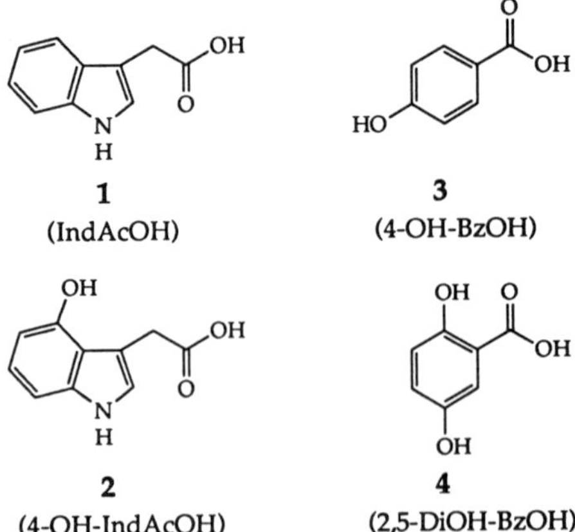

FIG. 3. Aromatic carboxylic acids found as head components of Agelenidae toxins.

fact, 5 of the 17 Agelenidae toxins were discovered in *A. aperta* as well as in *H. curta*.

C. ARANEIDAE TOXINS

The Araneidae polyamine toxins from several species of *Araneus, Argiope, Neoscona,* and *Nephila* are of a distinctively more complex construction than the Agelenidae toxins. All of them possess, in contrast to the Agelenidae toxins which consist only of the two mandatory parts A and C (Fig. 1), a peptide linker unit between the polyamine backbone and the lipophilic head (Fig. 1, part B). Some toxins of this class, like the already shown toxin Arg 636 (Fig. 1), are further acylated at the tail of the polyazaalkane moiety with a basic α-amino acid or a basic oligopeptide unit (Fig. 1, part D).

Besides the two acids **1** and **2**, already detected in Agelenidae toxins as acyl components of the head portion, the two compounds **5** and **6** were additionally claimed to be found in hydrolysates of Araneidae toxins (Fig. 4). Within this spider family, the acid moieties **1** and **5** have been described as head portions in the toxins of all *Argiope* species, as well as in the toxins of *Nephila clavata,* whereas the (4-hydroxyindol-3-yl)acetyl group (4-OH-IndAc, acyl part of **2**) is asserted to be found in *Argiope* toxins and the isomeric (6-hydroxyindol-3-yl)acetyl group (6-OH-IndAc, acyl part of **6**) only in *Nephila* toxins. It is conceivable, however, that there may have been mistakes with the structural assignment to **6** as the head portion of some NPTX toxins as a consequence of the very similar spectroscopic behavior that would be expected for such closely related components such as **2** and **6** (see also discussion below). In contrast to the compounds

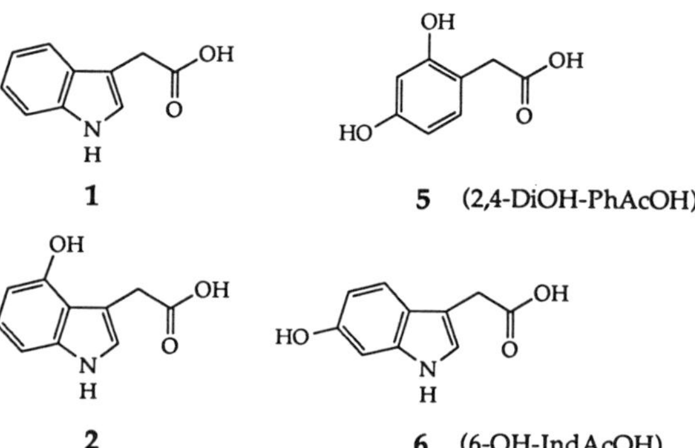

FIG. 4. Aromatic carboxylic acids found as head components of Araneidae toxins.

deduced to comprise **6** on the basis of spectroscopic evidence only, some structures containing **2** have also been verified by total synthesis.

The polyamine backbones of some Araneidae toxins differ characteristically from those of the Agelenidae-derived natural products. Whereas in most *Argiope* toxins the backbone is still a simple linear polyamine containing two to four nitrogen atoms, which are sometimes methylated, some *Argiope* toxins (Arg 728, 743, 744, and 759) and in particular most *Nephila* toxins (all but JSTX-1, JSTX-4, NPTX-9, and NPTX-11) contain unusual ω-amino acid units, like **7** (in *Argiope* toxins only), **8,** and **9** (in *Nephila* toxins only), as building blocks of the polyazaalkane chain (Fig. 5). Oxidation of the amino groups to hydroxylamino functionalities has not been found in Araneidae toxins.

The linkage between the polyamine portion C and the aromatic head moiety A (see Fig. 1) in *Argiope* toxins, as well as in *Nephila* toxins, is made in most cases by asparagine. Only Arg 630 with ω-methyllysine and some NPTX toxins with a Lys-Asn group [NPTX-3 (molecular weight 819) and NPTX-4 (molecular weight 929)] or a dipeptide Lys-Asp [NPTX-7 (molecular weight 772)] involve additional α-amino acids. Finally, the tail modification of the most complex Araneidae toxins consists of an amide connection of the terminal polyazaalkyl amino group to a basic α-amino acid [most often arginine, rarely ornithine (NPTX-11)] or in some cases a basic di- or oligopeptide (Orn-Arg in JSTX-4, NPTX-2, and NPTX-9 or Ala-Gly-Arg in clavamine).

D. Miscellaneous Spider and Wasp Toxins

Only four compounds of the remaining six miscellaneous spider and wasp toxins summarized in Table III are structurally fully elucidated (Fig.

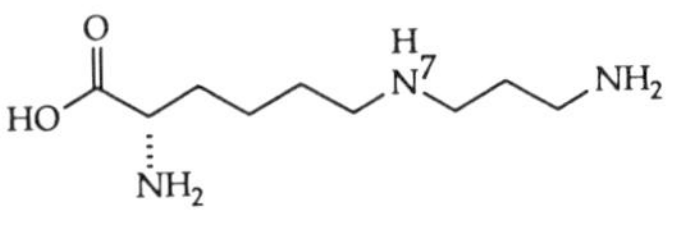

further modified at the N^7 atom
by methylation or permethylation

7 [ω-(3-aminopropyl)lysine]

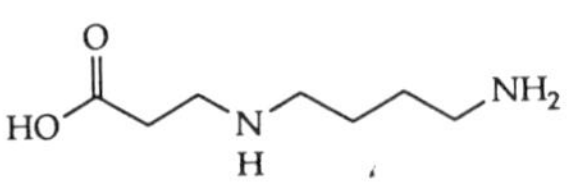

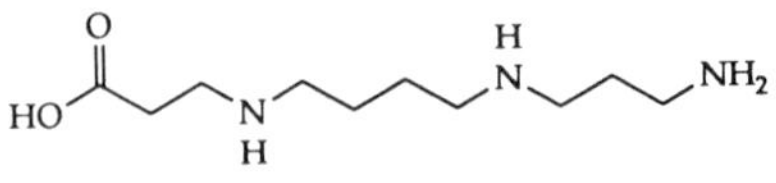

8 [putreanine, Pta]   **9** [ω-(3-aminopropyl)putreanine, AP-Pta]

FIG. 5. Unusual ω-amino acid building blocks found in the polyazaalkane portion of Araneidae toxins.

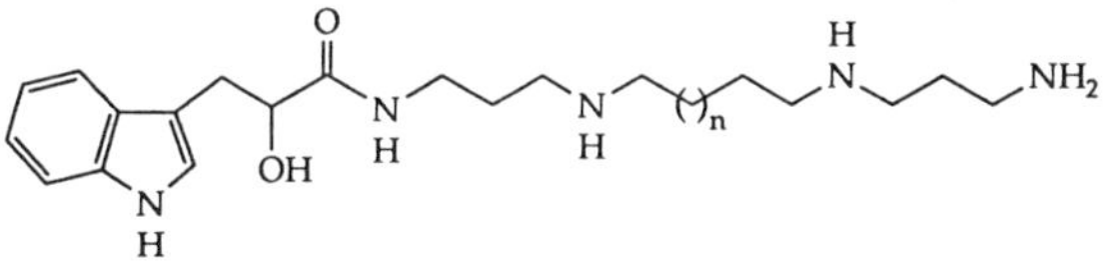

n = 1 **Het 389**
n = 2 **Het 403**

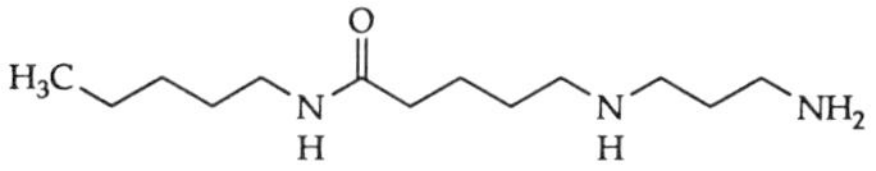

β-PhTX

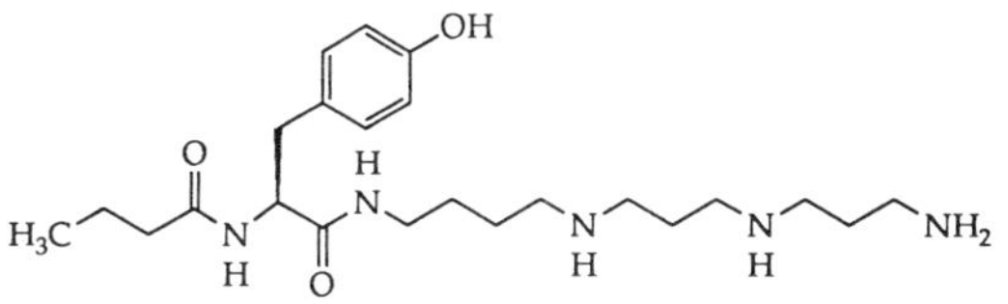

δ-PhTX

FIG. 6. Miscellaneous polyamine toxins.

6). The spider toxins Het 389 and Het 403 from *Hebestatis theveniti* and *Harpactirella* sp. are closely related to the Agelenidae toxins. They consist simply of a polyazaalkane chain, differing between Het 389 and Het 403 solely by one methylene unit, which is connected to 2-hydroxy-3-(indol-3-yl)propionic acid as the lipophilic head portion.

The two wasp toxins β-PhTX and δ-PhTX structurally resemble, despite the distant genealogical relationship of spiders and wasps, the Agelenidae as well as the Araneidae toxins. The valeric acid derivative β-PhTX possesses, like the Agelenidae toxins, a polyazaalkane backbone connected directly to a lipophilic head. In contrast to all other polyamine toxins known so far, however, the polyazaalkane backbone consists solely of a ω-amino acid portion and can therefore not be terminated at the head position by an acyl group. The head termination in β-PhTX is the pentyl group of a *N*-pentylamide.

The physiologically more active toxin δ-PhTX is more complex in structure than β-PhTX, and it is more closely related to the spider toxins mentioned above. It consists of a polyamine portion (PA334)* connected to tyrosine as the linker amino acid and to butanoic acid as the lipophilic head moiety. It correlates, therefore, with the most simple Araneidae toxins, differing only in the linking α-amino acid and the acyl moiety involved in the lipophilic head portion.

* This type of abbreviation is used in accordance with accepted polyamine naming conventions: PA334 indicates a linear polyamine (PA) possessing four nitrogen atoms that are separated by three, three, and four (334) methylene units.

III. Syntheses of Polyamine Toxins and Related Analogs

Considering the rather simple construction of polyamine toxins, even in the most complex forms comprising only four portions linked together as amides, the synthetic problems that could arise in the preparation of such compounds would appear to be reduced to difficulties with the syntheses of the respective separate fragments needed for a given target molecule. These fragments, suitably protected, should be easily connected to one another in a defined manner using well-elaborated methods from peptide chemistry. In fact, for the synthesis of a number of polyamine toxins exactly this simple strategy, or slight variations thereof, was followed. In the following discussion, we therefore do not focus primarily on detailed descriptions of complete syntheses of particular toxins. We wish more to highlight several specific problems arising in the construction of relevant polyamine toxin building blocks, especially in the development of suitably protected and modified polyamine portions.

A. AGELENIDAE AND *Hebestatis* TOXINS, AND RELATED ANALOGS

The Agelenidae and *Hebestatis* toxins (Tables I and III) consist, as already mentioned, solely of a polyamine backbone acylated on one side with an aromatic carboxylic acid. Seven Agelenidae toxins of the 17 compounds listed in Table I, the closely related Het 389 (Table III), as well as several toxin analogs of similar construction have been synthesized so far. Comparing the structures of the natural compounds given in Fig. 7, which share, except for Agel 489, a spermine (PA343) subunit on the tail side of the polyamine backbone, it is evident that their syntheses should start with a building block consisting essentially of this common section. Such a strategy would allow the economical preparation of a number of natural compounds using the same intermediates. In fact, the reported syntheses of all of the compounds in Fig. 7 (*1–4*), including that of Agel 489a, start with the construction of the polyamine portion from the tail side.

1. Syntheses of Selectively Protected Polyamines

For the spermine-containing materials, the synthesis of a tris-protected PA343 moiety of type **11** (Scheme 1) had to be accomplished initially. The selective protection of polyamines, however, is a rather laborious and uneconomical task. Several research groups working in the field of polyamine chemistry have been confronted with this problem, and advances in its solution have been reviewed (*5,6*). It is apparent that the

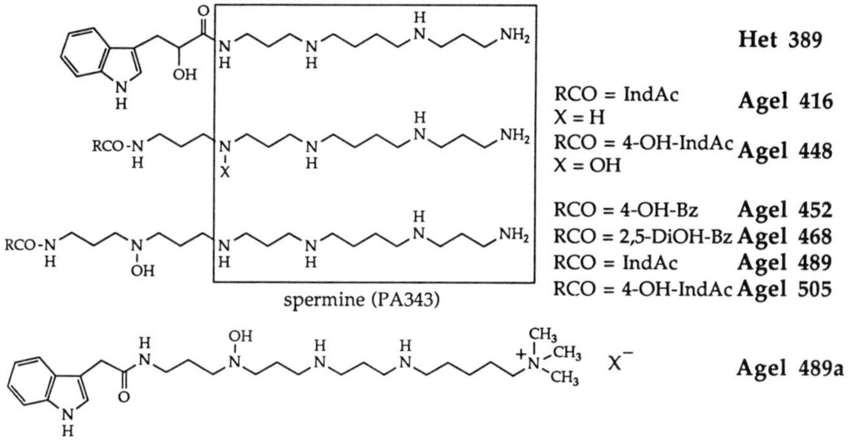

Het 389

RCO = IndAc **Agel 416**
X = H

RCO = 4-OH-IndAc **Agel 448**
X = OH

RCO = 4-OH-Bz **Agel 452**
RCO = 2,5-DiOH-Bz **Agel 468**
RCO = IndAc **Agel 489**
RCO = 4-OH-IndAc **Agel 505**

spermine (PA343)

Agel 489a

FIG. 7. Agelenidae (*1–3*) and *Hebestatis* (*4*) toxins synthesized thus far.

direct chemoselective reaction of three nitrogen atoms of a symmetrical tetramine like spermine (PA343, **10,** Scheme 1) is fundamentally difficult because chemical distinction of the two primary (or the two secondary) amino groups is almost impossible.* Therefore, indirect routes to compounds of type **11** had to be found. The problem, however, of selectively protecting one of the two chemically equivalent amino groups remains, even if the synthesis is begun with simpler diamines like putrescine (PA4, **12**) or other biogenic diamines.

Since 1964, several research groups have investigated the mono-derivatization of linear α,ω-diaminoalkanes (see Refs. *7–13* and citations

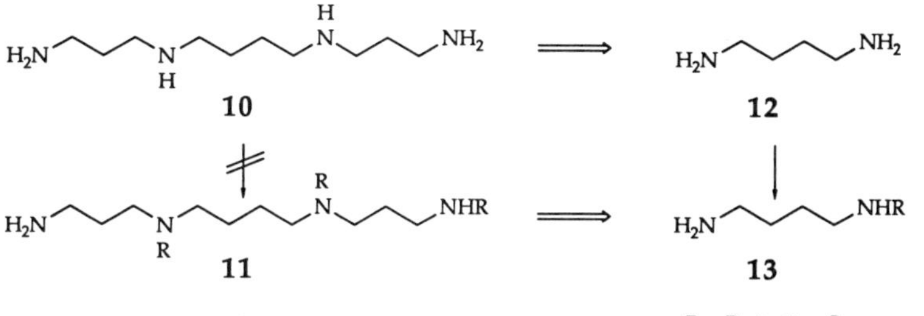

10

12

11

13

R = Protective Group

SCHEME 1. Synthetic approaches to selectively protected polyamines.

* The free amino group of a symmetrical, partially protected polyamine should in principle be chemically distinctive from an amino function of the nonreacted starting compound. However, if the transformations are carried out in homogeneous solution, the chemoselectivities were generally found to be rather small and, with more complex polyamines, nonexistent.

in Refs. *5* and *6*). Statistical alkylation or acylation of such starting materials was initially the only method at hand and gave at best about 60% of the desired mono-substituted products of type **13** together with unchanged starting materials and bis-derivatized side products. Mechanistic studies indicated that high dilution, a large excess of the diamine, and less reactive acylating agents were necessary to minimize the yield of bis-protected species (*12*). Because the mixtures arising from such nonselective reactions are generally easy to separate, especially with smaller diamino reactants, one was content with this solution to the problem until 1990, when Krapcho, and Kuell (*14*) presented an elegant way to obtain better yields of the desired products. They showed that the reaction of small diamines (PA2–PA6) with Boc_2O in dioxane (Boc is *tert*-butoxycarbonyl) gives up to 90% of the respective mono-protected derivatives. The reason for such high yields, which exceed the expected maximal possible yields achievable from a statistical reaction, is given by the low solubility in dioxane of the mono-Boc-protected compounds. Such materials precipitate from the reaction mixture, thus evading further conversion of the second amino group with Boc_2O. Analogously, the hydrolysis of bis-Boc-protected ethylenediamine in ethereal HCl solution, as performed by Geiger (*15*), takes advantage of the low solubility of the desired mono-reacted substrate, which precipitates almost quantitatively as the hydrochloride salt from the reaction mixture.

An entirely different, and less direct, strategy to obtain mono-protected diamines is to start the syntheses with suitably ω-functionalized mono-amines (or monoamine precursors) of type **14** (Scheme 2). Protection of the primary amine and substitution of the group Y by NH_2 would lead to the desired products. Several N-mono-protected linear diamines were obtained with this methodology beginning with different starting materials. The reaction of commercially available *N*-(4-bromobutyl)phthalimide (**15**) with sodium azide gave, by substitution of the bromine, the corresponding azido compound and, after catalytic hydrogenation, the mono-phthalimide-protected target molecule **16** in 40% overall yield (*16*). Similarly, but starting from 4-bromobutylamine hydrobromide (**17**) or several ω-aminoalcohols **19,** the syntheses of more N-mono-derivatized α,ω-diaminoalkanes were performed. Acylation of amine **17**, followed by substitution of the bromine with NaN_3 and reduction of the azido functionality to an amino group, as described above, delivered mono-acylated putrescines of type **18** in overall 45–58% yield (*17*).

Selective Boc protection of the amino functions in compound **19,** on the other hand, and mesylation of the alcohols, followed by subsequent substitution of the mesylates with NaN_3 and reduction of the azido compounds as aforementioned, gave the corresponding mono-Boc-protected diamines

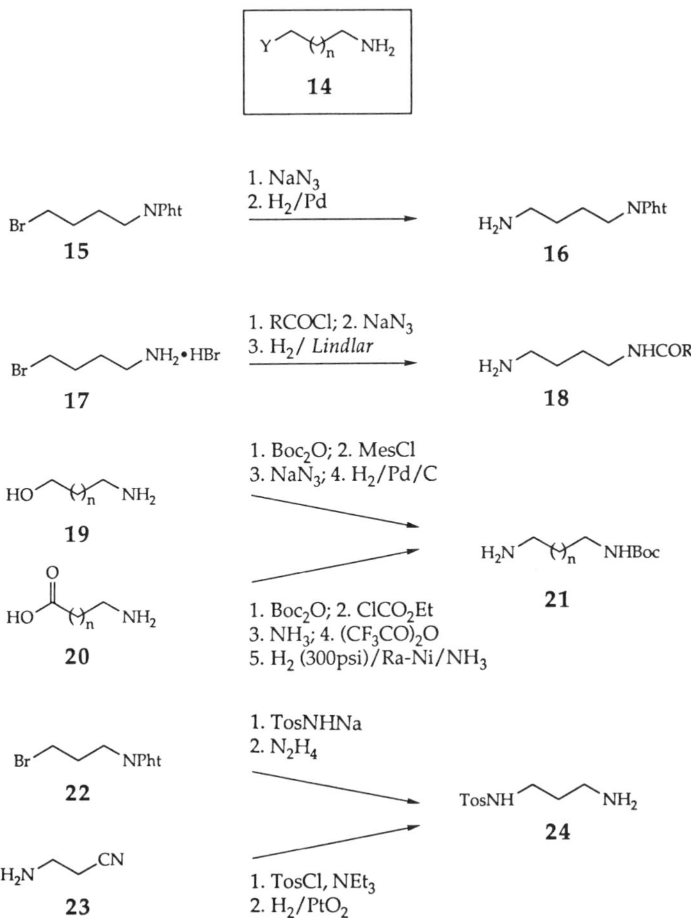

SCHEME 2. Indirect syntheses of N-monoprotected α,ω-diaminoalkanes.

21 with 66–88% yields (Scheme 2) (*18*). The latter compounds were also attained in less attractive yields from ω-amino acids (**20**) by Boc protection of the amino functions, conversion of the acids to the respective amides, dehydration of the latter to the corresponding nitriles, and hydrogenation of the cyano groups with H_2 at a pressure of 300 psi and in the presence of Raney nickel as the catalyst (23–46% yield) (*19*). The analogous toluenesulfonyl (Tos)-protected PA3 derivative **24** was obtained in approximately the same yield by substitution of the bromine in compound **22**, but this time with a tosylamino group instead of an azido group as above, followed by removal of the phthalimide group with hydrazine (*20*). Better yields of **24** were obtained by Tos protection of 3-aminopropanenitrile (**23**) followed by catalytic reduction of the cyano group (*21*).

Even though the indirect methods to obtain mono-protected diamines are generally higher yielding than their direct counterparts, the latter methodologies are frequently preferred. Starting from inexpensive materials and performing one single process is evidently superior to a multistep procedure beginning with more expensive reagents.

2. Syntheses of Spermine-Containing Agelenidae Toxins

Saccomano, Volkmann, and co-workers, the leading researcher collective studying the synthesis of Agelenidae toxins (*1–3,22*), as well as Skinner *et al.*, the authors who published the synthesis of Het 389 (*4*), started their syntheses of the respective protected spermine units required for the preparation of the desired natural products from an α,ω-diaminoalkane (putrescine, PA4, **12**). However, they did not use the aforementioned novel and high yielding N-mono protection methodology introduced by Krapcho *et al.* (*14*), but rather statistical transformations (Scheme 3).

For the synthesis of several Agel toxins, the tris-Boc-protected derivative **28** was prepared (*1–3*). Nonselective reaction of PA4 (**12**) with acrylonitrile gave, according to known procedures, the polyamino nitrile **25** in 58% yield (*7,23*). This polyamine was elongated by treatment with *N*-Boc-protected 3-bromopropylamine in the presence of neutral KF/

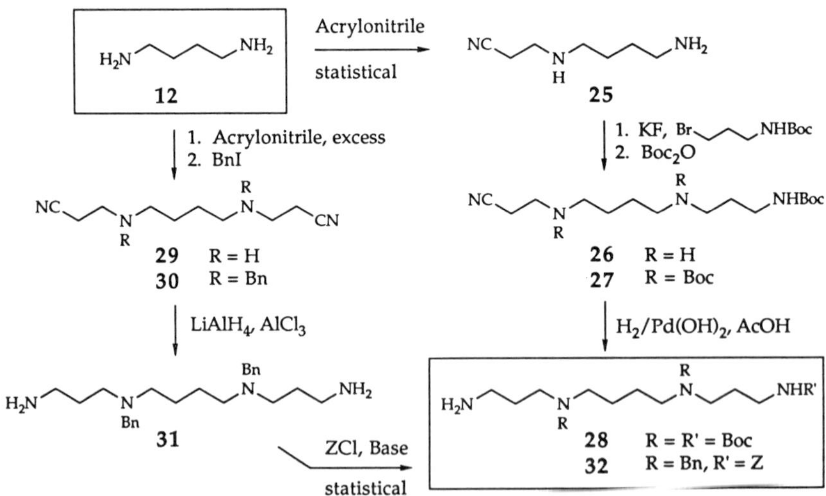

SCHEME 3. Syntheses of the spermine moieties used for Agel toxins (*1–3*) and Het 389 (*4*), respectively.

Celite as originally described by Ando and Yamawaki (*24*), furnishing in 60% yield the desired product **26,** together with a branched bis-alkylated side product. Protection of the free amino functionalities of compound **26** with Boc groups, followed by catalytic hydrogenation of the nitrile **27,** finally gave the spermine derivative **28** in 32% overall yield.

The corresponding compound **32,** utilized as the protected spermine moiety in the synthesis of Het 389, was also prepared starting from putrescine (*4*). Double Michael addition of the diamine **12** to excess acrylonitrile produced the dinitrile **29,** which was bis-benzylated to furnish compound **30.** Reduction of the two nitrile groups with LiAlH$_4$ in the presence of AlCl$_3$ provided the symmetrical spermine derivative **31** and finally, after nonselective reaction with benzyl chloroformate (ZCl, where Z is benzyloxycarbonyl) the suitably protected spermine **32.** The yields of these reactions were not reported; however, the overall yield of the sequence could be estimated not to exceed 40%. With the statistical and nonselective reaction occurring at the end of the tube, this preparation of a tris-protected spermine can hardly be regarded as economical.

With the appropriate polyamine portion at hand, the assembly of racemic Het 389 was completed in an unreported yield by acylation of the free primary amino group of compound **32** with commercially available (±)-3-(indol-3-yl)lactic acid (**33**) in the presence of dicyclohexyldicarbodiimide (DCC) and 1-hydroxybenzotriazole (HOBT). Simultaneous removal of all protecting groups by hydrogenolysis yielded Het 389 (Scheme 4).

Further modifications to the spermine portion **28** were necessary for the syntheses of the PA343 subunit-containing Agel toxins, namely, elongation of the polyamine by one or two additional aminopropyl units and, for Agel 448, 452, 468, 489, and 505, the preparation of the N^4 atom of the final

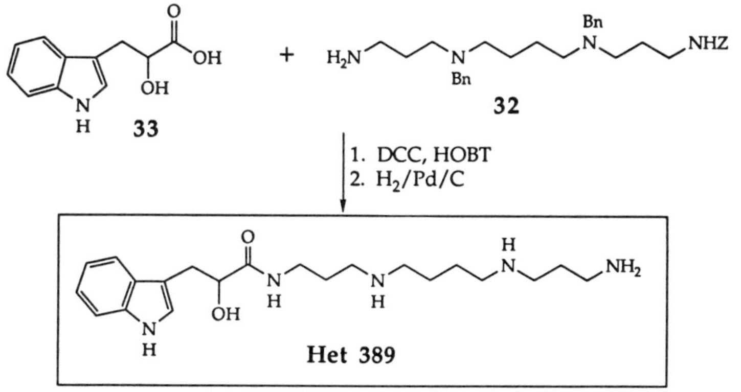

SCHEME 4. Completion of the synthesis of Het 389 (*4*).

polyamine for oxidation to a hydroxylamino functionality. Elongation of the chain was accomplished by cyanoethylation reactions using high yielding standard procedures (Scheme 5) (*1–3*). Addition of the spermine derivative **28** to acrylonitrile gave, almost quantitatively, the polyaminonitrile **34**. Compound **34** was protected at the free amino group by treatment with Boc_2O leading, after reduction of the nitrile, to the tetrakis-protected pentamine **36**, the ultimate polyamine precursor of Agel 416, in overall 81% yield.

Naturally, a different protective group other than Boc, for example, an allyloxy carbonyl group as used in the synthesis of Agel 489a (*22*), see below, could be introduced into molecule **34**, if necessary. This would lead, after reduction of the nitrile, to a tetrakis-protected pentamine similar to compound **36**, but with an N^4 atom free to be liberated independently from the other protected amines.

The selective deprotection of the N^4 atom in polyazaalkane **36**, or a subsequent product, would permit the regioselective modification of the polyamine at this particular center; for example, it would allow for the selective introduction of the desired hydroxylamino functionality at the

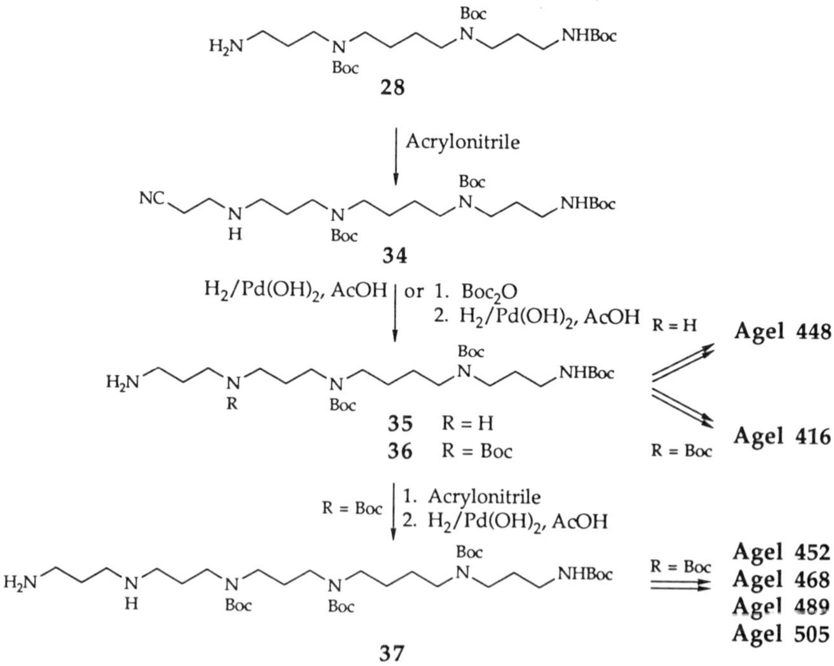

SCHEME 5. Syntheses of the protected polyamine portions for some Agel toxins (*1–3*).

correct position for Agel 448. Blocking of the free secondary amine in compound **34,** however, proved to be dispensable because the diamine **35,** obtained on reduction of **34** with H_2/Pd(OH)$_2$, as well as the tetrakis-protected hexamine **37,** obtained from **36** by an additional cyanoethylation/reduction sequence, did preferentially couple the primary amino functionalities with carboxylic acids in the presence of DCC and *N*-hydroxysuccinimide (NHS) (56–68% yield, Table V, entries 1 and 3–6). The secondary amino groups were left mostly untouched under such conditions. Coupling yields, however, could still be substantially improved if all of the internal amino functions were protected. This is exemplified by the reaction of the optimally protected polyamine **36** with acid **42,** leading to the corresponding amides in 88% yield (Table V, entry 2).

The syntheses of the PA343-containing Agel toxins were accomplished by acylation of the primary amino groups of the protected polyamine portions **35, 36,** and **37** with the respective, suitably derivatized, acids **41, 42, 43,** or **44** (for their synthesis, see Scheme 6) needed for a given target molecule, followed by oxidation of the free N^4 nitrogen to a hydroxylamino function, if necessary, and the simultaneous removal of all alcohol and amine protecting groups. The reaction conditions and chemical yields of the individual steps are summarized in Table V.

Several methods of performing the amine to hydroxylamine transformation were investigated. Oxidation of the respective secondary amines generated, in all cases, mixtures of the corresponding hydroxylamine and nitrone-containing compounds, which were treated with $NaCNBH_3$ to

TABLE V

COMPLETION OF SYNTHESES OF PA343-CONTAINING Agel TOXINS

Entry	Polyamine portion	Acyl portion[a]	Coupling[b] yield (%)	Oxidation yield (%)	Deprotection yield (%)	Final product	Ref.
1	**35**	**41**	68	77[c]	73[f]	Agel 448	*1,2*
2	**36**	**42**	88	No oxidation	70[g]	Agel 416	*3*
3	**37**	**41**	62	75[c]	70[f]	Agel 505	*1,2*
4	**37**	**42**	56	31[d]	95[g]	Agel 489	*1,2*
5	**37**	**43**	56	74[c]	96[g]	Agel 452	*1,2*
6	**37**	**44**	56	66[c], 33[e]	100[g]	Agel 468	*1,2*

[a] For structures of the acyl portions **41–44,** see Scheme 6.
[b] DCC, NHS.
[c] 2-(Phenylsulfonyl)-3-phenyloxaziridine (Davis reagent) followed by $NaCNBH_3$.
[d] MCPBA followed by $NaCNBH_3$.
[e] Dimethyldioxirane followed by $NaCNBH_3$.
[f] HCl/dioxane followed by water.
[g] TFA.

reduce the overoxidized nitrone by-products to the desired components. Use of the Davis reagent (*25*) as the oxidizing agent proved to be superior to 3-chloroperoxybenzoic acid (MCPBA) or dimethyldioxirane, the latter two reagents affording only approximately 30% yield (Table V, entries 4 and 6) compared with up to 77% yield of desired products employing the oxaziridine (Table V, entries 1, 3, 5, and 6).

Trifluoroacetic acid (TFA) treatment was determined to be a suitable method to deprotect the ultimate Agel toxin precursors. Alternatively, but proceeding in lower yield (Table V, entries 1 and 3), deprotection was performed by the action of dry HCl in dioxane, leading to the corresponding carbamic acid, which could be isolated or directly hydrolyzed by water.

The four aromatic carboxylic acid derivatives (**41–44**) to be coupled with the respective polyamine portions were obtained from commercially available starting materials in a few standard steps (*2*) analogously to literature procedures (*26,27*) (Scheme 6). Methoxymethyl (MOM)-protected 4-OH-IndAcOH **39** was prepared from 4-hydroxyindole (**38**) in four steps, and in 14% overall yield, by consecutive protection of the phenolic hydroxyl function as a MOM ether, Mannich-type dimethylaminomethylation of the protected indole derivative along with exhaustive methylation, and substitution of the introduced dimethylamino group by cyanide, followed by basic hydrolysis of the resulting nitrile. The synthesized aromatic carboxylic acid **39** and the commercially available acid **40** were then likewise

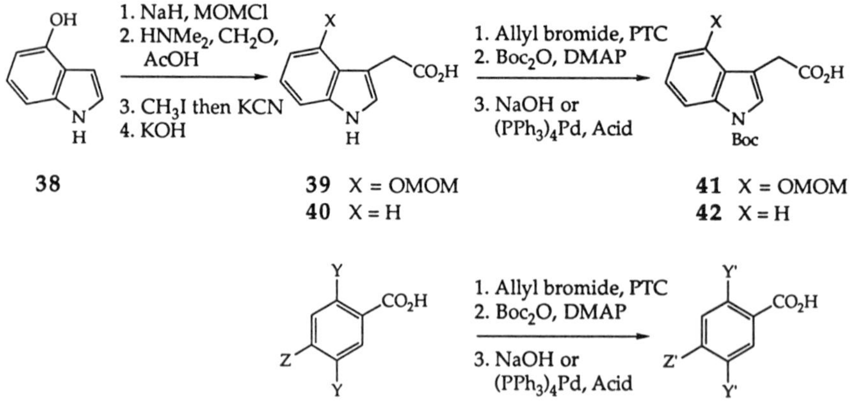

38

39 X = OMOM
40 X = H

41 X = OMOM
42 X = H

3 Y = H, Z = OH
4 Y = OH, Z = H

43 Y = H, Z = OMOM
44 Y = OMOM, Z = H

SCHEME 6. Syntheses of the carboxylic acid portions used for the preparation of Agel toxins (*2*).

transformed into the *N*-Boc-protected derivatives **41** (61% yield) and **42** (27% yield), respectively, by allylation of the acid groups under phase transfer catalysis (PTC) conditions, protection of the nitrogen functionalities with Boc, and conversion of the allylic esters to the free acids by saponification (high yielding reaction) or by $(PPh_3)_4Pd$-catalyzed transesterification (low yielding reaction). Analogously, the benzoic acid derivatives **43** (33% yield) and **44** (21% yield) were prepared from precursors **3** and **4** via the corresponding allylic esters, which were reacted with MOMCl to protect the phenolic hydroxyl groups and hydrolyzed as above to regenerate the required acid functions.

3. Synthesis of Agel 489a

The synthesis of Agel 489a, particularly of the polyamine portion, requires a completely different strategy than that for the other Agel toxins. The unsymmetrical polyamine backbone of Agel 489a not only differs in the type and arrangement of the diamine subunits, but, in addition to the head derivatization, is also modified at the terminal amino group. This demands a more refined protecting group technique than was required for the preparation of the polyamine portions **35–37.**

Concerned over the stability of the terminal quaternary ammonium salt and the internal hydroxylamine moiety, Jasys *et al.* decided to incorporate these functionalities in a late stage of the synthesis (*22*). This implied the construction of a polyamine portion of type **45** (Scheme 7), possessing three independently removable groups R^1, R^2, and R^3, which could be coupled with the already known carboxylic acid **42.** Allyloxycarbonyl (Aoc), Boc, and 2,2,2-trichloroethoxycarbonyl (Troc) functions were chosen as the three groups R^1, R^2, and R^3, respectively. The strategy to build up the corresponding derivatized compound **48,** lacking for the present the Aoc group at N^4, parallels earlier polyamine synthetic efforts by the same research group (*1,28,29*), with the contiguous polyamine components being assembled by a repetitive alkylation/amine protection process (*22*).

The synthesis of **48** opens with the alkylation of mono-Troc-protected cadaverine **46,** obtained from the corresponding mono-Boc analog by reaction of the free amino function with TrocCl and hydrolytic removal of the Boc moiety. Treatment of compound **46** with *N*-(3-bromopropyl) phthalimide (**22**) in the presence of KF/Celite, analogously to the preparation of **26** (Scheme 3), followed by treatment with Boc_2O and hydrazine gave **47,** but in fairly poor yield (25%). Two further sequences of that kind, omitting the Boc protection step in the last cycle, afforded **48** in 20% yield, the lowest yielding steps in all cycles being the alkylation reactions. It is unclear why the authors abandoned the well-known cyanoethylation/

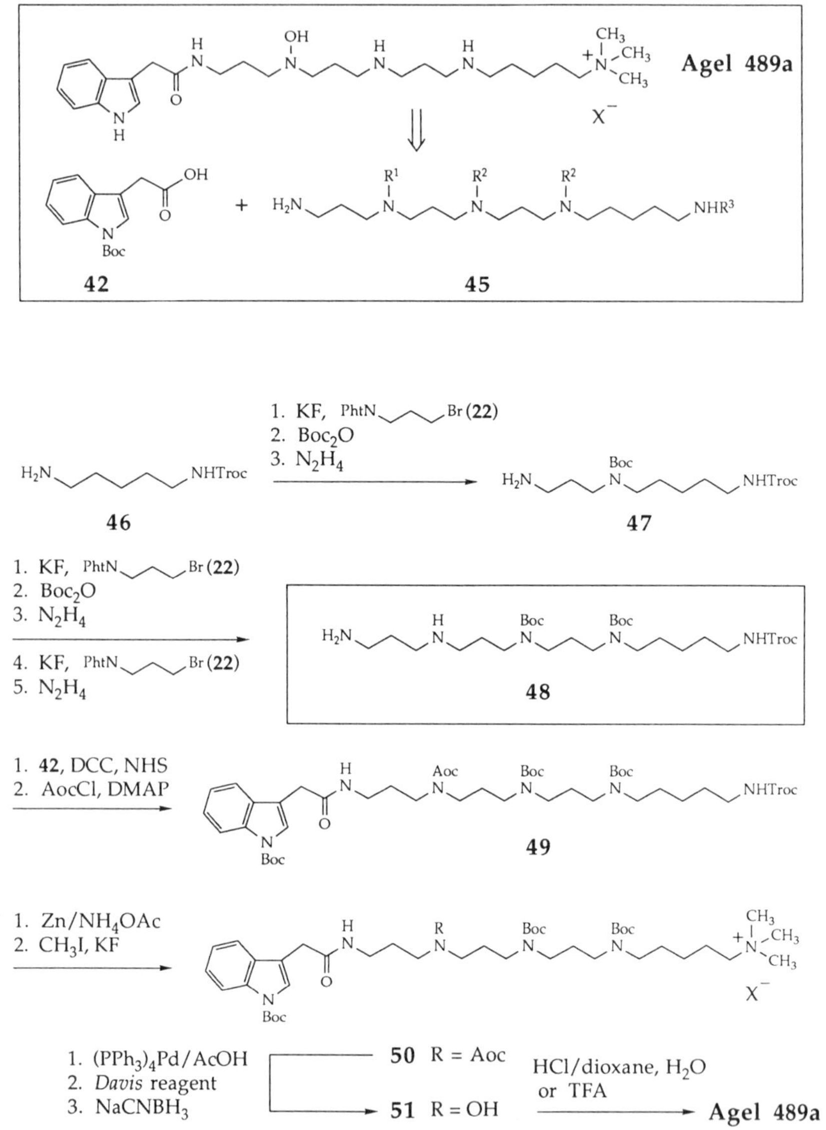

SCHEME 7. Synthesis of Agel 489a (*22*).

reduction sequence employed successfully in the syntheses of the other
Agel toxins. The Troc function possibly does not survive the acidic reaction conditions generally used for the catalytic reduction of a nitrile
[AcOH, H$_2$/Pd(OH)$_2$], even though this protective group should be stable
under such conditions according to reactivity charts (*30*).

Standard coupling of tris-protected pentamine **48** with acid **42** (DCC, NHS, 75%) yielded a compound that was immediately transformed into carbamate **49** by treatment with allyl chloroformate (AocCl, 71% yield). At this stage, the question arises why the authors omitted to protect the internal amine in compound **48,** that is, before the polyamine was coupled with acid **42.** Such an alteration in the coupling/protection sequence should lead to improved yields, as the authors themselves mention (*1*). Nevertheless, selective removal of the Troc protective group by treatment of compound **49** with zinc/ammonium acetate, followed by quaternization of the released amine terminus with excess methyl iodide, provided the ammonium salt **50** (70% yield). The Aoc group of **50** was then split off using $(PPh_3)_4Pd$ in the presence of acetic acid (to prevent N-allylation), and the liberated amine was oxidized using the Davis reagent and $NaCNBH_3$ to generate the N^4-hydroxylamino function. Treatment of the resulting unstable compound **51** directly with TFA finally provided a product identical with natural Agel 489a. The isolated yield of the toxin following reversed-phase HPLC was low (<20% from **50**), but the amount of pure compound obtained was sufficient to fully characterize the sample by spectroscopic means and to confirm the revised structure proposal, which contradicted earlier assignments (*31*).

4. Syntheses of Agelenidae-Type Toxin Analogs

Two patents from Usherwood and collaborators (*32,33*)* and three papers from Asami *et al.* (*34–36*) describe the biological activity and syntheses of therapeutic polyamine amides of type **52** (Fig. 8), which are synthetic toxin analogs of the Agelenidae (and *Hebestatis*) type. Most of these compounds derive from symmetrical spermine (PA343) and thermine (PA333) units. In contrast to the approaches mentioned before, no protecting group strategy was employed for their synthesis. The unprotected polyamines were directly coupled at the primary amino group, in the presence of DCC, with a number of equally unprotected phenolic or other substituted aromatic carboxylic acids to give mixtures of mono- and bis-acylated compounds, which were then separated by column chromatography. Such procedures typically delivered the respective toxin analogs in approximately 30% yield; in one example a 64% yield was attained.

Even though the statistical reactions above are rather low yielding, the yields are still high enough to compete with the overall yields obtained

---

* These compounds are denoted as hybrid toxins by the authors because they possess partial structures of either Arg 636 or δ-PhTX. As the α-amino acid linkage, which is mandatory in Araneidae toxins or δ-PhTX, is missing in these structures, however, they are more related to the Agelenidae toxins. More compounds of this type have been recently prepared by Goldin *et al.* (*33a*).

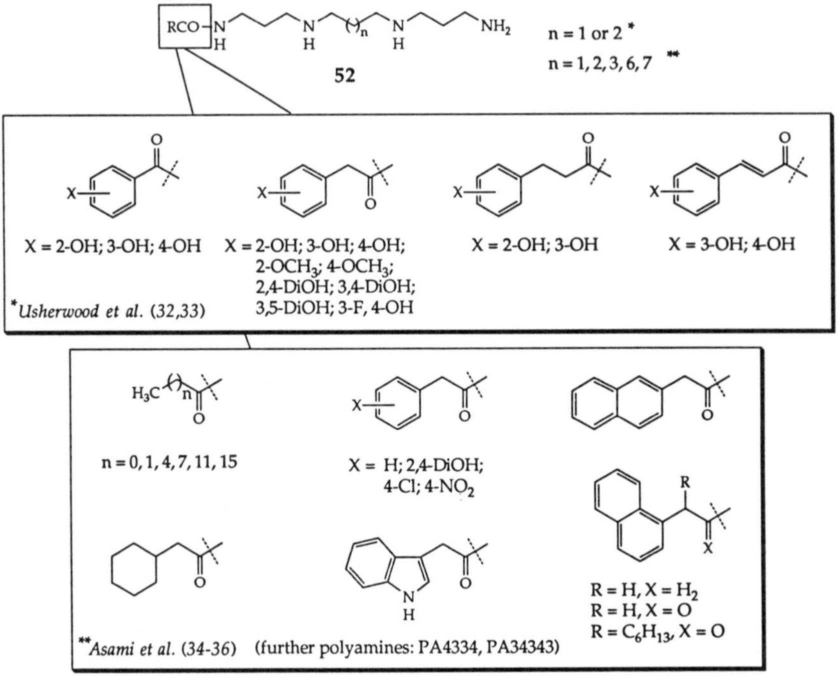

FIG. 8. Toxin analogs of the Agelenidae type synthesized by Usherwood *et al.* (*32,33*) and Asami *et al.* (*34–36*).

using protecting group strategies. However, only a limited class of toxins or toxin analogs can be efficiently prepared this way. The direct acylation method, which consists essentially of the nonselective reaction of a symmetrical tetramine with an activated carboxylic acid, dismissing all of the related problems discussed earlier (Section III, A,1), is limited exclusively to the syntheses of acylpolyamines possessing *symmetrical* polyazaalkane moieties. The method is not suited at all to prepare toxins or toxin analogs possessing *nonsymmetrical* polyamine backbones, like the Agel toxins outlined earlier (Sections III, A,2 and III, A,3). Statistics would favor the desired products too little (one out of four different terminal and internal amino groups must be acylated selectively), and, additionally, separation of the reaction mixtures would also be expected to be substantially more laborious.

B. Araneidae Toxins, δ-PhTX, and Related Analogs

The Araneidae toxins and δ-PhTX (cf. Tables II and III) have a characteristically more complex structure than the Agelenidae toxins (see Table I and discussion in Section III,A). The Araneidae toxins and δ-PhTX possess a mandatory α-amino acid linkage between the polyamine backbone and the acyl head portion. An optional further modification to the terminal amino group might also be present. Their synthesis, however, turned out to be easier than expected. In fact, the structurally most simple members of the class, namely, tail-unmodified toxins possessing an ordinary polyamine backbone only, like δ-PhTX, Arg 373, or JSTX-1, should be attained using similar strategies as outlined in the previous section and differing only in the last steps, where the derivatization of the primary amino moiety at the head end of the polyamine has to involve an N-acylated amino acid instead of an aromatic carboxylic acid as before. Such a derivatization could be performed directly, by introduction of the head and linker portion already assembled, or stepwise using standard peptide chemistry. Interestingly, only δ-PhTX and analogs, but no Araneidae toxins of this most simple type, have been synthesized so far.

1. Synthesis of δ-PhTX and Related Analogs

The construction of δ-PhTX by Nakanishi and co-workers begins with the preparation of the suitably protected polyamine portion **56** and the carboxylic acid coupling component **54** [Scheme 8 (*37–40*); see synthetic details in Ref. *41*]. The latter was obtained in three steps from commercially available protected tyrosine **53** by conversion to the 4-nitrophenyl (NP) ester, followed by hydrolytic Boc deprotection of the amine and acylation of the liberated nitrogen atom with butyryl chloride (56% overall yield). Amine **56** was prepared using standard procedures. The starting tris-Boc-protected thermospermine **55** was gained in overall 24% yield (six steps) from PA4 (**12**) in a customary fashion by a double sequence of cyanoethylation, Boc protection, and nitrile reduction (cf. synthesis of Agel toxins, Scheme 5). Compound **55** reacted with ZCl to afford the corresponding fully blocked tetramine in 94% yield, and it gave the desired partially protected amine coupling component **56** after exposure to TFA (no yields were given for the deprotection step). Reacting the two compounds **54** and **56** together in methanolic solution provided the PA433 product acylated at the head end and, finally, after simultaneous removal of the two different protecting groups by hydrogenolysis, a compound identical to natural δ-PhTX (11% yield). The coupling step was low yielding (23%), as would be anticipated, because the two unprotected internal amino groups could lead to side reactions as discussed earlier. However, acylation of the maximally protected amine **55** with the same activated acid **54** unexpectedly generated the corresponding coupling product in just fairly better yields (39%). This might be due to the nonoptimal nature of the acid activating group (4-nitrophenyl ester). Coupling a fully protected PA433 unit with the acid corresponding to **54** under DCC activation provided the desired product in 90% yield (*42*).

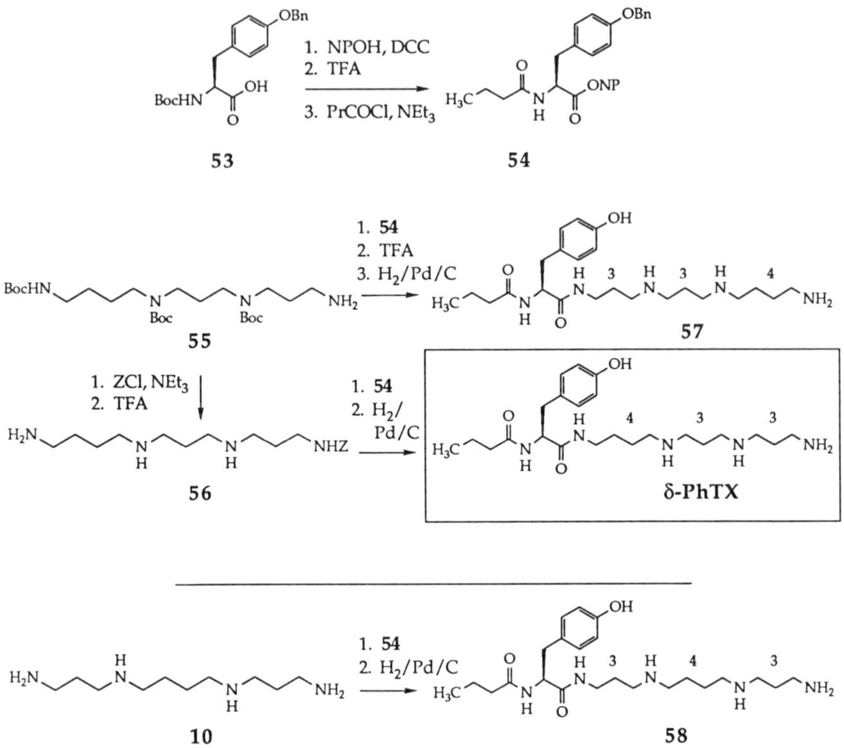

SCHEME 8. Synthesis of δ-PhTX and isomeric analogs (*41*).

The coupling product obtained from **54** and **55** was transformed into the toxin analog **57** by the consecutive action of TFA and H_2/Pd/C to remove the Boc and benzyl protecting groups, respectively. The compound obtained by this procedure differs from the natural toxin simply by the sequence of the methylene groups between the nitrogen atoms of the tetramine backbone (3-3-4 instead of 4-3-3). A further toxin analog, which is also isomeric with δ-PhTX, namely, compound **58**, possessing PA343 instead of PA334 as the polyamine portion, was synthesized directly by coupling of the unprotected symmetrical spermine **10** with ester **54** (38% yield, cf. syntheses of Agel toxin analogs above and Refs. *32* and *33* as well as *43*). The synthesis of all three isomeric compounds, δ-PhTX, **57**, and **58**, was necessary to determine unambiguously the structure of the natural product, which was not possible by spectroscopic means at that time because of the relatively limited amount of material available.

The easy access to synthetic δ-PhTX as well as the toxin analogs **57** and especially **58**, together with the interesting glutamate receptor antagonizing effect of these compounds (*50*), motivated Nakanishi and co-workers to prepare a number of additional toxin analogs using the same strategies and intermediates (*37,38,40,41,44,45*). Taking the most easily available *Philanthus* toxin analog **58** as the base structure, and showing great endurance, they varied the acyl head portion, the α-amino acid moiety, and the polyamine part as schematically outlined in Fig. 9. Even terminal acylated compounds, which more resemble some Araneidae toxins, have been prepared. Similar and some identical compounds as those outlined in Fig. 9 were synthesized by Piek and co-workers to investigate the structure–activity characteristics of this class of toxins (*46–49*); however, no synthetic details have been reported. Detailed descriptions of the experiments leading to some spermidine-containing toxin analogs of the same type are given by Fiedler *et al.* (*5,51*). Their synthetic compounds possess a racemic alanine linker unit and, in two cases, the less ordinary acyl components **59** and **60** (Fig. 10).

Details are not available as to how the terminal acylated compounds in Fig. 9 were acquired by Nakanishi and collaborators.* Probably, chain N-deprotected toxin analogs were mono-acylated with the corresponding activated acids and the desired terminal derivatized materials separated from undesired internal acylation products. An example of a sequence leading to this conclusion is the synthesis of toxin analog **64** (Scheme 9), where the precursor **61** was coupled at the tail amino group with the protected L-lysine **62** to produce intermediate **63** (*41*).

2. Syntheses of Nephila *Toxins and Related Analogs*

The synthesis of **64** (Scheme 9) shows that the construction of α,ω-bis-acylated polyamines is basically unproblematic, if the two terminal amino groups can selectively be subjected to reaction one at a time. Such an undertaking is not too exhausting, as just described, if only diamines are to be incorporated into a toxin or toxin analog. Often, statistical acylation of unprotected starting compounds already delivers mono-derivatized products in satisfactory yields. Otherwise, several less direct routes to mono-acylated diamines have been developed (cf. Scheme 2 and related discussion).

A number of ω-derivatized *Nephila* toxins that only contain cadaverine (PA5) as the polyamine subunit in their structures have been synthesized (Fig. 11). The complete polyazaalkane backbone of some members of this toxin class comprise, in addition to PA5, unusual ω-amino acid units, which are attached to the diamine as amides. The syntheses of compounds

---

* *Addendum:* The synthesis of these and further δ-PhTX analogs is described in full details in the latest communication of Choi *et al.* (*45a*).

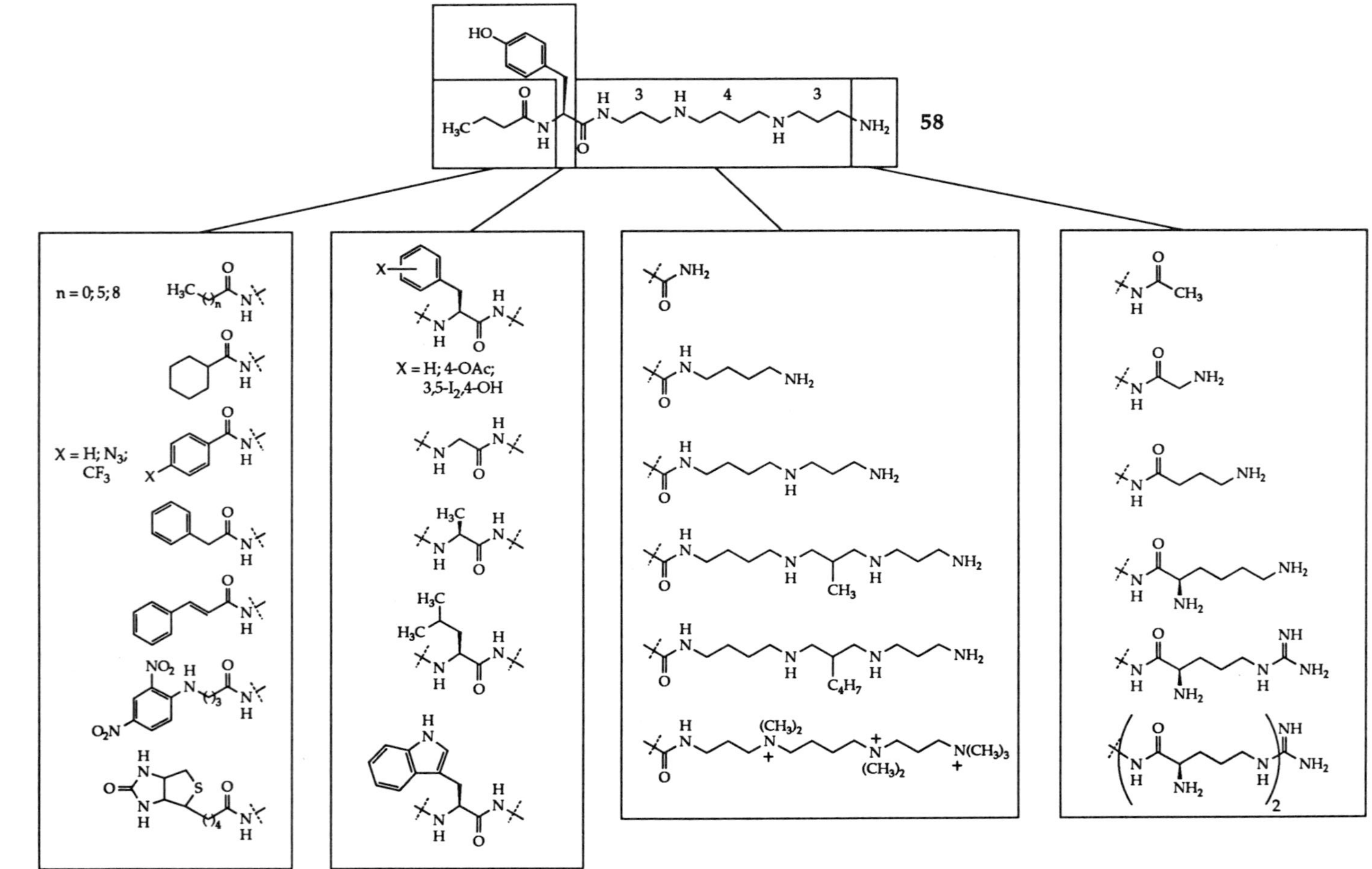

58
n = 0; 5; 8
X = H; N3; CF3
X = H; 4-OAc; 3,5-I2,4-OH

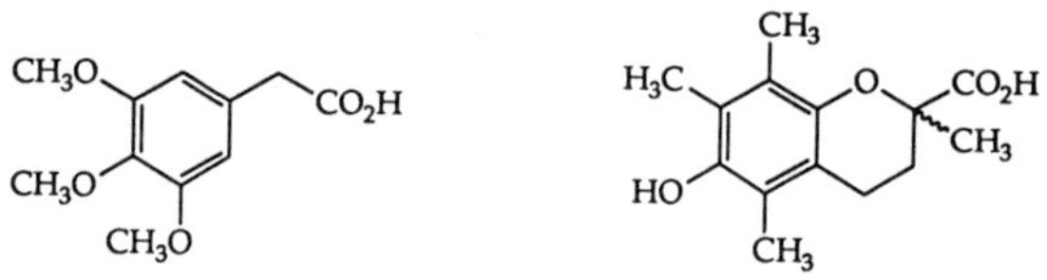

FIG. 10. Acyl head components used in toxin analogs by Fiedler *et al.* (*5,51*).

containing such uncommon amino acid building blocks, however, should not differ in essence from the syntheses of toxins that are derivatized at the two terminal amino functions with more familiar amino acids or oligopeptide portions.

Because the collection of NPTX toxins in Fig. 11 and the remaining *Nephila* toxins JSTX-3, NSTX-3, and clavamine comprise a common head portion, construction of the ubiquitous building blocks **68** and **71**, respec-

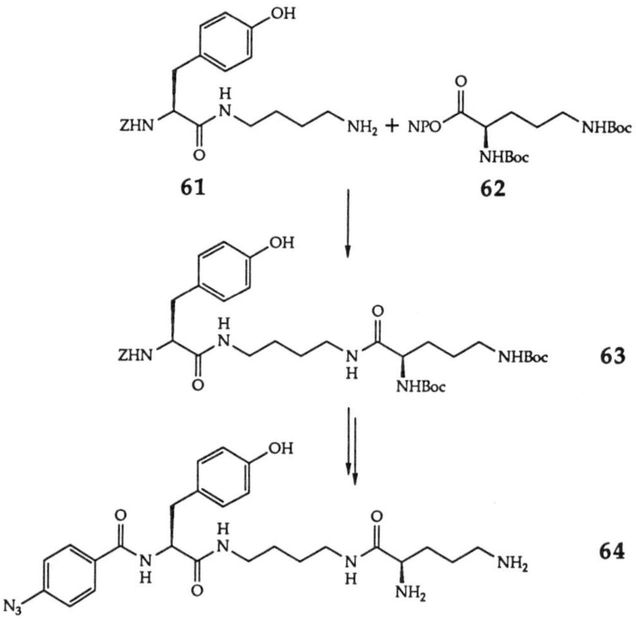

SCHEME 9. Synthesis of a tail-acylated δ-PhTX analog (*41*).

FIG. 9. Synthetic δ-PhTX analogs prepared by Nakanishi *et al.* (*37,38,40,41,44,45*) [see also toxin analogs synthesized by Piek *et al.* (*46–49*)]. Further analogs were obtained by simultaneous variation of more than one of the assigned portions above.

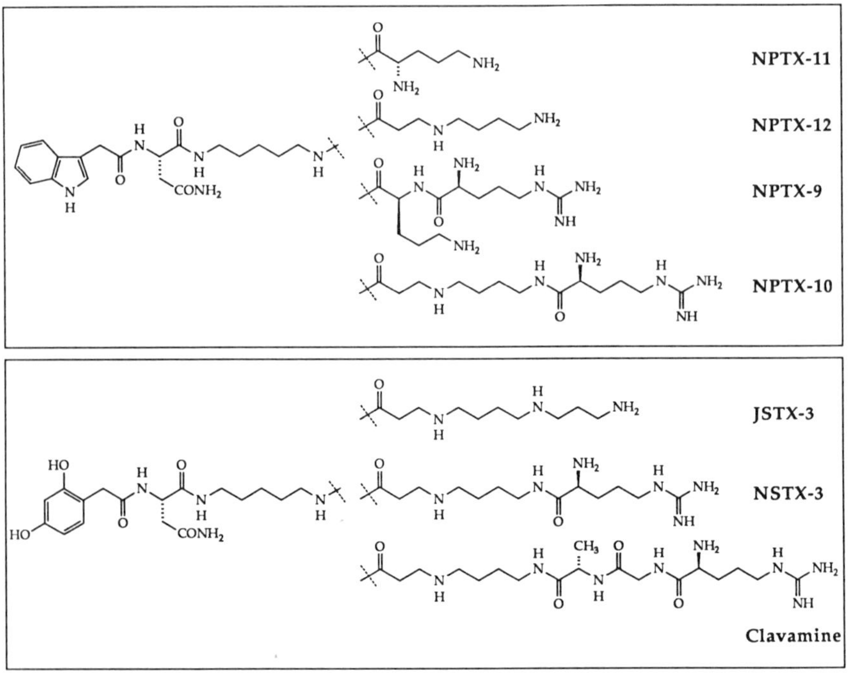

FIG. 11. Synthesized Araneidae toxins.

tively, was performed by some research groups as a first stage (Scheme 10). Miyashita *et al.* (*52,53*) started their preparation of **68** with azido-protected cadaverine **65,** which was obtained according to previously described procedures (*18*). Standard coupling with asparagine derivative **66** and activated (indol-3-yl)acetic acid **67** followed by catalytic hydrogenation of the azido group provided the desired NPTX toxin precursor **68** in 52% yield. The corresponding analog **71** was prepared similarly by Nason *et al.* (*29*). Two successive coupling reactions of Boc-PA5 (**69**) with asparagine and 2,4-dibenzyloxyphenylacetic acid *N*-succinimidyl ester (2,4-DiOBn-PhAcONSu, **70**), using standard peptide chemistry, and final removal of the Boc group by the action of TFA delivered the expected product **71** in 48% overall yield. In another approach, intermediate **71** was obtained directly from PA5 by its mono-acylation with preformed asparagine derivative **72** in an unexpectedly high 92% yield (*34,54,55*).

The syntheses of the NPTX toxins were completed employing standard protocols for peptide syntheses (Scheme 11). Reaction of **68** with activated ornithine derivative **73** provided an α,ω-acylated cadaverine intermediate (89% yield), which delivered, on complete deprotection (TFA followed by

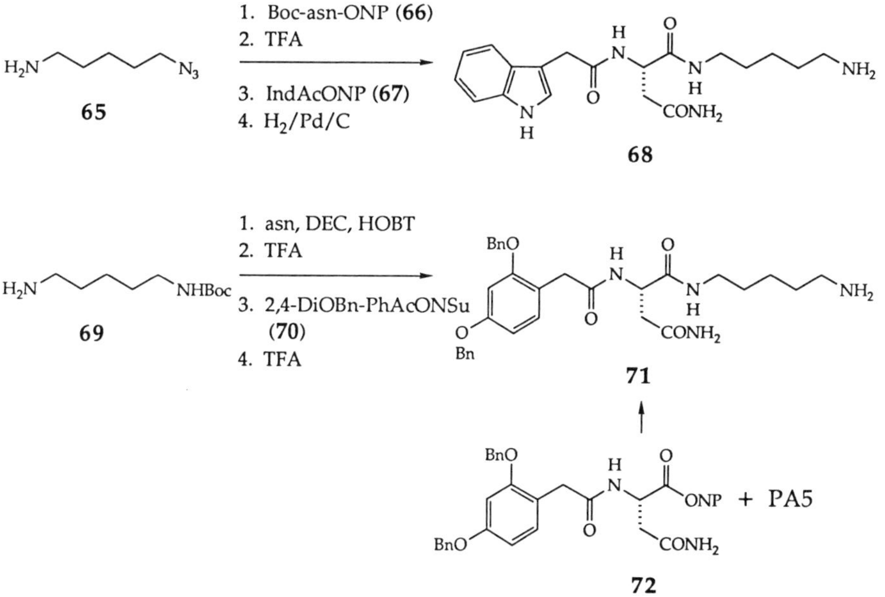

SCHEME 10. Syntheses of the relevant *Nephila* toxin head portions **68** (*52,53*) and **71** (*29,34,54,55*).

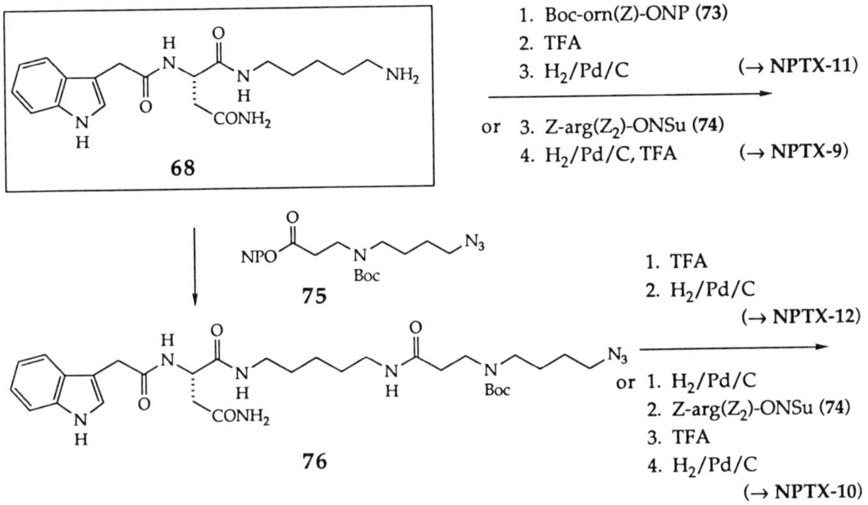

SCHEME 11. Completion of the syntheses of NPTX toxins according to Miyashita *et al.* (*52,53*) (for toxin structures, see Fig. 11).

H$_2$/Pd/C), the first required toxin NPTX-11 in 51% yield (*52,53*). Partial deprotection of the ornithine coupling intermediate by treatment with TFA alone and subsequent coupling of the released amino function with the arginine derivative **74** gave, after simultaneous hydrogenolytic removal of the Z and Bn groups, the second toxin NPTX-9. The remaining NPTX toxins, NPTX-10 and NPTX-12, were attained similarly. The versatile precursor **68** reacted with activated acid **75** to afford the ω-azido compound **76** (76% yield), which was either directly deprotected and hydrogenated to produce a compound identical to natural NPTX-12 (55% yield) or was further elongated by acylation of the hydrogenation product with activated ester **74** leading, after splitting off the protective groups, to synthetic toxin NPTX-10 in 75% yield.

Equally simple as the formation of the NPTX toxins described above was the preparation of JSTX-3 and NSTX-3 beginning with **71** as the

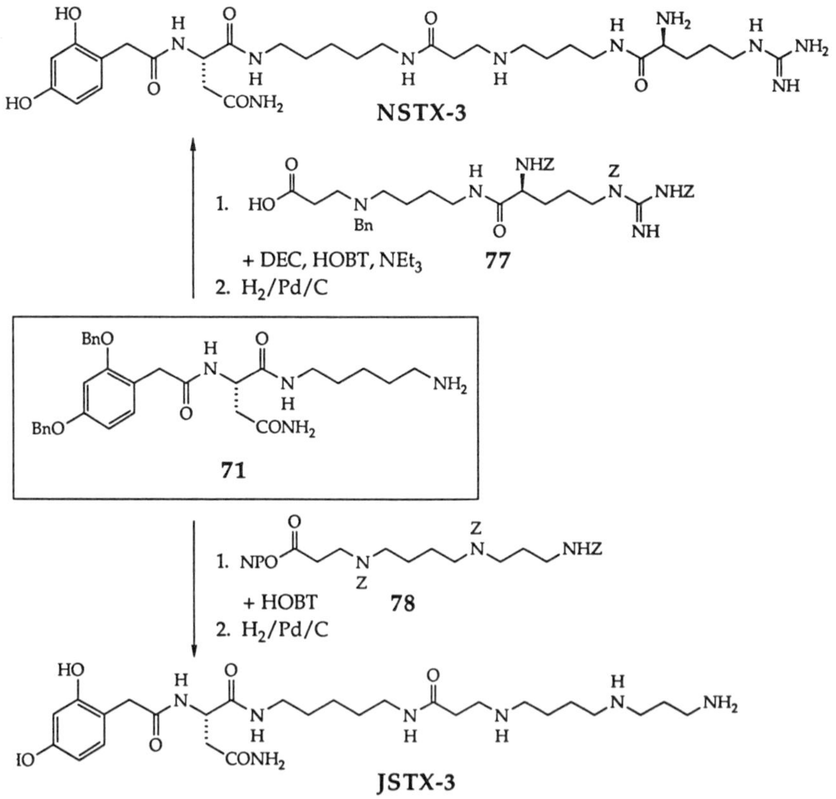

SCHEME 12. Completion of the syntheses of NSTX-3 and JSTX-3 according to Nason *et al.* (*29*) and Hashimoto *et al.* (*34,54,55*), respectively.

starting material (Scheme 12). Acylation of the primary amino group of **71** with putreanine derivative **77** (*29*) or ω-(3-aminopropyl)putreanine derivative **78** (*34,54,55*), followed by catalytic hydrogenation, produced NSTX-3 and JSTX-3, respectively, in 29 and 46% yields.

The unusual acid moieties **75, 77,** and **78** (Schemes 11 and 12) utilized in the coupling reaction with **68** and **71** were prepared according to procedures outlined in Scheme 13. The putreanine portion **75** was attained starting with 4-aminobutanol **79** (*53*). Michael addition of the amine to methyl acrylate, followed by Boc protection of the resulting secondary amine, yielded an alcohol intermediate, which was further transformed into the corresponding azide **80** by the consecutive action of methanesulfonyl chloride (MesCl) in the presence of base and NaN_3 (71% yield). Saponification of the ester and treatment of the corresponding acid with NPOH in the presence of DCC furnished the desired product **75** in 78% yield.

Reduction of the azide group in compound **80** and coupling with a protected arginine moiety should lead to **77**. Nason *et al.*, however, chose a different path for the synthesis of **77** (*29*). Intermediate nitrile **82** was constructed by reaction of 3-benzylamino propanenitrile **81** (obtained from benzylamine and acrylonitrile by Michael addition) in a typical manner with *N*-(4-bromobutyl)phthalimide (**15**) in the presence of KF. Acid treatment, resulting in simultaneous hydrolysis of the nitrile and the phthalimide groups, provided an ω-amino acid, which was coupled with an arginine portion **74** to yield the desired acyl coupling component **77**. The

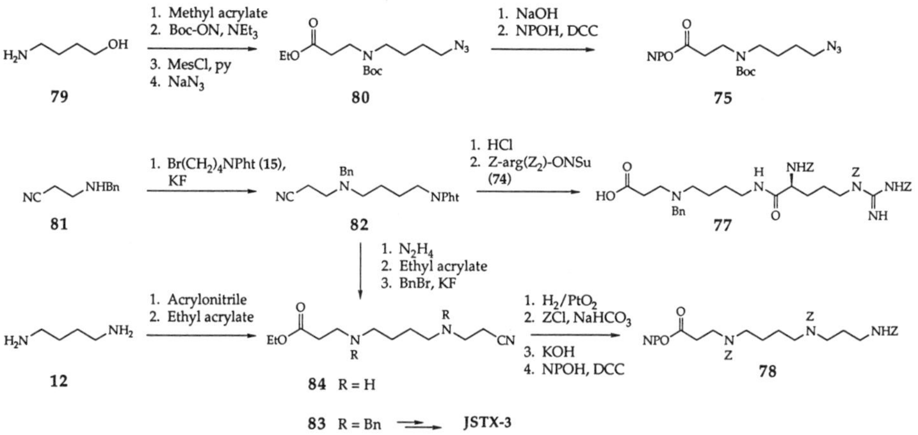

SCHEME 13. Preparation of the unusual amino acid moieties **75** (*53*), **77** (*29*), and **78** (*34,54,55*).

authors utilized nitrile **82** in an innovative way for the construction of an ω-(3-aminopropyl)putreanine (PA-Pta) moiety as well. Selective removal of the phthalimide protecting group by hydrazinolysis and Michael addition of the resulting amine to ethyl acrylate provided ester **83.** However, this product was not directly connected in a convergent way with **71** (cf. Scheme 12) to yield, in a few steps, JSTX-3, but rather was stepwise elongated at the ester end with a PA5, an asparagine, and a 2,4-DiOH-PhAc moiety (*29*) (this synthesis of JSTX-3 is not separately outlined). The ω-(3-aminopropyl)putreanine moiety **78,** which was actually used in the coupling reaction mentioned above, was synthesized similarly to compound **83** via the α,ω-cyanoester **84.** Mono-cyanoethylation of putrescine (PA4, **12**), according to a procedure published by Yamamoto and Maruoka (*23*), followed by Michael addition of the resulting product to ethyl acrylate, gave rise to intermediate **84** in reasonable yield. Catalytic hydrogenation of the cyano group, protection of the free amines with Z groups, and subsequent activation of the carboxylic acid as the NP ester finally furnished the desired derivative **78** in 14% overall yield (from PA4).

The putreanine portion **87** employed by Teshima *et al.* in the syntheses of NSTX-3 (*56–58*) and clavamine (*59*) was built up with a sparsely utilized reductive amination step (Scheme 14). Conversion of the phthalimido aldehyde **86,** obtained in two steps from aminoacetal **85** by protection of the amine and hydrolysis of the acetal (95% yield), with β-alanine in the presence of NaCNBH$_3$, followed by reaction with ZCl, afforded the putreanine derivative **87** (71% yield). This product was coupled with the PA5 unit **69,** the phthalimido protective group at the terminal amino function of the putreanine subunit was exchanged for a Troc group, and the Boc group at the PA5 unit was removed by hydrolysis. The intermediate **88** obtained this way was then extended stepwise at the head end with the required acid portions. The action of zinc/acetic acid to remove selectively the introduced Troc group afforded compound **89** (31% yield), the common intermediate in the preparation of both NSTX-3 and clavamine. When the free amino group of **89** was acylated with protected arginine, synthetic NSTX-3 was obtained in 43% yield after release of the hitherto blocked hydroxyl and amino functions by catalytic hydrogenation. On the other hand, stepwise attachment of an alanine, a glycine, and an arginine moiety to intermediate **89** provided, after removal of the protecting groups, clavamine (14% yield).

Compound **89** was not only used as an intermediate for the preparation of the two natural toxins mentioned above. It was also employed as a precursor of *Nephila* toxin analogs (*60*) (Fig. 12). Direct catalytic hydrogenation of **89** (see Scheme 14 for structure), or preliminary acylation of the terminal amino group with several basic, acidic, or neutral groups, de-

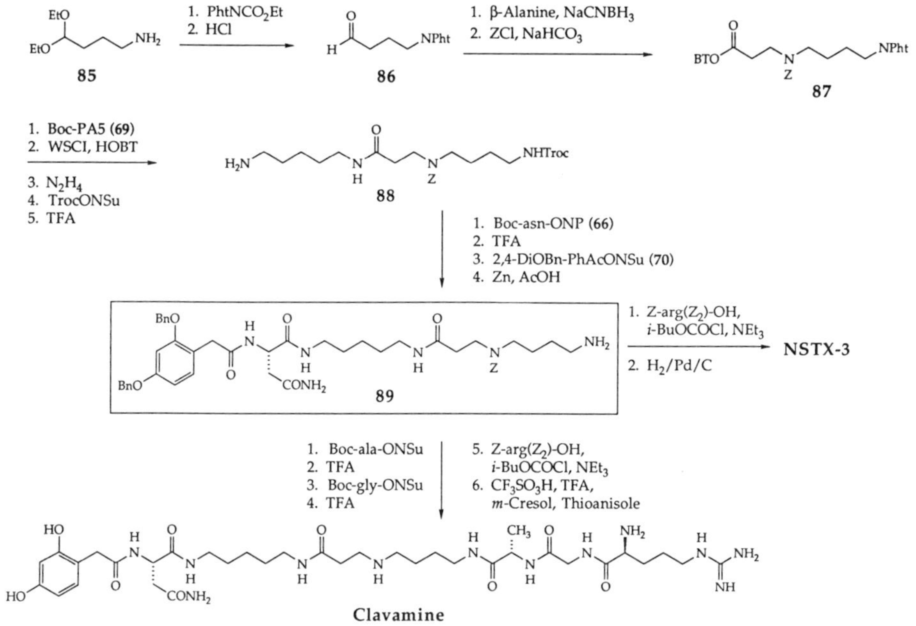

SCHEME 14. Syntheses of NSTX-3 (*56–58*) and clavamine (*59*) according to Teshima *et al.*

livered the putreanine-containing toxin analogs presented in Fig. 12. A number of toxin analogs possessing the 2,4-DiOH-PhAc head moiety were also similarly prepared by Hashimoto and collaborators starting from intermediate **71** or **72** (see Scheme 10 for structures) (*34,54,55*). In this case, the terminal amino groups were not acylated, which is the reason why these compounds are termed analogs of JSTX-3.

More systematic variations of the subunits contained in the original toxins were performed by Goto and Nakajima (*61*), who secured a patent for the production and use of amide compounds of the general structure **90**, also given in Fig. 12. The syntheses of these compounds do not differ in principle from previously described preparations of similar materials.

Synthetic access to NPTX-type toxin analogs has been investigated by our group in Zürich (*62*). The 10 compounds delineated in Fig. 13 were prepared using standard protocols for peptide syntheses. Of special note are the structures **92** and **93,** which differ in the sequence or installation mode of the subunits, with compound **92** having the linking α-amino acid unit located between the tail amino group of the polyazaalkane moiety and the terminal lysine modification and **93** containing an alanine portion, which is inserted between the lipophilic head and the polyamine backbone in a reversed manner, thus presupposing an ω-amino acid and an arylal-

ANDREA SCHÄFER *ET AL.*

89

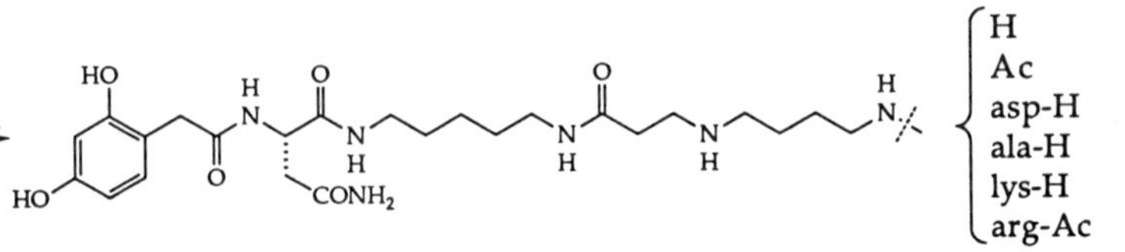

71

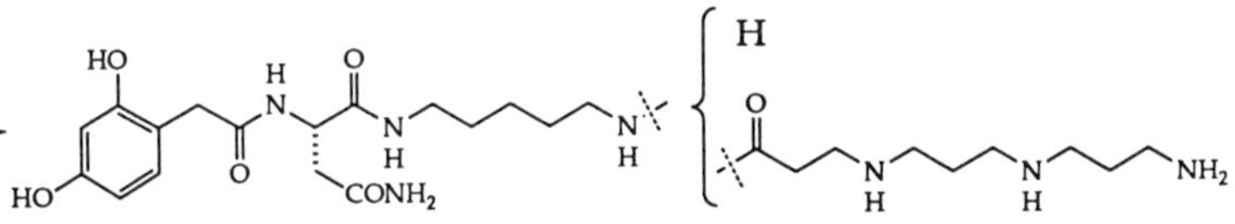

72

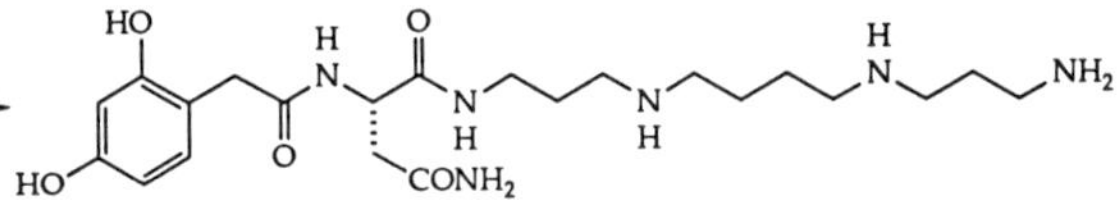

90

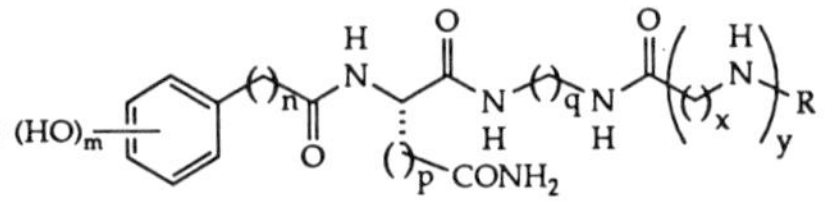

m = 1 - 3   n = 1 - 4
p = 1 or 2   q = 1 - 6
x = 2 - 6   y = 1 - 3

FIG. 12. Analogs of toxins NSTX-3 and JSTX-3 (*34,54,55,60,61*).

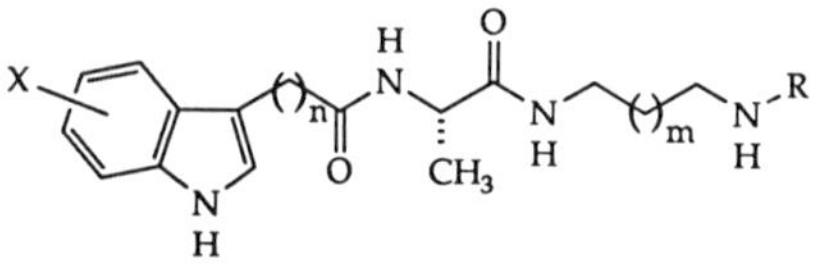

91a-h

n	m	X	R
1	1	H	Ac
1	4	H	Ac
1	4	H	lys
1	4	4-OH	Ac
1	4	4-OH	lys
1	4	5-OH	Ac
1	4	5-OH	lys
2	4	H	lys

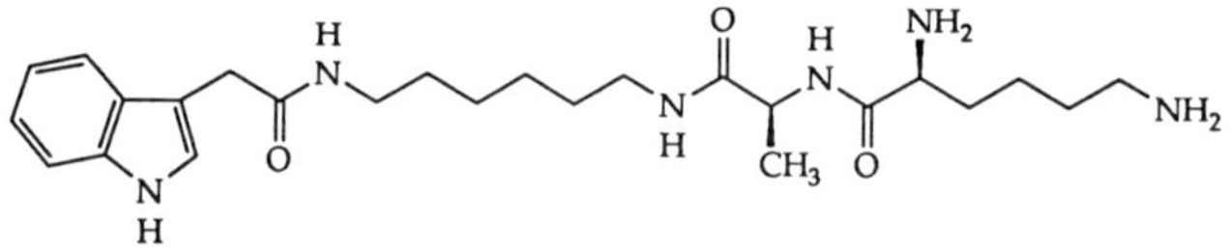

92

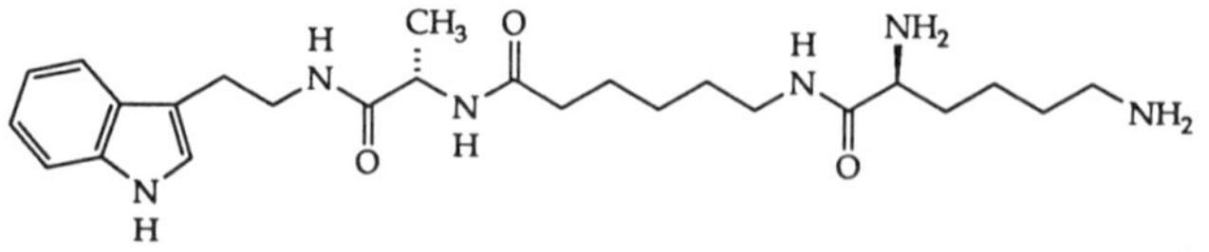

93

FIG. 13. Analogs of the NPTX toxin type (*62*).

kylamino unit as the polyazaalkane and the aromatic head portions, respectively. With these structural features, compound **93** resembles β-PhTX, the synthesis of which is described in a later section.

3. Syntheses of Argiope Toxins and Related Analogs

Until now, only the α,ω-bis-acylation of diamines has been discussed in detail. We have not yet looked into the preparation of head and tail modified spider toxins possessing more complex polyamine portions. Selective derivatization of the two terminal amino groups of such polyamines, however, is not as laborious as might be anticipated. We have already described how chemoselective coupling of the free terminal amino group of an α,ω-diaminopolyazaalkane moiety with mild acylating agents can be performed with satisfactory yields without protection of the internal amino functionalities. The blocking group, preventing the opposite amino function from reacting, might be a classic protecting group as discussed earlier or another acyl group, for example, a group directly involved as a part of a given target molecule. This means essentially that toxin precursors of the aforementioned δ-PhTX type (see Section III,B,2) are ready to be elongated at the opposite end using standard peptide synthesis procedures, and that therefore the same procedures as exploited for the construction of simpler toxins can also be used for the erection of α,ω-modified polyamines. Because free secondary amino groups always react to some extent with acylating reagents during coupling reactions of solely terminal mono-protected polyazaalkanes, it might still be advisable to block the internal amino functions of such portions before performing coupling reactions, in order to increase yields. Both paths, with or without protection of the internal amines prior to the coupling reaction, have been followed in four independent syntheses of spider toxin Arg 636, which is of the α,ω-bis-acylated polyamine type.

Shih *et al.* (*63*) started their synthesis of Arg 636 with the tris-Boc-protected polyamine **94** (Scheme 15), which was obtained by two cyanoethylation/Boc protection/reduction sequences beginning with the mono-Boc-protected cadaverine (Boc-PA5), as described earlier (see Scheme 5). These authors omitted blocking the internal amino groups independently from the terminal amino function, which would have been an easy task. Their assumption that a primary amine would react with enough preference over an internal secondary amine with mildly activated carboxylic acids was in this case proved true, as can be recognized later in the synthesis. Prior to that, the free amine of **94** was coupled with the activated tris-Z-protected arginine **95** to give the tail-modified polyamine portion **96** in excellent (95%) yield. All Boc groups were then removed by treatment with TFA to give **97,** which, in fact, delivered in the crucial step

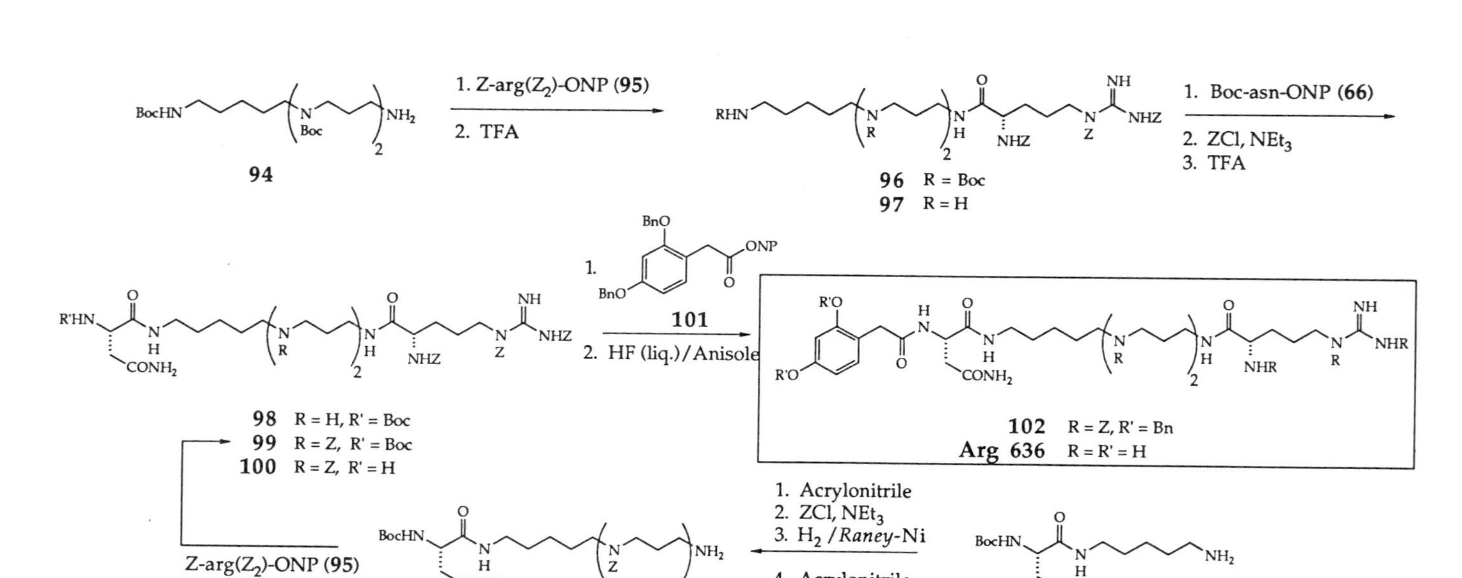

SCHEME 15. Syntheses of Arg 636 by Shih *et al.* (63).

by treatment with asparagine derivative **66** the coupling product **98** in high yield (>74%). The final transformation of **98** to Arg 636 was performed in a few steps using standard peptide chemistry. The secondary amino groups of the polyazaalkane chain were Z-protected (**99**), the primary amino functionality of the asparagine moiety was released, and the intermediate **100** obtained in this way was coupled with activated carboxylic acid **101** to afford the fully protected toxin precursor **102** in 59% overall yield (from **97**, four steps). Finally, toxin Arg 636 was obtained by removal of all the benzyl functions, which was according to the authors best achieved in liquid HF containing approximately 10% anisole.

The same authors synthesized intermediate **99** in 15% overall yield according to a different scheme (*63*). They assembled compound **99** from the asparagine end starting with cadaverine derivative **103** (Scheme 15). In contrast to the procedure discussed above, the optimal protected polyamine intermediate **104** prepared over several steps, was utilized as the coupling partner of **95,** also affording **99** in good yield. It is interesting that in the preparation of the partially protected building block **104,** this product was obtained following cyanoethylation/Z-protection/reduction sequences using catalytic hydrogenation processes (in the presence of sponge Raney nickel) for transformation of the cyano group to the primary amino function. The success of this route depended strongly on the resistance of the Z groups toward the reductive conditions, which, although anticipated by the authors, is rather surprising.

Another synthesis of Arg 636, which is, particularly in the final assemblage of the toxin, similar to the synthesis described above in the first part of Scheme 15, was published by Yelin *et al.* (Scheme 16) (*64*). Despite the analogy to the method of Shih *et al.*, this synthesis is nevertheless, interesting, because of the construction in the first stage of the mono-triphenylmethyl (trityl, Tr)-protected PA533 unit **109,** using a method differing from all previously described protocols to obtain comparable compounds. The polyamine portion was built up starting with the trityl-protected β-alanine derivative **105.** This ester was saponified and treated with β-alanine methyl ester (**106**) in presence of DCC and *N*-hydroxysuccinimide (HONSu) to furnish diazaalkane derivative **107** in 68% yield. Subjection of **107** to excess cadaverine (PA5) afforded diamide **108** (70% yield), which was reduced with $LiAlH_4$ to the mono-protected tetramine **109** in 47% yield. Except for the low yield at the final step, such a sequence would present a remarkable alternative to other methods.

Polyamine portion **109** was coupled without further protection of the internal amino groups on the head side with asparagine derivative **110** to furnish compound **111** (39% yield). This intermediate afforded the target

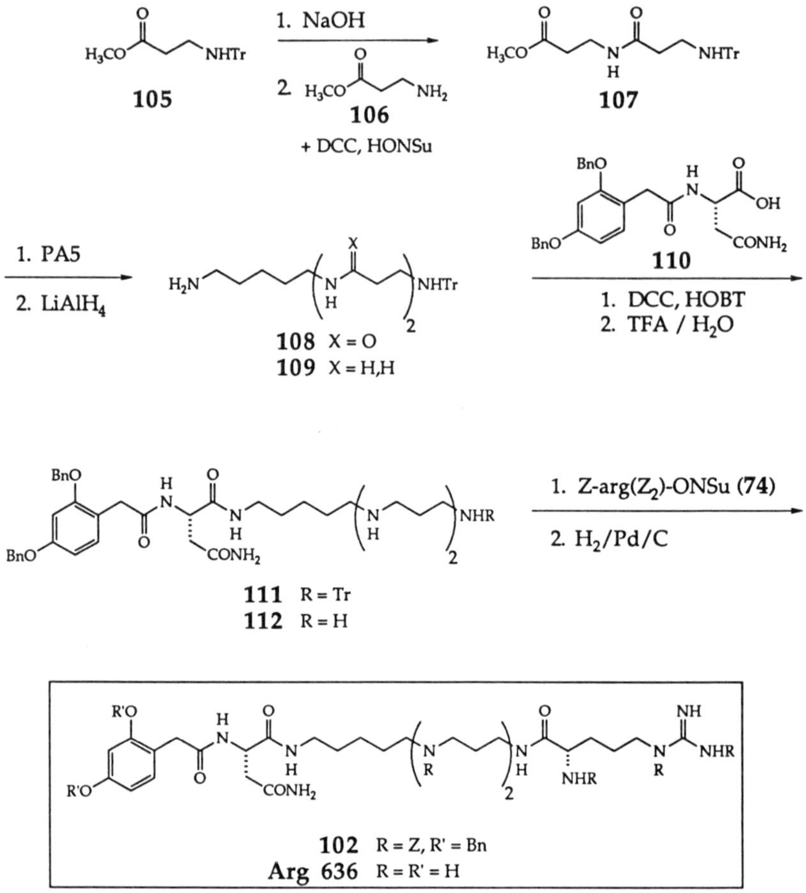

SCHEME 16. Synthesis of Arg 636 by Yelin *et al.* (*64*).

molecule Arg 636 after hydrolytic removal of the trityl group, coupling of the resulting primary amine **112** with Z-protected arginine *N*-succinimidyl (NSu) ester **74,** and complete deprotection of **102** using H_2/Pd/C in TFA/methanol solution (overall yield of 72%).

Two more groups synthesized Arg 636, together with the related toxin Arg 659, which differs from Arg 636 only in the aromatic acyl portion. Both groups used polyamine building blocks that were protected at the internal amines, in contrast to the intermediates employed in the syntheses discussed before (Scheme 17). Whereas Adams *et al.* (*65*) started directly with mono-Boc-bis-diphenylmethyl-protected **113**, without mentioning its source, Jasys *et al.* (*28*) first described the synthesis of the corresponding mono-Boc-bis-Bn derivative **114**. This compound was obtained from

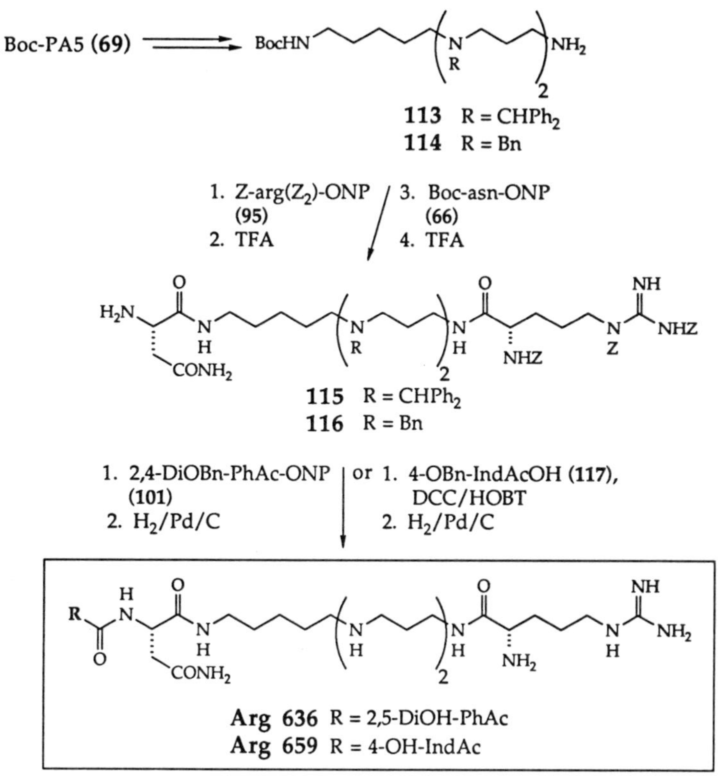

Boc-PA5 (**69**) $\Longrightarrow$ BocHN ... N(R) ... NH$_2$

113 R = CHPh$_2$
114 R = Bn

1. Z-arg(Z$_2$)-ONP (**95**)
2. TFA
3. Boc-asn-ONP (**66**)
4. TFA

115 R = CHPh$_2$
116 R = Bn

1. 2,4-DiOBn-PhAc-ONP (**101**)
2. H$_2$/Pd/C

or

1. 4-OBn-IndAcOH (**117**), DCC/HOBT
2. H$_2$/Pd/C

Arg 636 R = 2,5-DiOH-PhAc
Arg 659 R = 4-OH-IndAc

SCHEME 17. Syntheses of Arg 636 and Arg 659 by Adams *et al.* (*65*) and Jasys *et al.* (*28*).

mono-Boc-cadaverine (Boc-PA5, **69**) following two alkylation/protection/ deprotection sequences analogous to those discussed earlier (see syntheses of spermine moieties, Scheme 3, or synthesis of Agel 489a, Scheme 7). The benzyl protective groups were introduced in good yield by reductive amination of benzaldehyde with the internal secondary amines of the polyamine precursors.

The syntheses of Arg 636 were completed by both groups, beginning with building block **113,** as well as proceeding with its analog **114,** in the same way, differing only in small experimental details. First, the arginine moiety was attached to the tail end of the polyamine portion; then, after releasing the opposite primary amino group, the head parts were added stepwise. Intermediate **115** or **116** was coupled in the last assembling step not only with **101,** but also with **117,** delivering, after catalytic hydrogenation, in addition to Arg 636 a compound identical with natural toxin Arg 659.

The Araneidae toxin Arg 673, another compound related to Arg 636, was prepared by Jasys *et al.* (*28*) employing similar reactions as already shown (Scheme 18). However, the entire polyamine portion of this compound was not synthesized to start. In the first stage, only one of the two required aminopropyl portions was attached to the mono-Boc-protected *N,N'*-dimethylputrescine (**118,** obtained in 48% yield from *N,N'*-dimethylputrescine). Reaction of compound **118** with *N*-(3-bromopropyl)phthalimide (**22**), followed by hydrazinolysis, delivered the triamine **119** in 51% yield. This intermediate was acylated with activated tris-Z-protected arginine **74** to furnish, after splitting off the Boc protective group, derivative **120** (85% yield), having the complete tail portion of Arg 673. Reaction of the released amine with *N*-Boc-3-bromopropylamine in the presence of base finished the construction of the desired polyamine backbone. Compound **121** ob-

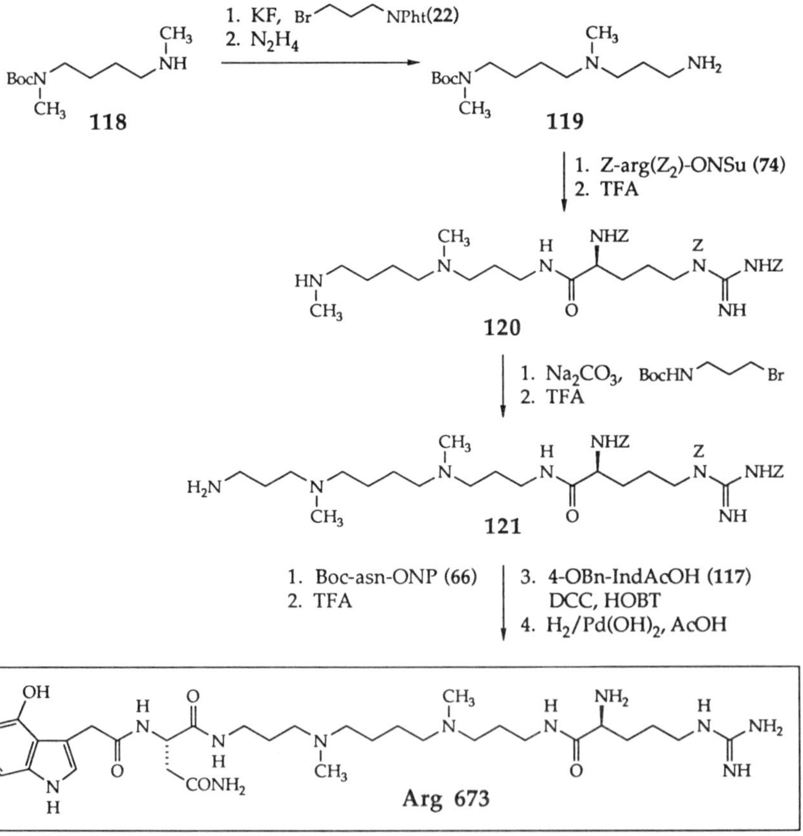

SCHEME 18. Synthesis of Arg 673 according to Jasys *et al.* (*28*).

tained this way yielded, after stepwise coupling with **66** and **117** using standard peptide synthesis protocols, and after concurrent hydrogenolytic removal of the Z and Bn groups, synthetic spider toxin Arg 673.

At this stage of the discussion, we wish to append the description of the synthesis of the acid **122** used for the preparation of Arg 636 (Scheme 19). Like the syntheses of the other aromatic acyl portions discussed previously, the preparation of **122** is not very dramatic. Three approaches targeting this compound have been described. The first approach started with bis-Bn-protected resorcinol **123,** which was formylated by treatment with the Vilsmeyer reagent (70% yield) (*64*). The aldehyde **124** was then converted to acid **122** in two steps, which were not further explained (47% yield). Probably an Arndt–Eistert procedure was followed as in the second approach, where β-resorcylic acid was the starting material (*63*). Benzylation of the phenolic hydroxyl functions afforded acid **125,** which was treated successively with oxalyl chloride, diazomethane, and silver benzoate to provide the desired product **122** in 57% overall yield. In the last approach, resorcinol was mono-allylated to **126** (50% yield) and gave, after Claisen rearrangement and treatment of the deprotonated intermediate with BnBr bis-protected allylresorcinol, compound **127,** in 41% yield (*56–58*). Oxidative degradation of the olefin also provided **122.**

A number of *Argiope* toxin analogs possessing a 1-naphthylacetyl (NphAc) group as the head acyl moiety have been prepared by Toki *et al.* (*66*). Most interesting are the molecules of this type outlined in Scheme 20 that contain as part of their structures a macrocyclic polyazaalkane

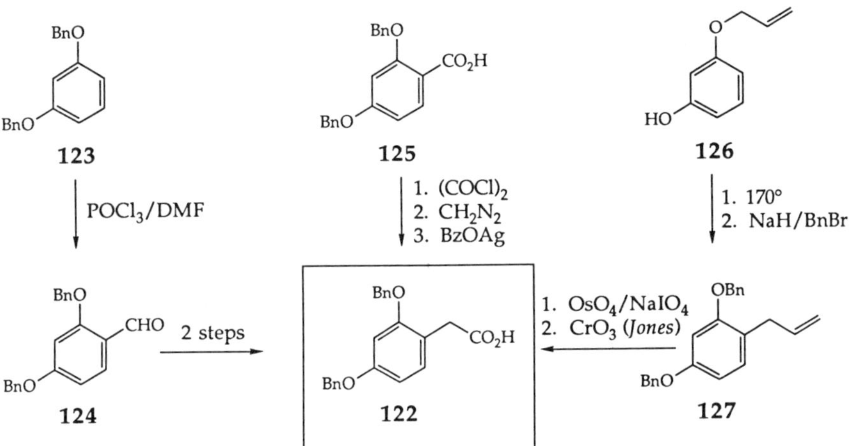

SCHEME 19. Synthesis of acyl coupling component **122** (*56,57,60,63,64*).

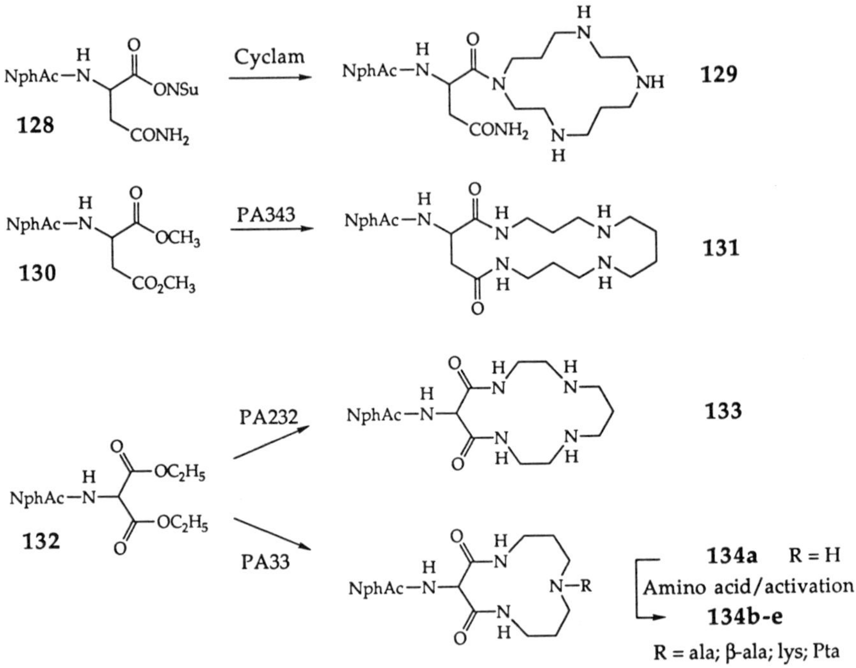

SCHEME 20. Toxin analogs possessing macrocyclic polyazaalkane portions synthesized by Tabushi *et al.* (*67*).

portion instead of the linear polyamine backbone. These compounds are easily obtained from simple precursors. Reaction of asparagine derivative **128** with cyclam (1,4,8,11-tetraazacyclotetradecane) afforded **129** in a well-known manner. The other compounds were synthesized according to procedures published by Tabushi *et al.* (*67*). The *N*-NphAc-derivatized precursors **130** and **132** were treated with PA343, PA232, or PA33 to deliver, in low yield, the corresponding macrocyclic products **131**, **133**, and **134a**, respectively. Acylation of the free secondary amine of **134a** with different amino acids provided the remaining toxin analogs **134b–e.**

C. β-PhTX AND RELATED ANALOGS

In contrast to all other spider and wasp toxins found so far, only the toxin β-PhTX possesses an unusual ω-amino acid and no real polyamine subunit as the polyazaalkane backbone. Consequently, problematic chemoselective mono-derivatization of polyamine precursors is unnecessary for its preparation, which simplified the synthetic access of this type of compounds enormously. The naturally occurring β-PhTX was obtained twice by Karst *et al.* (*68*) by different synthetic routes as outlined in Scheme 21. In the first approach, the 5-(3-aminopropylamino)valeric acid derivative **136** was prepared from 5-aminovaleric acid (**135**) by a cyanoethylation/reduction/protection sequence, as in earlier syntheses. The intermediate **136** was then coupled with pentylamine in the presence of DCC, and the resulting product was deprotected by the action of TFA to afford the desired toxin β-PhTX, although in low and not precisely quantified yields. The second approach provided compound β-PhTX in "much better" yields (the exact yields were also not given for this procedure). Reductive amination of 2,4-dimethoxybenzaldehyde (**137**) with pentylamine and reaction of the obtained secondary amine with ω-bromovaleric acid chloride furnished the amides **138** as a 1:1 mixture. The halogen atoms of the intermediates were substituted by an aminopropyl moiety (reaction with **139**), affording a fully protected toxin precursor. Simultaneous removal of the Boc and 2,4-dimethoxybenzyl groups by treatment with TFA, followed by splitting off the *N*-benzyl group by hydrogenolysis, yielded β-PhTX.

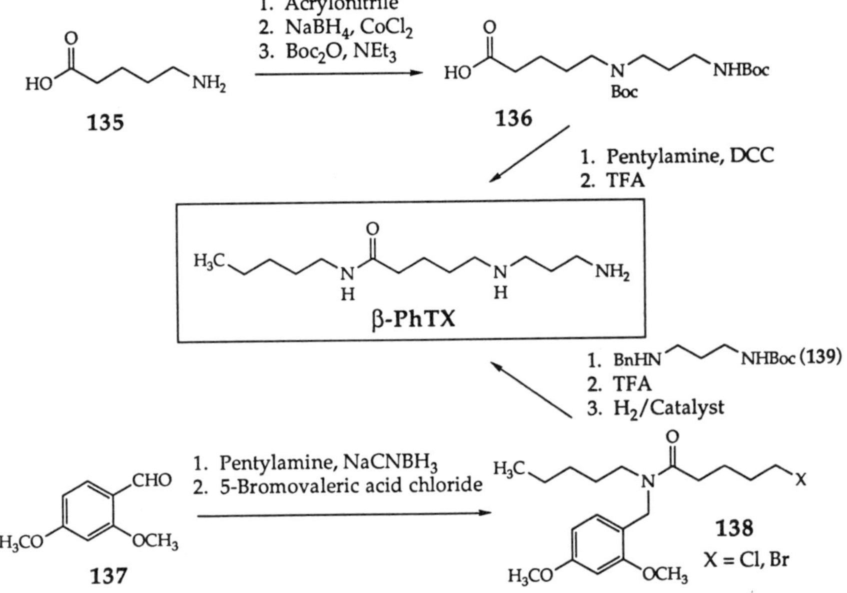

Scheme 21. Syntheses of β-PhTX by Karst *et al.* (*68*).

Until now, little effort has been made to synthesize toxin analogs of the β-PhTX type. However, our group in Zürich has prepared some compounds of this class exhibiting structural resemblance not only to β-PhTX, but, at least for some examples, to Araneidae toxins as well (Scheme 22) (*21,69*). The N-(3-aminopropyl)-β-alanine-derived compounds were prepared employing well-known protocols for peptide synthesis, as shown in Scheme 22. Bis-protected diamino acid **140** was activated and coupled, either with a simple alkylamine or with an alkylamide of alanine, lysine, or glutamine, to provide the respective toxin analogs in approximately 50–60% yield after deprotection of the hitherto blocked amino functionalities.

Compound **93,** an α,ω-bis-derivatized β-PhTX toxin analog shown earlier in connection with NPTX toxin analogs (see Fig. 13), was synthesized similarly to the N-(3-aminopropyl)-β-alanine-derived compounds above (*62*) (Scheme 23). The ω-amino acid **141** was initially coupled with bis-Z-protected lysine, then activated by treatment with 4-nitrophenyl trifluoroacetate to provide ester **142** (30% yield). This reactive intermediate was attached to N-alanyltryptamine (**143**), and the obtained product was deprotected by hydrogenolysis. Compound **93** was prepared by this convergent procedure in 22% overall yield, starting from amino acid **141.**

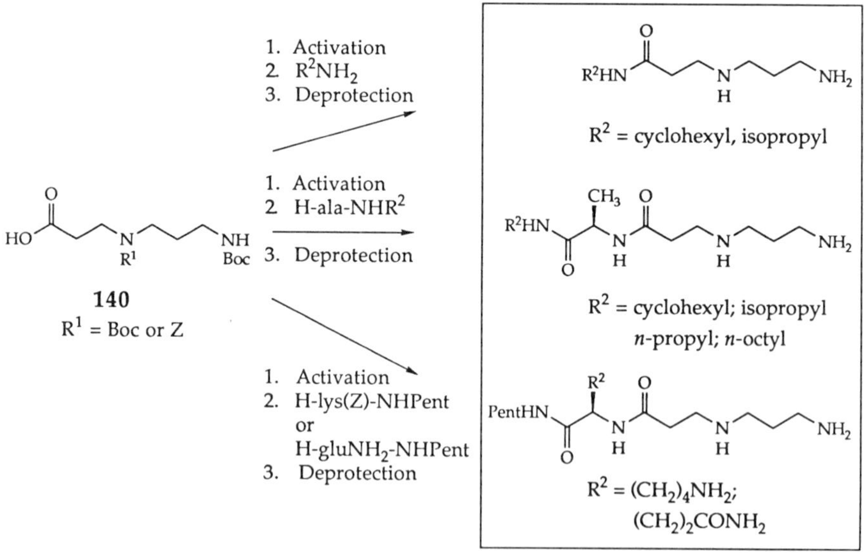

SCHEME 22. Syntheses of β-PhTX toxin analogs (*21,69*).

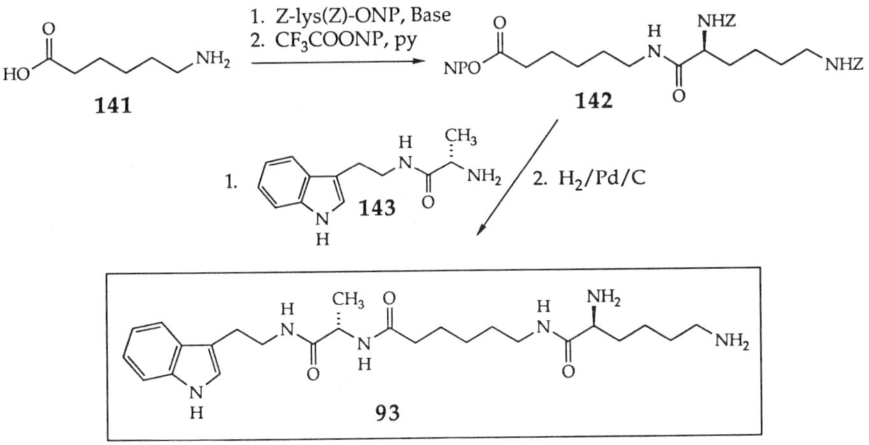

SCHEME 23. Synthesis of an α,ω-bis-derivatized β-PhTX toxin analog (*62*).

IV. Isolation and Purification of Spider and Wasp Toxins

A. ISOLATION

The difficulty of obtaining significant amounts of spider or wasp toxins is a major reason that favored interest in snake venoms over spider, wasp, scorpion, or ant venoms. Isolation, purification, and characterization of small amounts of biologically active compounds require particular and specialized strategies for success. A variety of methods first had to be developed. Many of the early studies on spider toxins were carried out with tarantulas (Othognata/Mygalomorphae, Theraphosidae) or black widow spiders (Labidognatha/Araneomorphae, *Latrodectus* species). Thus, the development and evaluation of methods discussed in the literature often concern these types of spiders. They represent some of the better known families, but many uncertainties and unanswered questions remain concerning them. Research on low molecular weight toxins began with tarantula spiders (*70*). There are no low molecular weight polyamine spider toxins known from widow spiders. However, the general difficulties involved in the isolation and purification of venom components are always the same, and therefore some examples referring to widow spiders are also cited.

A complication when dealing with spider venoms is the true identity of the spiders. Spider taxonomy is sometimes not standardized, and the determination of species is often not an easy task. For the widow spiders

(*Latrodectus* sp.) it has been stated that all widow venoms have similar biological effects (*71*), and, therefore, the venoms are often referred to as just "black widow spider venom," although there seem to be some differences at least in the venom composition between the Mediterranean black widow (*Latrodectus mactans mactans*) and the North American black widow (*Latrodectus mactans tredecimguttatus*) (*72*). The point is that the phrase "black widow spider venom" is not exact. As long as studies on the venom of spiders of all widow species are incomplete, and the venoms are not proved to be identical in composition, one must state exactly the species from which the venom discussed was obtained (*73*). The same criteria are applicable to other spider families, for example, the ones discussed in this chapter.

Another question is whether there is a difference between the venom of male and female spiders of the same species. In 1936 D'Amour *et al.* (*74*) reported that male black widow venom is nontoxic. The extract of twelve pairs of venom glands from male spiders, when injected into a rat, did not evoke any evident alteration of behavior (*75*). From then on only the venom from female black widow spiders has been studied. This was probably also due to the larger size of the female specimen. A different situation is reported for the venom of the Sydney funnel-web spider *Atrax robustus*. This time the male venom was established as lethal to humans (*76*), but here the venom of the females was also usually studied, owing to the more ready availability of females of the species. In fact, for many spiders males of the species are not easily identifiable until fairly late in life. Thus, the question arises whether the reported comparison between venoms from male and female specimens is actually a comparison between male and a combination of female and (male) late nymph spiders (*73*).

Within the same species of spiders as well as snakes, the venom can be affected by feeding, seasonal variations, regional variations, and maturation. For most spiders, the relevant data are not available or are very incomplete. Contributions from several factors towards venom potency were tested for the funnel-web spider *Atrax infensus* (*77*). It could be shown that the potency of the venom of adult female *Atrax infensus* increases as the spiders pass from the winter hibernated state to the active early summer state, and then decreases in midsummer. Over a period of 15 days, the potency of the venom of unfed spiders increased, while that of fed spiders decreased. This fact may explain the rise in venom potency with the commencement of summer, which is followed by a decline in midsummer, when the opportunities to feed are normally the best. The amount of venom obtained from adult female spiders, was substantially greater than that from the corresponding immature group, but there was no difference in potency observed. The venom of adult males was found to be statistically far more potent than that of adult females (*77*).

The chemical composition of the venom is related to the methods used to obtain the venom itself. The venom toxicity may also vary according to the method applied. Bettini and Maroli (*75*) described five methods for obtaining venom from black widow spiders (*Latrodectus* sp.), which are equally applicable to other spiders: (1) extraction of whole homogenized glands, (2) piercing the gland sac with a capillary, (3) electrical milking, (4) collecting the venom at the tip of the fangs through a capillary pipette, and (5) absorbing the venom on cotton plugs. For the first two processes, the complete glands are surgically dissected from the respective spider or wasp as a first step. The glands are then either homogenized and extracted with aqueous acetonitrile, or the intact gland sacs are pierced with a fine glass capillary to obtain the venom from the gland lumen. Whereas the extraction technique often results in forcible contamination of the venom with impurities, for example, deriving from foreign tissue material, this, although still observed, is only a minor problem with the second method.

Even fewer problems with contamination arise with the "milking" methods. In a typical procedure of electrical milking (method 3), a spider is anesthetized by exposure to dry ice and placed in a restraining device under a microscope. A suction tube is placed into the mouth of the spider to prevent contamination of the venom with foreign substances from the alimentary canal, and the ejection of the venom, which is collected from the fangs by suction with small tubes [~0.2–0.5 μl per milking for, e.g., *Argiope trifasciata, A. florida,* and *Agelenopsis aperta* (*1,78*)], is stimulated by passing an alternating current across the head. Spiders treated in this fashion recover fully and can be repetitively milked at 1- to 3-week intervals. In similar procedures, spider venom is obtained by placing the distal portions of the fangs in contact with a capillary pipette, which directly draws up the venom released by the irritated spider (method 4), or the venom is absorbed on a small cotton plug (method 5). Particularly in the last case, as with electrical milking, thorough care must be taken to avoid contamination with digestive secretions.

Generally, the electrically stimulated milking techniques seem to be most useful with orthognath spiders (Mygalomorphae, tarantulas and allies). Such methods were used routinely in the 1980s to obtain pure venom from the American tarantula, *Dugesiella hentzi* (*73,79*). With smaller spiders the dissection method was still favored at that time (*73*). For example, in 1978 Bettini and Maroli (*75*) commented that milking was not the method of choice for *Latrodectus* spiders (Labidognatha), owing to contamination of the venom obtained with stomach contents. The jaws (chelicerae) of these spiders are tipped with fangs, which connect to the venom glands. Since spiders practice external digestion (*80*), one can imagine that electrical stimulation elicits not only venom release, but also secretion of digestive juices (*73*).

Preparation by extraction yields far more venom per spider than does milking. A comparison of the amount of crude venom obtained by milking and by extraction from dissected glands of the western black widow (*Latrodectus mactans hesperus*) showed that the yield by extraction was 10 times higher (*73*). The LD_{50} of the respective venom was similar for both venom preparations. The squeezing or piercing of dissected venom glands with glass capillaries requires that the venom glands are hollow sacs filled with venom. In fact, the venom glands contain secretory epithelial cells (*81,82*). When a spider bites, the muscles surrounding the glands contract and disrupt the secretory cells. Spiders can control the number of cells disrupted by the nervous control of the musculature.

The squeezing and piercing techniques obviously yield less venom than extraction procedures. In 1978 the opinion was, "The choice of methods obviously depends on the use to be made of the venom. If work is to be carried out on separated venom fractions, whole gland extracts from large numbers of spiders may be justified" (*75*). In the following years, the trend for smaller spiders has still been to dissect the venom apparatus and to prepare soluble extracts (*73*). Over the last few years, however, the milking methods have improved, and for a number of different types of spiders special milking procedures have been developed (Ref. *83* and literature cited therein). In 1990, Quicke and Usherwood (*83*) stated, "Without doubt the ideal starting material for studies of spider toxins is pure, milked venom, that can readily be obtained by teasing the chelicerae from anesthetized spiders. However, this is not always available and instead many groups have used extracts from venom glands or whole venom apparati. Nevertheless, with a little practice it is quite possible to milk spiders (even small ones) at almost the same rate as their glands could be extricated." For *Argiope* spiders a milking procedure is described that should avoid contamination of the venom with regurgitate (*83*).

Extraction of the glands delivers a crude venom that can differ markedly from the naturally excharged venom of the spider. Even the mixture obtained by milking a spider does not have exactly the same composition as the naturally expelled venom. The electrically induced stimulation of the cephalothorax muscles may unnaturally squeeze out venom fractions from the gland that would naturally be retained (*84*).

The venom obtained with one of the methods cited above is pooled (5–1000 glands or milkings) and frozen at $-80°C$ until further use. At ambient or slightly elevated temperature the venom rapidly loses biological activity. Heating the venom for a short time in a boiling water bath prevents such an alteration, indicating that the toxic components of the mixture are in fact heat-stable compounds, which decompose only in the presence of an active protease or peptidase that contaminates the raw venom (85). For instant use and when checking for activity, glands or venoms are homogenized and diluted in physiological saline (65).

Two different methods are used in general to obtain crude venom from wasps, a small-scale preparation (method a) and a larger scale preparation (method b) (86): (a) Dissection and extraction of the venom reservoirs yields about 16–26 mg per 100 venom reservoirs. (b) The wasps are shaken carefully with liquid nitrogen in a vessel with a perforated cover. Because the wasps become brittle at −192°C, the abdomens will float on the nitrogen and can easily be separated from the rest of the body parts. The abdomens are then homogenized and extracted. This procedure yields about 8 g of crude venom per 2500 abdomens. From the gland of a single solitary digger wasp (*Philanthus triangulum*) about 1.1 μg of the pure toxin δ-PhTX can be obtained (40).

For the investigation of spider toxins it might be important to know how, and under what conditions, a particular venom and its toxins were obtained. As indicated above, besides the method of isolation, feeding, seasonal, and regional factors are important for the quality of the venom. In addition, the venom composition can differ in male, female, and immature arthropods. Unfortunately, it is not possible to survey the significance of all these factors on the composition of the venom, particularly on the content and distribution of polyamine toxin components, for all investigated spiders and wasps in an unequivocal manner; research in the various laboratories differs too much with respect to tools and goals.

B. PURIFICATION

The basic structural elements of polyamine spider and wasp toxins are very similar. Purification and structure elucidation or identification of the different components are therefore not easy tasks. Considering only the polyamine part of the general structure given in Fig. 1 (see Section I), there are many variations realized in nature. One can find sequence isomers, constitution isomers, homologs, and derivatives with *N*-methyl, *N*-hydroxyl, or oxo groups. This often results in minor differences in the chemical and physical properties of such compounds, which make isolation and purification, as well as definite assignment of the structure, a challenge.

In 1957 Fischer and Bohn (70) noted that the venom of seven species of so-called tarantula spiders (Theraphosidae) contain both spermine and aromatic chromophoric residues. Gilbo and Coles (*76*) subsequently found in 1964 that the venom of the female Sydney funnel-web spider *Atrax robustus* contains the polyamine spermine, associated with a phenolic carboxylic acid and γ-aminobutyric acid.

The paper strip electrophoresis and chromatography techniques employed in these early studies were dependent on the availability of large quantities of venom, which is presumably the reason why some of the largest spiders known to man were used. Some of the *Grammostola* species have a body length of up to 10 cm and a span between the first and fourth pair of legs of 35 cm. Their glands reach 1 cm and can eject 4 mg of venom (dry weight) at one bite (*70*). The average "garden variety" spider discussed in this chapter cannot be compared to these monsters. Fractionation of the venom from smaller spiders has been made possible by the subsequent development of high-performance liquid chromatography (HPLC) and gel filtration chromatography (*87*). A summary of the applications of these techniques for the isolation of polyamine toxins from both spider and wasp venoms has been published (*83*).

V. Characterization of Spider and Wasp Venoms

A. Araneidae Toxins from *Argiope* Species and *Araneus gemma*

Neuroscientific investigations on the interaction of spider and wasp venoms with transmitter receptors started several decades ago. Toxins isolated from different sources have been used for the investigation of the properties of macromolecules in nerve and muscle cell membranes. Until the mid 1980s it was generally accepted that most neurotoxins contained in spider and wasp venoms were molecules of relatively high molecular weight, for example, peptides, polypeptides, or proteins, somewhat like the proteinaceous toxins found in snake venom. The discovery of polyamine amides with molecular weights less than 1000 in certain spider and wasp venoms (*37,88,89*) produced a dramatic change in this viewpoint. Interest in the structure of these low molecular weight toxins rose especially when it was shown that these compounds are highly potent antagonists of certain transmitter receptor macromolecules (e.g., the L-glutamic acid/glutamate receptor) (*87*).

The low molecular weight spider toxins were originally thought to be polypeptides (*89*), before it became evident that they may be composed of peptides as well as polyamine moieties. Conventional chemical analysis for structure elucidation, such as amino acid sequencing, can fail entirely if the toxin is not a straightforward molecule such as simple peptide. Thus, in the case of the Araneidae toxins it has become necessary to apply more elaborate analytical procedures. Fast atom bombardment–mass spectrometry (FAB-MS) and various NMR experiments have proved to be particularly useful (*83*). The structures proposed from sophisticated NMR techniques, mass spectrometry, and other analytical methods were so unusual that it was not until the completion of the total synthesis of these compounds that these polyamines were recognized as a new family of toxins (*87*).

In 1984, the first report appeared describing the occurrence of glutamate receptor antagonists in orb-weaver spider venom with molecular weights of less than 1000 (*85,88*). Kawai *et al.* (*90*) and Abe *et al.* (*91*) had previously shown that venom from spiders of the genus *Nephila* antagonizes lobster muscle quisqualate receptors (QUIS-R). No earlier than 1984, Kawai *et al.* assumed that the active principle in this venom was a toxin with a molecular weight of about 500 (*92*). The same year, the active principles in the venoms of *Argiope trifasciata* and *Araneus gemma*, which antagonize the postjunctional glutamate receptor complex, were isolated. Concerning the structure of the active toxin Usherwood *et al.* (*88*) stated, "Preliminary attempts biochemically to characterize the toxin responsible for this antagonism indicate that it is a hydrophilic molecule of low (probably <1000 Da) molecular weight [*sic*]."

1. Toxin Arg 636

The most abundant and pharmacologically potent toxin from spiders of the genus *Argiope* is Arg 636 (refer to Section II for nomenclature). From 1985 to 1988 Arg 636 was isolated by several research groups from five different species, namely, from four *Argiope* and one *Araneus* species (*Argiope lobata, A. trifasciata, A. florida, A. aurantia,* and *Araneus gemma*) (*65,78,85,93–95*), but in none of the cases was the structure determination completed with all necessary spectroscopic or chemical analyses. The toxin Arg 636 is a good example to point out the difficulties involved in attempts to isolate, identify, and structurally elucidate a natural compound unambiguously. Although a considerable number of people were involved in its structure elucidation, still not all of the necessary data are available for its unequivocal structure determination. For example, the absolute configuration of the amino acids contained in Arg 636 has not yet been defined, which is surprising, for one would expect this to be a minor problem and obligatory for a complete structure determination.

One research group in particular (*78,85,95*), which contributed major achievements in the investigations of spider venoms and which studied the three species *Argiope trifasciata, A. florida,* and *Araneus gemma,* describes that the isolation and purification procedures for the toxin Arg 636 from the three species are identical. However, a clear statement concerning the structural identity of the toxin Arg 636 itself obtained from the three different species is never given. In this regard it is mentioned, at one point, that substances with "similar" UV characteristics were obtained from the venom of *Argiope florida, A. trifasciata,* and *Araneus gemma* (78). By using the word "similar" the authors indirectly say that from their experience the word "identical" is not justified. In the same publication (78) one can find the statement, "It appears that isomers of the major *Argiope* toxin species exist. For example two species of the mass 636 were observed, which were well separated on HPLC indicating fundamental structural differences between the molecules" (78). An artifact might have been observed here, as in the literature a second toxin of mass 636 was never mentioned again. The only clear statement that can be given concerning the comparison of the toxin Arg 636 isolated from different species is that the FAB mass spectrum shows an ion at *m/z* 637, which is interpreted in each case as the $(M + H)^+$ ion. An identical molecular weight, however, is definitely not sufficient proof for structural identity. Among the Agelenidae toxins two examples (Agel 489a and Agel 505a) are known where the signals in the mass spectra were misunderstood, leading, in these cases, initially to incorrect molecular weights and finally to incorrect structure assignments (31,96,97). The ions observed for the two compounds by FAB-MS at *m/z* 489 and 505, respectively, correspond, owing to the quaternary ammonium ion structure of the toxins, to the M^+ (22) and not, as initially interpreted, to the $(M + H)^+$ ions.* For several years this had not been recognized.

In 1990 a review article on spider toxins by Quicke and Usherwood (83), who were both involved in the structure elucidation of Arg 636 from *Argiope trifasciata, A. florida,* and *Araneus gemma,* briefly addressed the problem of the identity of the toxin(s) Arg 636 by stating that Grishin *et al.* (93) published the structure "for a related compound from the venom of *Argiope lobata.* This compound appears to be particularly similar to the 637 molecular ion (argiotoxin 636) which we have identified in the venom of *Argiope trifasciata, A. florida,* and *Araneus gemma* (98), and they may, in fact, be identical compounds."

The amount and purity of the toxin available in the different research groups varied. As a consequence, the quality and completeness of the data

---

* In Ref. 22 the structures given for Agel 488 and Agel 504 (31,96,97) were revised, and the toxins were renamed Agel 489a and Agel 505a.

relevant for a definite structural assignment vary. Some of the publications concerning the structure of Arg 636 appear within a relatively short time interval. Therefore, one might assume that the isolation and characterization in these cases was done independently.

From the known physical and chemical data of the toxins from different *Argiope* species, it is most likely that the toxins with identical molecular weights are identical. Therefore, there is no distinction made in the nomenclature of the *Argiope* toxins originating from different species. However, one has to keep in mind that, as long as the physical and chemical data are not known in complete detail, a definite proof to verify their identity still has to be given.

All of the circumstances described above compel the reviewer, who can only compare published data, to discuss the structure elucidations done by different research groups separately. The problems described for Arg 636 also relate to all other toxins with identical molecular weights discussed in this chapter.

a. Toxin Arg 636 from Argiope lobata. The first isolation and structural characterization of a polyamine amide spider toxin, which was called Arg 636 (argiopine) and obtained from *Argiope lobata,* was published in 1986 by Grishin and co-workers (*93,99*) (Fig. 14). The crude venom from *A. lobata* was fractionated by reversed-phase HPLC with an acetonitrile/ water gradient. Studies involving UV, ^{1}H-NMR, nuclear Overhauser effect (NOE) NMR, and ^{13}C-NMR spectroscopies, microanalysis, mass spectrometry, and amino acid analysis were accomplished to establish the structure of Arg 636. Figure 14 shows that Arg 636 (argiopine) is composed of six structural units.

Hydrolysis and amino acid analysis yielded equimolar amounts of arginine and asparagine or aspartic acid. ^{1}H-NMR analysis showed an amidation of the β-carboxyl group of fragment II, and could thus define

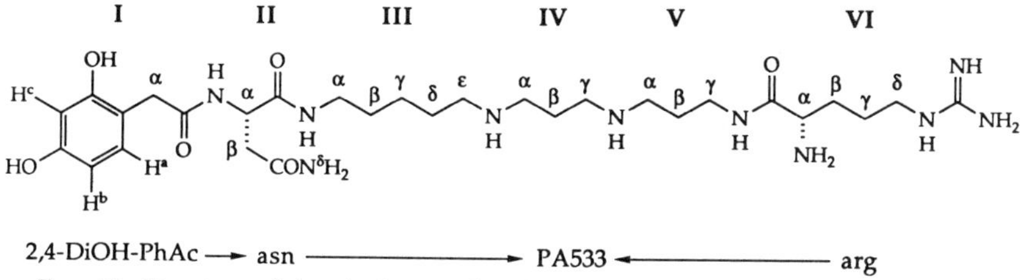

FIG. 14. Structure of the *Argiope* toxin with a molecular weight of 636 (Arg 636, argiopine).

asparagine. The absolute configuration of the amino acids, however, was not mentioned.

The ^{1}H-NMR spectrum determined, in addition, that the α-amino and β-carboxyl group of the asparagine each are parts of a peptide bond. Concerning the arginine unit, ^{1}H-NMR and Edman degradation determined the presence of a free N-terminus. In the UV spectrum, a maximum was detected at 279 nm ($\varepsilon = 2827$), indicating an aromatic system. In the ^{1}H-NMR spectrum, the signals of three aromatic protons were observed, H^a in the ortho and H^c in the meta position relative to H^b ($J_{H^aH^b} = 8.8$, $J_{H^aH^c} \approx 0$, $J_{H^bH^c} = 2.6$ Hz). From the NOE between the H^a protons and the two protons of the acyl group of fragment I, it could be concluded that this CH_2 group is in the ortho position to H^a. The NOE observed between the NH proton of the asparagine and the acyl protons of fragment I showed that the asparagine was linked by an amide bond to the carboxyl group of fragment I.

For further investigations, Arg 636 was methylated with diazomethane. By comparison of the electron impact (EI) mass spectrum of the methylated and the natural product, fragment I was identified as a diphenol. This result was in good accord with the pH dependence of the chemical shifts of the aromatic protons in the ^{1}H-NMR spectrum. The pK values of the two ionogenic groups were determined as 9.3 and 11.2, demonstrating that fragment I contains two HO groups.

Mass spectrometric data determined the molecular weight to be 636. Taking into account the structures of the fragments I (2,4-DiOH-PhAc), II (asparagine), and VI (arginine) and the results of the elemental analysis, the empirical formula $C_{29}H_{52}N_{10}O_2$ was proposed for Arg 636. This means that the fragments III, IV, and V have a net molecular weight of 214 and the empirical composition $C_{11}H_{26}N_4$. In the ^{1}H,^{1}H-correlated spectroscopy (COSY) NMR spectrum of Arg 636 the spin systems of the fragments III, IV, and V could be identified as $NH(CH_2)_5$ (III), $(CH_2)_3$ (IV), and $NH(CH_2)_3$ (V). The proton signals of the fragments III and V proved that the NH groups of these units are parts of amides. The structures of the fragments were confirmed by ^{13}C-NMR data. The signals relate to 18 aliphatic, 6 aromatic, and 4 carboxylic carbon atoms, besides one guanidine carbon. The NOE observed between the NH proton of fragment V and the $C^\alpha H$ and $C^\beta H_2$ of the arginine residue establishes a peptide bond. A peptide bond is also detected between fragment III and the asparagine (II). This left only fragment IV to be localized. The ^{1}H-NMR spectrum shows that IV is linked to two NH groups, which belong to the fragments III and V. Comparison of the ^{1}H-NMR spectra at pH 3 and pH 8 showed at pH 8 a high-field shift of the $C^\delta H_2$ and $C^\varepsilon H_2$ of fragment III and $C^\beta H_2$ and $C^\gamma H_2$ of fragment V, as well as of all the protons of fragment IV, owing to the deprotonation of the basic NH groups.

With the data given above, Grishin *et al.* (*93*) determined the structure of Arg 636 to be as given in Fig. 14. The arrangement of two amino acids linked via a polyamine and terminating as a (2,4-dihydroxyphenyl)acetamide was considered quite unusual. The only aspect of a complete structure determination that was not addressed by the research collective concerns the configurations of the two amino acids.

b. Toxin Arg 636 from Argiope trifasciata *and* Araneus gemma. In 1985, the major components of the venom of *Argiope trifasciata* and *Araneus gemma* were isolated. The FAB mass spectrum of the venom of *Argiope trifasciata* showed a signal at m/z 637. The toxins from both species were not fully characterized, but they appeared to be polypeptides with molecular weights between 500 and 1000 (*85*).

A detailed characterization of Arg 636 from *Argiope trifasciata* was given in 1987. The procedures described are stated to be applicable under identical conditions to the venom of *Araneus gemma* (*95*). The actual structure elucidation was done with the toxin Arg 636 obtained from *A. trifasciata*. Scheme 24 gives an example of the isolation and purification procedures for venom from *Argiope trifasciata* and *Araneus gemma*.

The UV spectrum of Arg 636 measured in 0.1 M HCl showed a λ_{max} at 278 nm (in methanol at 280 nm) with a shoulder at approximately 284 nm, which is characteristic of an aromatic chromophore. In base (0.1 M NaOH) the maximum was shifted bathochromically by 20 nm (λ_{max} at 298 nm). This is consistent with a phenol/phenolate system. In the ^{1}H-NMR spectrum of Arg 636 from *Argiope trifasciata,* taken in CD_3OD, two sets of signals were observed in the aromatic region, a doublet at 6.97 and 6.94 ppm ($J = 8$ Hz, 1 H) and a complex multiplet between 6.33 and 6.27 ppm (2 H), corresponding to a 2,4-dihydroxyphenyl system (resorcinol) substituted in position 1 by an alkyl residue (cf. the same chromophore is found in spiders of the genus *Nephila,* JSTX-1 to JSTX-4, NSTX-1 to NSTX-4) (*95*). The FAB mass spectrum defined the molecular weight as 636. Thus, Arg 636 has to possess an even number of nitrogen atoms.

The acetylation of Arg 636 under mild conditions ($MeOH/Ac_2O$) afforded a bis-acetylated product. Using more forcing conditions [AcOH, $(CF_3CO)_2O$], two compounds, a hexa-acetylated derivative of Arg 636 and its formal dehydration product, were detected by FAB-MS [$(M + H)^+$; m/z 889 and 872, respectively]. These results can be reconciled with the structure of Arg 636 (Scheme 25) in that with methanol/acetic anhydride the arginine portion of the natural product is bis-acetylated, whereas the remaining nitrogen groups stay unaffected. In a model reaction using the same conditions, arginine itself was bis-acetylated. Under more drastic conditions, however, the remaining amine and hydroxyl functions of Arg 636 were acetylated as well, leading mainly to the compound **144**

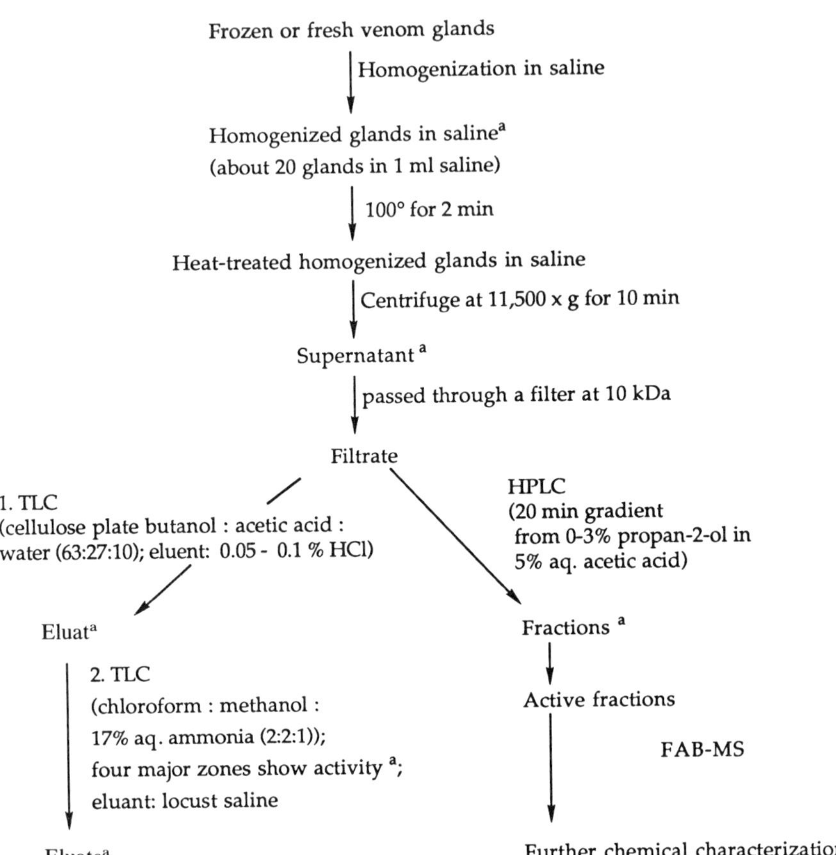

[a] Bioassay step; full biological activity has to be maintained after each step
of purification (saline: 140 mM NaCl, 5 mM KCl, 0.75 mM CaCl$_2$,
1 mM MgCl$_2$, 4 mM NaHCO$_3$, 5 mM HEPES, pH 7.2)

SCHEME 24. Isolation and purification of toxins from the venom of *Argiope trifasciata* and *Araneus gemma* (*83*).

(molecular weight 888). The FAB-MS signal at *m/z* 872 can derive from nitrile **145** (molecular weight 871), which was possibly formed under the forcing conditions by dehydration of the terminal N-unsubstituted amide of **144.** An example of an amide to nitrile transformation using trifluoroacetic anhydride is discussed by Campagna *et al.* (*100*).

Amino acid analysis indicated the presence of equimolar amounts of arginine and asparagine or aspartic acid. Concerning the *Argiope* toxins,

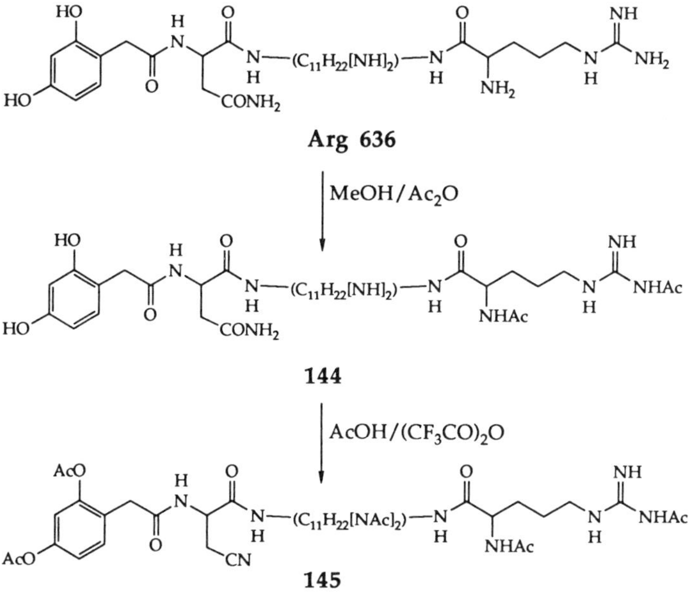

Arg 636

MeOH/Ac$_2$O

144

AcOH/(CF$_3$CO)$_2$O

145

SCHEME 25. Acetylation of Arg 636.

the differentiation between asparagine and aspartic acid is often not properly performed. The *Nephila* toxins (NPTX-1 to NPTX-12, see Figs. 29 and 30) are a good example that such structural options have to be thoroughly investigated. One of twelve *Nephila* toxins (NPTX-7) carries aspartic acid and not asparagine (*101*). Edman degradation of Arg 636 resulted in the formation of PTH (phenylthiohydantoin)-arginine. No other PTH amino acid was observed, indicating that the second amino acid, asparagine, was N-terminally blocked.

The structure proposed from these data (*95*) for Arg 636 from *Argiope trifasciata* and *Araneus gemma* is identical to that of Arg 636 from *Argiope lobata* (Fig. 14), isolated and characterized by Grishin *et al.* (*93*).

c. Toxin Arg 636 from Argiope trifasciata *and* Argiope florida. In 1988, an additional *Argiope* species was examined. Budd *et al.* (*78*) isolated Arg 636 from *Argiope florida* and *A. trifasciata,* but no distinction was made or seemed to be made between the toxins of *A. florida* and *A. trifasciata*. The venom was obtained by milking mainly female spiders. The venom from about 1000 milkings was pooled (*78*).

The UV characteristics for Arg 636 from *Argiope trifasciata* were described as "similar" to the ones known from *A. florida* and *A. gemma*. Both methanolysis and mild hydrolysis of Arg 636 liberated a compound containing the chromophore of the toxin, which subsequently could structurally be determined by EI-MS as a dihydroxyphenylacetic acid. In the case of methanolysis (methanolic HCl, 18 hr, 80°C) the chromophore-containing fraction could be readily separated by reversed-phase HPLC from an Arg 636 fragment carrying both amino acids. In a separate experiment, the acid hydrolysis product containing the chromophore part was esterified with an HCl-saturated 1:1 mixture of CH_3OH and CD_3OH and analyzed by EI-MS. The mass spectrum showed a characteristic pattern of unlabeled and labeled esterified compounds. This result showed that the hydrolysis product mentioned above contains only one carboxylic acid unit that can be methylated. Therefore, the chromophore unit could be interpreted as a dihydroxyphenylacetic acid (*78*).

About 100 μg of Arg 636 (obtained from 1500 spiders of *Argiope trifasciata* and/or *A. florida*) was prepared for ^{1}H-NMR analysis (Fig. 22△, see p. 86). After collecting data for 24 hr in CD_3OD, the 1-alkylated 2,4-dihydroxyphenyl part was recognized, but the small amount of sample complicated the interpretation. The low-field region of the spectrum was "obscured" by solvent impurity peaks. Proton decoupling and ^{1}H,^{1}H-COSY NMR experiments could not settle the remaining questions concerning the structure of the polyamine part (*78*), and the amount of toxin available was not sufficient for ^{13}C-NMR analysis. However, taking into account the NMR, MS, and UV data on the natural product and on the hydrolysis products, Budd *et al.* (*78*) concluded that Arg 636 from *Argiope trifasciata* and *A. florida* contains a 2,4-dihydroxyphenylacetyl chromophore.

Furthermore, amino acid hydrolysis of Arg 636 from *Argiope trifasciata* and *A. florida* yielded equimolar amounts of arginine and asparagine. The presence of the N-terminal arginine was defined by Edman degradation and dansylation. Edman degradation led to the results already described (*78*). Dansylation of Arg 636 and TLC analysis of the fully hydrolyzed product showed the presence of bis-dansylarginine as the only dansylated amino acid (*78*). These results indicated an N-terminal arginine and an asparagine residue concealed within the molecule. Together with the ^{1}H-NMR data, this gave evidence that the polyamine is linked through the carbonyl function of the asparagine, which itself is N-terminally blocked by the chromophore, 2,4-dihydroxyphenylacetic acid. There was no evidence, however, to rule out the possibility that the polyamine was linked by an amide bond to the asparagine (*78*). Grishin *et al.* (*93*) seemed to be able to determine, on the basis of NOE data, that the polyamine is bound to the α-carbonyl group. Concerning further structural definition, Budd *et al.* (*78*) explained that, "Because of the small amount of each *Argiope* toxin available for analysis and the requirement for lengthy data acquisition many high field signals were obscured by solvent and column derived impurities, which were also present in a control blank. Some tentative assignment could however be made despite the poor spectral quality, but the data so obtained were insufficient for full structural definition." The elemental composition of the compound was known, but the full structure remained to be defined. The structure proposed by the authors is "in broad agreement" (78) with the structure given for Arg 636 from *Argiope lobata* by Grishin *et al.* (93).

d. Toxin Arg 636 from Argiope aurantia. Investigations of the venom of *Argiope aurantia* again identified one of the toxins obtained as Arg 636 (65). The compound was mainly analyzed by NMR techniques. ^{1}H-NMR, nuclear Overhauser effect spectrosocpy (NOESY), two-dimensional (2D) ^{13}C-NMR, and distortionless enhancement by polarization transfer (DEPT) spectra allowed the assignment of all the proton and carbon-13 resonances and the determination of the structure of Arg 636 from *A. aurantia*. As discussed above (78,85,95), it was also shown that Arg 636 from *Argiope aurantia* contains two functional groups, which can rapidly be acetylated using a reagent that acetylates amino groups.

e. Final Remarks Concerning Toxin Arg 636. In the structure proposed for Arg 636 by different authors, the L form of the amino acids is shown. As mentioned above, to our knowledge no proof for this absolute configuration of the amino acids has yet been given. Arg 636 has two chiral centers, owing to the two amino acids (see Fig. 14). The total synthesis of Arg 636 to confirm its structure was done by Shih *et al.* (63), Yelin *et al.* (64), and Jasys *et al.* (28). The aim was, besides the verification of the structure, to develop synthetic strategies for achieving larger quantities of the toxin, required for detailed pharmacological screening. For the synthesis, a decision concerning the chirality of the amino acids had to be made. Shih *et al.* (63) concluded, "Since a definite X-ray structure was not available and since the amino acid stereochemistry does not appear to have been defined, the proposed structure was synthesized incorporating the L-form of asparagine and arginine." The synthetic toxin was identified by ^{1}H-,^{13}C-NMR, and FAB-MS data, which were reported to be consistent with those of the natural compound. Additionally, similar biological profiles were obtained for synthetic and natural Arg 636 (63).

The *Argiope* toxins were the first low molecular weight spider toxins investigated in detail. The fact that Arg 636 was the first structurally characterized *Arigope* toxin, and because of its high potency and high abundance in the venom, it is so far the most studied polyamine spider toxin. In all likelihood, the toxin Arg 636 is identical from the different spider species discussed.

2. *Further* Argiope *Toxins from* Argiope lobata

By 1988, from *Argiope lobata* nine *Argiope* toxins had been isolated and identified from 500 mg of lyophilized venom (*94,102*). The separation, performed by HPLC, is shown in Fig. 15.

The structure elucidation of the compounds was done mainly by NMR spectroscopy. Mass spectrometry (FAB-MS), UV data, and amino acid analysis supported the interpretation of the NMR data. The results obtained from the UV and NMR data enabled classification of the toxins, according to the nature of the chromophores, into three categories (Fig. 16). The argiopine type (Arg 636) contains 2,4-dihydroxyphenylacetic acid, the pseudoargiopinines (Arg 373, Arg 728, Arg 743) possess an (indol-3-yl)acetic acid, and the argiopinines (Arg 630, Arg 658, Arg 659, Arg 744, Arg 759) carry (4-hydroxyindol-3-yl)acetic acid as the chromophore (*94,102*). The structures of the *Argiope lobata* toxins are given in Fig. 17.

a. Toxin Arg 659 from Argiope lobata.   Detailed NMR data for Arg 659 (Fig. 17) are given by Grishin *et al.* (*94,102*). Arg 659 is the best investigated of the *Argiope* toxins which carry an indolic chromophore. It is

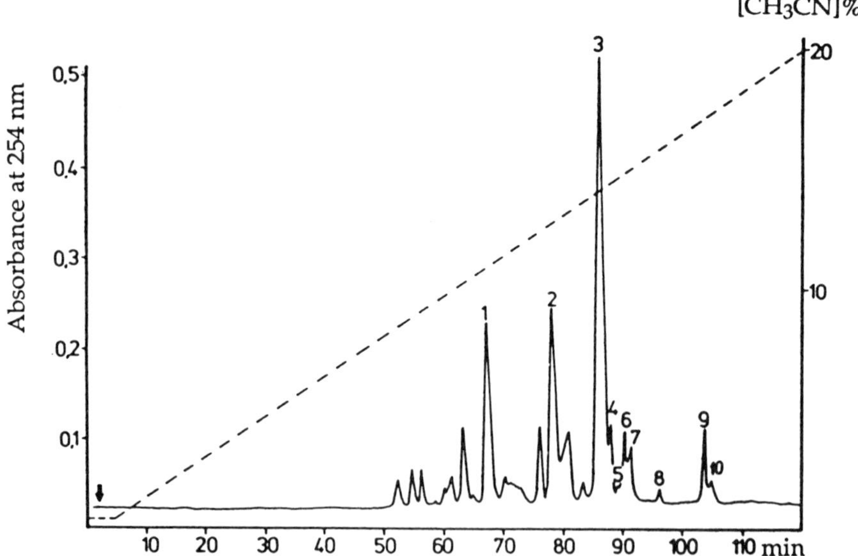

FIG. 15. Separation of components of the venom of *Argiope lobata* by reversed-phase HPLC with a linear gradient of water/acetonitrile (0–20%) containing 0.1% trifluoroacetic acid. (From Ref. *102*, with permission.) The identities of peaks 1–10 are given in Table VI.

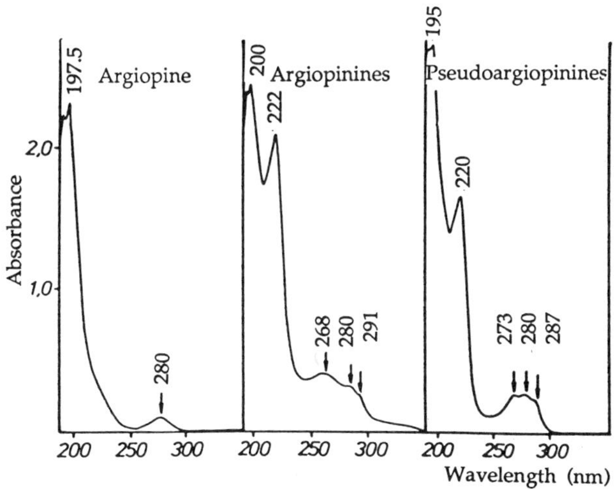

FIG. 16. UV spectra of argiopine (2,4-DiOH-PhAc chromophore), argiopinines (4-OH-IndAc chromophore), and pseudoargiopinines (IndAC chromophore); the solvent is presumably methanol. (From Ref. *102*, with permission.)

highly abundant in venom (Fig. 15, peak 3), and its biological potency is reported to be similar to that of Arg 636 (*78*).

The basic feature of the approach employed for the structure elucidation of Arg 659 from *Argiope lobata* was the use of 2D NMR spectroscopy of the double quantum filtered (DQF)-COSY type for the identification of proton spin systems and NOE difference spectroscopy to determine the sequence of the proton spin systems in the compounds studied. In addition, information concerning the pH dependence of the chemical shifts of proton signals was used to analyze the nature and position of ionogenic groups in the molecule (*94,102*). Figure 18 shows the DQF-COSY spectrum of Arg 659 in water, with the assignment of the cross-peaks of the protons coupled by scalar spin–spin interaction through two or three chemical bonds. It was possible to detect all of the proton spin systems of Arg 659. The chemical shifts of the protons participating in the spin systems are given in Table VII.

The down-field region of the one-dimensional ^{1}H-NMR spectrum of Arg 659 (Fig. 19) gives additional information. It is possible to detect the signals of an amide group (7.52 and 6.81 ppm), the broad signals of amino groups (8.5 ppm, ~3*H*; 8.22 ppm, 4–5*H*; 6.75 ppm, ~4*H*), and the signals of aromatic NH protons (10.21 ppm, 1*H*) as well as those of the aromatic CH protons (7.26 ppm, 1*H*; 7.14 ppm, 2*H*; 6.61 ppm, 1*H*). According to the DQF-COSY spectrum, spin–spin coupling is observed between the

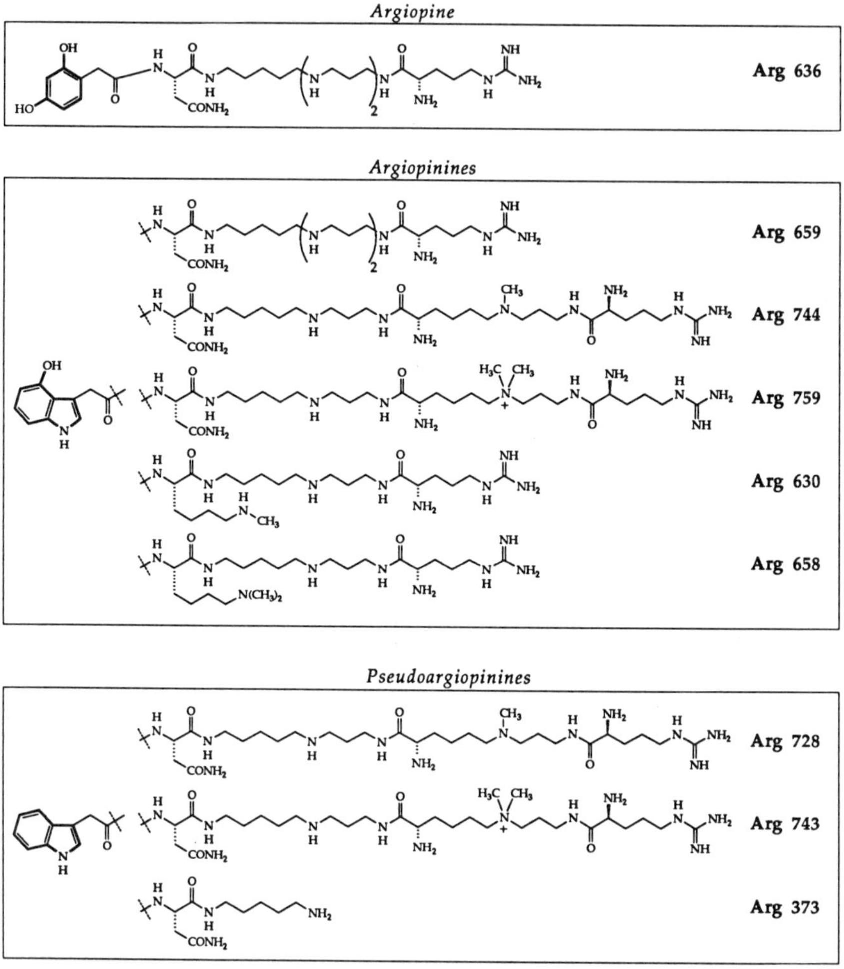

FIG. 17. Toxins of *Argiope lobata*.

aromatic NH (10.21 ppm) and CH (7.26 ppm) protons, and also between the CH protons at 6.61 and 7.14 ppm.

Because, on the basis of the results of the amino acid analysis, the molecule Arg 659 includes either an asparagine or an aspartic acid residue, spin system 1 (see Table VII) was assigned to this residue. The results of a study on the pH dependence of the chemical shifts of the proton signals showed the absence of carboxyl groups. Consequently, Arg 659 contains an asparagine residue, the NH signals of the amide group of which were observed at 7.52 and 6.81 ppm.

TABLE VI

CHARACTERISTICS OF COMPONENTS OF *Argiope lobata* VENOM[a]

HPLC fraction[b]	Compound	Molecular weight	Chromophore	Amino acid analysis data[c]
	Argiopine			
1	Arg 636	636	2,4-DiOH-PhAc	Asx, Arg
	Argiopinines			
6	Arg 759	759[d]	4-OH-IndAc	Asx, Arg
7	Arg 744	744	4-OH-IndAc	Asx, Arg
3	Arg 659	659	4-OH-IndAc	Asx, Arg
4	Arg 630	630	4-OH-IndAc	Asx
5	Arg 658	658	4-OH-IndAc	Asx
	Pseudoargiopinines			
9	Arg 743	743[d]	IndAc	Asx, Arg
10	Arg 728	728	IndAc	Asx, Arg
8	Arg 373	373	IndAc	Asx

[a] From Ref. *94*, with permission.
[b] Fractions correspond to the HPLC profile shown in Fig. 15.
[c] Asx, Asparagine or aspartic acid; Arg, arginine.
[d] These compounds contain one quaternary nitrogen atom.

The signals of spin system 5 [—$C^{\alpha}H$—$(CH_2)_3$—$N^{\varepsilon}H$—] and a broad, six-proton signal (6.75 ppm) belong to the guanidine group of the arginine residue. A broad signal at 8.50 ppm integrating for three protons probably relates to the α-amino group of the arginine residue. Arginine was identified by amino acid analysis and by Edman degradation to be the N-terminal amino acid. The same conclusion was drawn from a study of the pH dependence of the chemical shift of the signal due to the $C^{\alpha}H$ proton of the arginine (pK value of this amino group 7.1) (*94,102*).

Alteration of the pH of the medium from pH 8.57 to pH 12.02 resulted in changes of the chemical shift and multiplicities of the signals. The aromatic region of the spectrum of Arg 659 at various pH values is shown in Fig. 20. Analysis of the spectrum at pH 12.02 led to the conclusion that there were three protons in the aromatic system under investigation coupled to one another successively by scalar spin–spin interactions. The signal of the NH proton at 10.21 ppm (see Fig. 19) is considered characteristic for an indole group, and the spin–spin coupling between the NH proton (10.21 ppm) and a CH proton (7.26 ppm) was recorded in the DQF-COSY spectrum. Thus, the existence of a disubstituted indole group in Arg 659 was proposed. An NOE was observed between the NH proton at 10.21 ppm and a proton at 7.14 ppm. Thus, one could assume the presence of a disubstituted indole residue in Arg 659, in position 4 of which there is

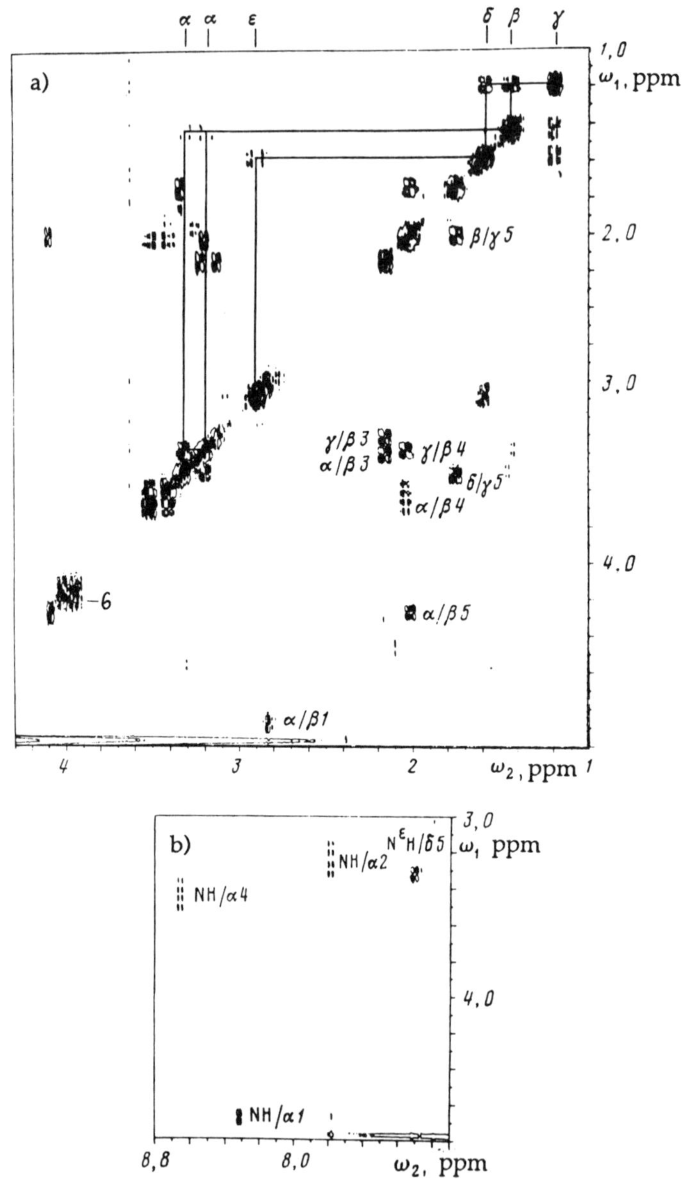

FIG. 18. Double quantum filtered COSY spectrum of Arg 659 (6.5 mM in D$_2$O, pH 3; 30°C): (a) aliphatic region of the spectrum (ω_1 = 4.8–1.0 ppm); (b) region of the spectrum containing cross-peaks with the participation of NH amide protons (ω_1 = 4.8–3.0 ppm; ω_2 = 8.8–7.1 ppm). The figures show the spin systems of the protons detected in the spectrum (see Table VII). (From Ref. *94*, with permission.)

TABLE VII

CHEMICAL SHIFTS OF PROTON SIGNALS OF Arg 659[a]

	Spin System	NH	$C^\alpha H$	$C^\beta H$	Others
1	$- NH - C^\alpha H - C^\beta H_2 -$	8.28	4.65	2.80	$N^\delta H_2{}^+$: 6.81; 7.52
2	$- NH - (CH_2)_5 -$	7.74	3.26; 3.14	1.41	$C^\gamma H_2$: 1.16; $C^\delta H_2$: 1.55; $C^\varepsilon H_2$: 2.84; $N^\delta H_2{}^+$: 8.22
3	$- (CH_2)_3 -$		3.18; 3.18	2.11	$C^\gamma H_2$: 3.08
4	$- NH - (CH_2)_3 -$	8.62	3.46; 3.36	2.00	$C^\gamma H_2$: 3.15; $N^\delta H_2{}^+$: 8.22
5	$- C^\alpha H - (CH_2)_3 - N^\varepsilon H -$		4.04	1.98	$C^\gamma H_2$: 1.71; $C^\delta H_2$: 3.29; $N^\varepsilon H$: 7.27; $C(NH_2)_2{}^+$: 6.75
6	$- CH_2 -$		3.97; 3.89		
	4-Hydroxyindol-3-yl				N(1)H: 10.21; C(2)H: 7.26; C(5)H: 6.61; C(6)H: 7.14; C(7)H: 7.14

[a] Spectra were collected in D_2O, pH 3.0, 30°C. Chemical shifts (δ) are given in ppm. (From Ref. *94*, with permission.)

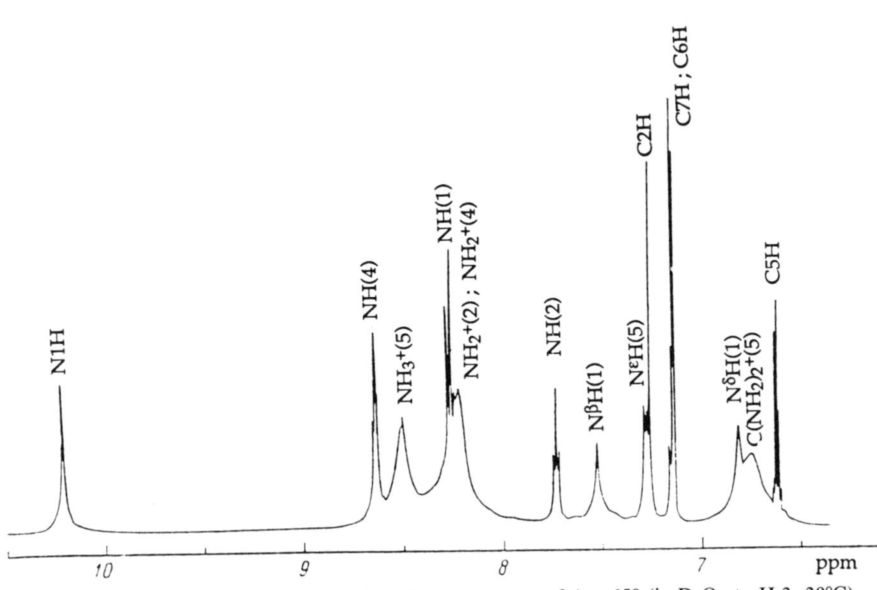

FIG. 19. Down-field region of the ^{1}H-NMR spectrum of Arg 659 (in D_2O at pH 3, 30°C). (From Ref. *102*, with permission.)

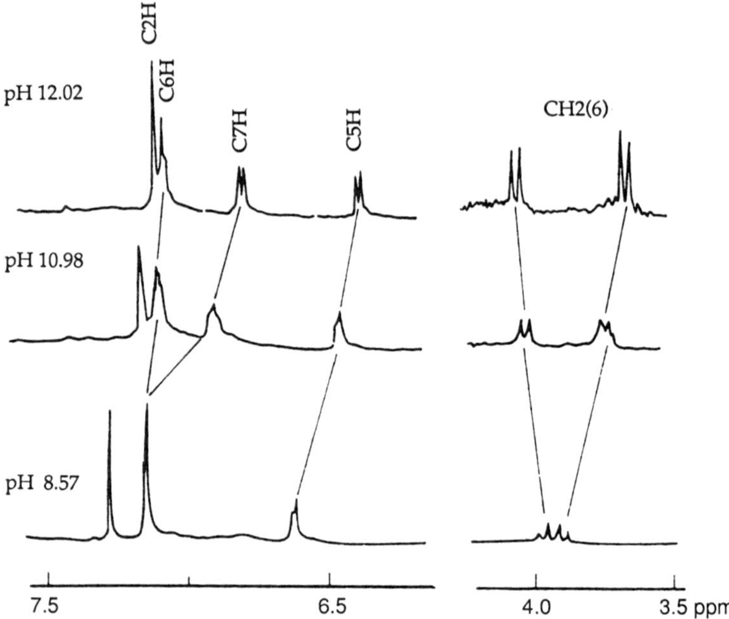

FIG. 20. ^{1}H-NMR spectrum of Arg 659 (in D_2O at different pH values, 30°C), showing the effect of pH changes on the aromatic proton signals and the signals of the methylene protons (proton spin system 6, Table VII). (From Ref. *102*, with permission.)

an ionogenic group (the pK value was determined as 10.6) and a methylene group in position 3. The results obtained permit the conclusion that Arg 659 includes a (4-hydroxyindol-3-yl)acetic acid residue [cf. chromophore of the *Nephila* toxins NPTX-1 to NPTX-6, which carry a (6-hydroxyindol-3-yl)acetic acid moiety]. For the complete determination of the molecular structure of Arg 659, NOE difference spectroscopy was employed. It was found that the NH proton of spin system 4 [—HN—$(CH_2)_3$—] and the C^αH and $C^\beta H_2$ protons of the arginine residue are spatially close (Fig. 21). Therefore, these fragments presumably are linked to one another by a peptide bond.

Similarly, an NOE was observed between the asparagine backbone NH proton and the protons of the CH_2 group of the (4-hydroxyindol-3-yl)acetic acid moiety. Because Arg 659 lacks a free carboxy group, the asparagine residue and the spin system 2 [—HN—$(CH_2)_5$—] are also linked via a peptide bond between the NH group of spin system 2 and the carbonyl group of the asparagine residue. In this case, the use of NOESY did not appear to be possible because of the masking of the C^αH proton signal by water. Nevertheless, on the basis of all the facts given above, Grishin *et al.*

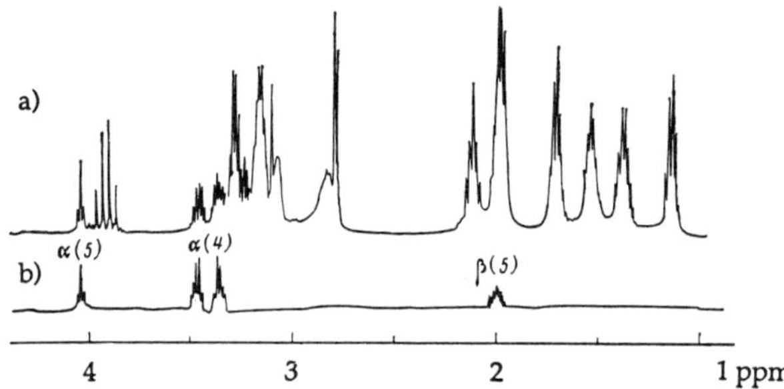

FIG. 21. (a) Aliphatic region (1.0–4.3 ppm) of the ¹H-NMR spectrum of Arg 659 (in D_2O, pH 3, 30°C). (b) NOE spectrum obtained on irradiation of the NH proton (8.62 ppm) of spin system 4 [— HN — $(CH_2)_3$ —] (Table VII). (From Ref. *94*, with permission.)

(*94,102*) declare their structure elucidation for Arg 659 (see Fig. 17) as "unequivocal."

b. Argiopinines Arg 630, 658, 744, and 759 from Argiope lobata. The structures of Arg 744 and Arg 759 possess considerable similarities to Arg 659 (Fig. 17). The proton spin systems of —NH—$(CH_2)_5$—NH— and $C^{\alpha}H$—$(CH_2)_4$— were identified in the DQF-COSY spectrum of these compounds instead of the proton spin systems 2 [—HN—$(CH_2)_5$—] and 3 [—$CH_2)_3$—] of Arg 659, respectively. In addition, signals typical for *N*-methyl groups were observed in the ¹H-NMR spectra of these compounds. The structures of Arg 744 and Arg 759 were deduced only from the above results, namely, the data on the pH dependence of the chemical shifts of the proton signals and the molecular masses determined by FAB-MS (*102*). Compound Arg 759 is a quaternary ammonium salt, and FAB-MS therefore shows the signal of the M^+ ion at *m/z* 759, and not the $(M + H^+)$ ion.

The presence of *N*-methyllysine and *N,N*-dimethyllysine is characteristic of the structures of Arg 630 and Arg 658 (Fig. 17). All other argiopinines contain an asparagine residue instead of the modified lysines. The DQF-COSY spectra of Arg 630 and Arg 658 contain the spin system —NH—$C^{\alpha}H$—$(CH_2)_4$—, which is typical for lysine residues. The remaining spin systems identified in the DQF-COSY spectra are identical for Arg 659, Arg 630, and Arg 658. One *N*-methyl group was identified for Arg 630. The amino acid analysis of this compound showed no lysine; therefore, an *N*-methyllysine residue was assumed. Thus, the structure of the entire Arg 630 molecule was determined. Signals of three *N*-methyl groups were identified in the ¹H-NMR spectrum of Arg 658, which, together with the molecular mass, led to the structure proposal.

c. Pseudoargiopinines Arg 373, Arg 728, and Arg 743 from Argiope lobata. The pseudoargiopinines (Arg 373, Arg 728, Arg 743) contain (indol-3-yl)acetic acid instead of the (4-hydroxyindol-3-yl)acetic acid found in the argiopinines (Arg 630, Arg 658, Arg 659, Arg 744, and Arg 759). This is reflected by the difference in molecular weight of 16 amu between Arg 759 and Arg 743 and between Arg 744 and Arg 728. DQF-COSY and NOE spectra were taken. The proton spin systems of Arg 759 and Arg 743, as well as those of Arg 744 and Arg 728, were identical for each pair of compounds, except for the signals in the aromatic region of the spectra [derived from the (indol-3-yl)acetyl and the (4-hydroxyindol-3-yl)acetyl group, respectively]. These facts allowed the structural determination of Arg 743 and Arg 728.

In the ^{1}H-NMR spectrum of Arg 373, the proton spin system of cadaverine was identified, in addition to an (indol-3-yl)acetic acid and an asparagine residue. Besides the proton spin systems, a broad signal of an amino proton was observed. Because a free carboxyl group was absent, the asparagine was assumed to be linked via a peptide bond to the (indol-3-yl)acetyl residue and to the cadaverine unit.

As described above, the toxins of *Argiope lobata* (Fig. 17) show high structural similarity. All possess a chromophore, an aliphatic polyamine moiety, and, except for Arg 373, an arginine residue. The chromophore is linked to the polyamine chain by an asparagine or modified lysine unit (*94,102*).

3. *Further* Argiope *Toxins from* Argiope trifasciata, Argiope florida, Argiope aurantia, *and* Araneus gemma

In 1987/1988 further low molecular weight toxins were isolated from *Argiope trifasciata, A. florida, A. aurantia,* and *Araneus gemma* (*65,78,95,96*). The characterization was done by UV, ^{1}H, ^{13}C, NOE, and COSY NMR, EI-MS, FAB-MS, and micro-chemical derivatization (acetylation and esterification), ^{2}H-labeling, amino acid analysis, dansylation, and Edman degradation. Data have been published mainly for toxins of the venoms of *Argiope trifasciata* (*78,85,95*) and *A. aurantia* (*65,103*).

Similar to the treatment of toxins of Argiope lobata by Grishin *et al.* (*94,102*), the toxins from *Argiope trifasciata* were divided into two distinct families according to different UV spectral characteristics (*78*). Members of both families of *Argiope* toxins have a terminal arginine linked via a polyamine to asparagine, which, in turn, is coupled to a phenol in one family (argiopine) or to an indole moiety in the other (argiopinine).

The first family possesses a chromophore identical to that of argiopine [Arg 636 from *Argiope lobata* with a (2,4-dihydroxyphenyl)acetic acid chromophore] and shows UV spectra in methanol with λ_{max} at 280 nm and a shoulder at 286 nm, which shifts bathochromically by 20 nm in base, suggesting the presence of a phenolic hydroxyl group. The second family, the members of which are called argiopinines by Grishin *et al.* (*102*) (e.g., Arg 659 from *Argiope lobata*), possesses a 4-hydroxyindole-like spectrum (λ_{max} at 268 nm and at 292 nm) (see UV spectra in Fig. 16 and ^{1}H-NMR spectra in Fig. 22).

As already mentioned, the amount of venom available limited the experiments of Budd *et al.* (*78*). The ^{1}H-NMR spectrum of about 100 μg of the indolic species Arg 659 was taken, but, as in the case of Arg 636, the spectrum was complicated by the presence of background signals derived from so-called solvent impurities. Thus, only the aromatic region of the spectra could be interpreted (Fig. 22). Signals were observed at 7.08 (1 *H*,s), 6.95 (2 *H*, d), and 6.45 ppm (1 *H*, m). The chemical shifts are in agreement with those expected for indolic protons, and the coupling constants suggested an indole molecule substituted in the 5 or 6 position, although a definite assignment was impossible (*78*). As described above, Grishin *et al.* (*94,102*), who had a larger amount of toxin available, were able to define a (4-hydroxyindol-3-yl)acetic acid moiety in Arg 659 from *Argiope lobata*.

Amino acid analysis indicated, as already known for toxins from *Argiope lobata*, that both families of toxins from *A. trifasciata* contain equimolar amounts of arginine and asparagine. Analysis by FAB-MS gave the molecular weights of three compounds belonging to the first group, the argiopines, namely, Arg 636 and its lower and higher homologs Arg 622 and Arg 650. The best known species in the second group, possessing an indolic chromophore, is Arg 659. The higher and lower homologs of Arg 659, namely, Arg 673 and Arg 645, which were not detected in *A. lobata*, were present in *A. trifasciata* and *A. florida*. The relative abundance of each homolog was reported to vary with the venom source. For example, Arg 673 was the most abundant in the venom of *Argiope florida* (*78*).

The "mass difference" between the two families (best represented by the toxins Arg 636 and Arg 659) corresponds to the mass difference between a tyrosine and a tryptophan residue, a result which is in agreement with the UV data. From both families of toxins, carrying a (2,4-dihydroxyphenyl)acetic acid or a (4-hydroxyindol-3-yl)acetic acid chromophore, four different compounds could be isolated that carry no arginyl residue: Arg 480, Arg 494 and Arg 503, Arg 517, respectively. Again, the "mass difference" of the two groups corresponds to the mass difference between a tyrosine and a tryptophan residue.

The toxin Arg 622 was isolated and characterized in the same fashion as Arg 636 (see above). Analysis by NMR indicated that Arg 622 contains the same chromophore as Arg 636. The proposed structure differs by one methylene group; nevertheless, the exact structure of the polyamine part is still unknown (see Section II, Table II) (*78*). The same isolation and

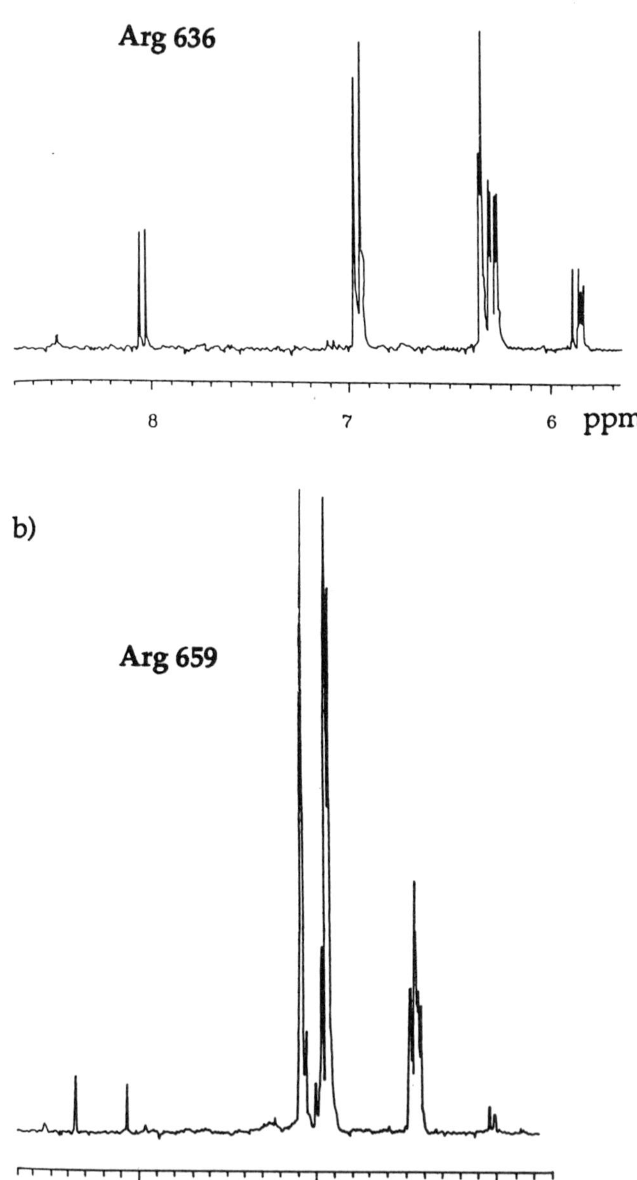

FIG. 22. Aromatic region of the ^{1}H-NMR spectrum (CD_3OD) of (a) Arg 636 and (b) Arg 659. (From Ref. *78*, with permission.)

characterization procedures were applied to the venom of *Araneus gemma* and gave similar results as in the case of Arg 622 from *Argiope florida,* within experimental error (*95*). The molecular weight of Arg 636 and its lower homolog Arg 622 were also determined by high-resolution FAB-MS (*78*).

The compounds Arg 636, Arg 659, and Arg 673 were also isolated from *Argiope aurantia* by Adams *et al.* (*65*) in 1987. All three are highly abundant constituents of the venom, occurring in essentially equal proportions at concentrations of 20 m*M* each (*65*). ^{1}H-NMR, ^{13}C-NMR, NOE, ^{1}H,^{1}H-, and ^{13}C,^{1}H-COSY data were reported. A chromophore similar to Arg 659 was detected for Arg 673. The ^{1}H-NMR spectrum showed the presence of two additional methyl groups, which were found to belong to 5,8-dimethylspermine. The FAB-MS fragmentation of the toxins is consistent with the NMR structural assignments. The FAB mass spectra were particularly valuable in assigning the location of the two methyl groups in Arg 673. The generalized structure elucidation follows the same principles already described.

Arg 636, Arg 659, and Arg 673 isolated by Adams *et al.* (*65,103*) seem to be identical to the *Argiope* toxins isolated from *Argiope lobata* (*93,94,99,102*). The pattern of hydroxylation observed in the aromatic moieties of some of the *Argiope* toxins is unconventional, for example, the hydroxyl substituent in the (4-hydroxyindol-3-yl)acetic acid moiety of Arg 659 and Arg 673. This seems to be unprecedented in a natural product (*65*). The Arg 659 and Arg 673 from *Argiope trifasciata, A. florida,* and *Araneus gemma* are most likely identical to Arg 659 and Arg 673 obtained from *Argiope aurantia* (*65,103*).

From the pharmacological behavior, it had been shown that most likely the major active compound of the venom in *Argiope trifasciata* and *Araneus gemma* is Arg 636 (*78*), and that this compound is also present in *Argiope florida, A. lobata,* and *A. aurantia.* Of all the compounds with an indolic chromophore, Arg 659, which was found in all species discussed above, was by far the most biologically active component (*133*). Identical pharmacological behavior is reported for the *Argiope* toxins with identical molecular weights. However, identical pharmacological behavior is no proof of structural identity among the toxins of identical molecular weight originating from different species.

B. Araneidae Toxins from *Nephila* Species

In 1982 Kawai *et al.* (*90*) were the first to isolate toxins from spiders of the genus *Nephila.* They reported the neurotoxic activities of the venom as a potent blocker of glutaminergic receptors. The venom of the Japanese spider Jorogumo (or Joro spider, *Nephila clavata*) is called Joro spider toxin (JSTX) (*104*). The dry weight of a single venom gland is 100–200 μg, and the content of the toxic principle JSTX seemed to constitute less than 10% of the gland. The molecular weight of Joro spider toxin was "tentatively" assumed to be 500 (*92*). At present, characterization of the venom leads to the conclusion that the toxic principle JSTX is a mixture of at least 17 different closely related compounds, named JSTX-1 to JSTX-4, NPTX-1 to NPTX-12, and clavamine (*1,59,101,105–107*).

In 1986, Aramaki *et al.* (*108*) isolated four toxins from *Nephila clavata* (JSTX-1 to JSTX-4) and three from *Nephila maculata* (NSTX-1 to NSTX-3). For some of the toxins, structures were proposed. The common structural unit was assumed to be a (2,4-dihydroxyphenyl)acetyl-asparaginyl-cadaverino-3-oxopropylamino moiety (see Fig. 23). Compounds of the NSTX type were thought to contain additional arginyl-cadaverine units, whereas JSTX toxins were composed of other kinds of polyamine moieties and/or amino acids.

The venom was obtained from homogenized glands of *Nephila clavata* and *N. maculata*. Based on an amount obtained from half of a venom gland, the separation conditions were established for the toxins from both spiders (Fig. 24). The crude venom corresponding to the glands of 200 spiders of *Nephila clavata* and 100 spiders of *Nephila maculata* was

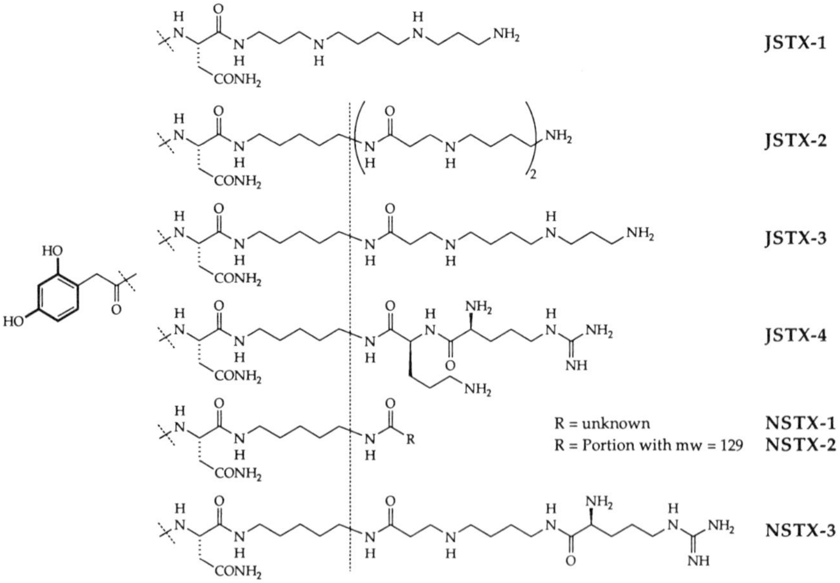

FIG. 23. Structures of JSTX-1 to JSTX-4 and NSTX-1 to NSTX-3.

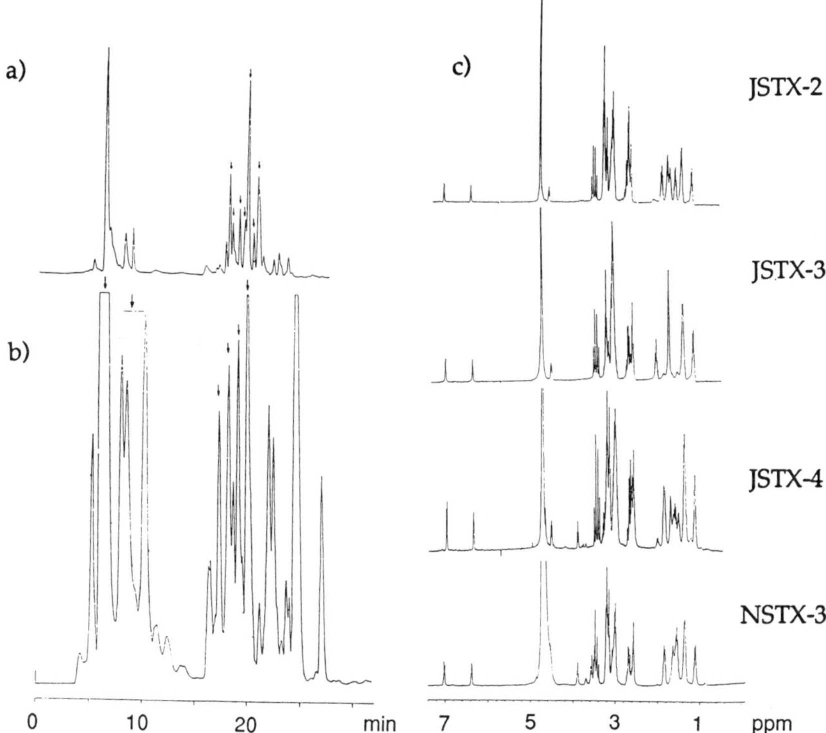

FIG. 24. HPLC chromatogram of the venom from (a) *Nephila clavata* and (b) *N. maculata*, obtained with a linear gradient from 0.02% aqueous HCl to acetonitrile/0.02% aqueous HCl (1:1) within 30 min at a flow rate of 0.5 min. (c) ^{1}H-NMR spectra of JSTX-2 to JSTX-4 and NSTX-3 in D_2O. (From Ref. *108*, with permission.)

separated preparatively using an identical elution system. The chromatogram was monitored with a UV detector at 210 nm. The toxins eluted were further purified by repeated chromatography applying a similar elution system until four JSTX toxins (JSTX-1 to JSTX-4) and three NSTX toxins (NSTX-1 to NSTX-3) from the respective venom extracts were isolated as fractions containing only a single compound.

All these spider toxins possess UV maxima at 210 and 280 nm and exhibit natural fluorescence at 310 nm when excited at 280 nm. The ^{1}H-NMR spectra (in D_2O) of some toxins showed a similar pattern and common significant signals in the high-field region around 6.5 and 7.0 ppm (Fig. 24). These signals were separated into three groups at about 6.1 (d), 6.3 (s), and 6.9 ppm (d) when the spectra were taken in DMSO-d_6. The structural element responsible for these signals was readily separated by hydrolysis (25 mM oxalic acid at 100°C for 16 hr) and extracted with ethyl acetate. Subsequent NMR measurements gave strong evidence for (2,4-dihydroxyphenyl)acetic acid. The same chromophore was found in some *Argiope* toxins, and its characterization was done in a fashion similar to the methods already discussed earlier in connection with the *Argiope* toxins (see Section V,A, referring to *Argiope* spiders, e.g., Arg 636).

Amino acid hydrolysis and time course studies for the partial acid hydrolysis of the toxins were performed. The configurations of the amino acids in the hydrolysate were determined by the chiral separation of amino acids derivatized by OPA (*o*-phthalaldehyde) and *N*-acetyl-L-cysteine, according to the method of Nimura and Kinoshita (*109*). In addition, NMR experiments were carried out with JSTX-2, JSTX-3, and JSTX-4. In terms of structural elucidation, the data obtained at that time were not sufficient for a definite structure assignment, but a raw interpretation was published (*108*). The JSTX toxins were chemically characterized a few years later (*104*).

Efforts were first focused on the structural determination of NSTX-1, NSTX-2, and, in particular, NSTX-3 (*105*) (Fig. 23). To obtain the toxin in a sufficient amount for structural analysis, the isolation started with about 200 mg of dried venom, which corresponds to the glands of about 1000–2000 spiders. Three toxins (NSTX-1 to NSTX-3) were obtained as chromatographically pure compounds. The amounts of pure toxins, roughly calculated by the UV absorption, were approximately 1 μmol of NSTX-1, 1.6 μmol of NSTX-2, and 3 μmol of NSTX-3. Analysis by FAB-MS of NSTX-1 did not give the molecular ion, the spectrum of NSTX-2 showed a signal at m/z 567 (later experiments proved that the value expected is m/z 523), and NSTX-3 gave a signal at m/z 665 for $(M + H)^+$. This enabled a mass assignment for this compound, which was confirmed by later measurements (*105*).

The active substances obtained by HPLC purification were hydrolyzed with 6N aqueous hydrochloric acid. Amino acids and polyamines in the hydrolysate were determined with an amino acid analyzer equipped with an ion-exchange column. To analyze both the amino acids and the polyamines, with only one type of column, a fourth buffer system (0.35 N sodium citrate, pH 5.28, containing 2 M NaCl and 20% methanol) was used following the conventional three buffer systems for usual amino acid analysis (*105*). The eluate from the column was mixed with a NaCl buffer, and then OPA was added. Both the column and the reaction temperature were maintained at 65°C. The OPA derivatives of the amino acids and the polyamines were identified fluorometrically with excitation above 410 nm (*105*).

The ^{1}H-NMR spectra (D_2O) of NSTX-1 to NSTX-3 share two character-istic signals in the aromatic proton region (6.4 and 7.1 ppm) for the protons of the 2,4-dihydroxyphenyl moiety, a structural element that is also present in the *Argiope* toxins (Fig. 25, e.g., Arg 636). Equimolar amounts of asparagine, cadaverine, and unknown substances were determined in the hydrolysates of NSTX-1 and NSTX-2. The complete structures of the toxins NSTX-1 and NSTX-2 are still not known. Equimolar amounts of asparagine, cadaverine, putreanine, and arginine were found for NSTX-3. Edman degradation revealed arginine. The degradation did not proceed to the next cycle, suggesting that arginine was located at a terminal end of NSTX-3 and that the α-amino group of arginine was not protected. Detailed ^{1}H-NMR, DQF-COSY, and NOE NMR experiments were conducted to determine the chromophore and how the structural units (2,4-dihydroxyphenyl)acetic acid, asparagine, cadaverine, putreanine,* and arginine were connected (*105*).

A ^{1}H,^{1}H-COSY NMR spectrum and FAB-MS measurement of NSTX-3 were observed. From these data, NSTX-3 was deduced to be *N*-[(2,4-dihydroxyphenyl)acetyl-asparaginyl]-*N'*-(arginyl-cadaverino-β-alanyl)-cadaverine (*109*). The authors slightly revised this structure based on improved NMR data a year later to *N*-[(2,4-dihydroxyphenyl)acetyl-L-asparaginyl]-*N'*-(L-arginyl-putreanyl)-cadaverine (*105,112*) (Fig. 25). This example shows the difficulty in interpreting NMR spectra obtained with small amounts of sample.

In 1987 Aramaki *et al.* (*106*) isolated JSTX-1 to JSTX-3 (Fig. 23), determined the structure of JSTX-3, and confirmed it by synthesis. The synthetic proof for the structure of JSTX-3 was also given by Hashimoto *et al.* (*54*) (refer to Section III,B). Comparison of the data from the ^{1}H-NMR and

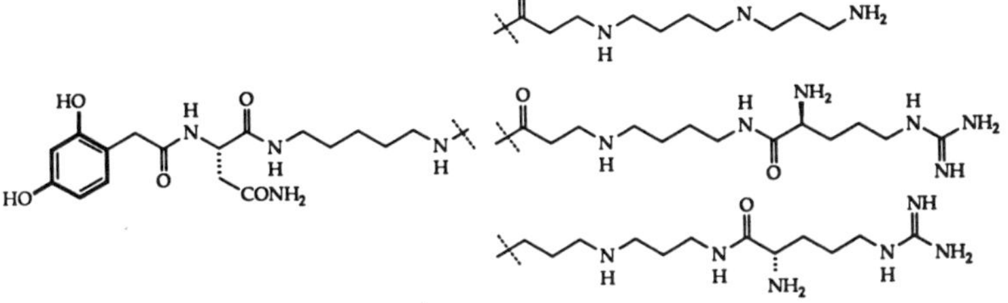

FIG. 25. Structures of the toxins JSTX-3, NSTX-3, and Arg 636.

* Putreanine was first reported by Kakimoto *et al.* (*110*) in mammalian brain and was synthesized by Shiba and Kaneko (*111*). Although the biological activity of putreanine is not yet clear, it might be of interest that this compound is incorporated into spider toxins.

UV spectra, the retention time on HPLC, the biological activities, and the diastereomeric derivatization of the synthetic compound with the data obtained from the natural JSTX-3 revealed that the structure of the two compounds, including the absolute configuration, was "strictly identical" (*54*).

Teshima *et al.* (*57*) confirmed the correctness of the structure of NSTX-3 proposed by Aramaki *et al.* (*105,106,108,112*) through preliminary synthetic studies in 1988, and finally by total synthesis in 1991 (*54,58*) (refer to Section III,B). Both NSTX-3 and JSTX-3 possess a (2,4-dihydroxyphenyl)acetyl-L-asparaginyl-cadaverine-putreanyl substructure, but they differ in that NSTX-3 terminates with an arginine and JSTX-3 with a propylamine residue (Fig. 25).

Among the JSTX and NSTX toxins, compounds JSTX-3 and NSTX-3 seem to be the most abundant in the venom and are therefore the most investigated. Their structures have been confirmed by synthesis. The basic characterization of JSTX-3 and NSTX-3 followed the same principles (*112*). A report concerning the structural characterization of the other JSTX toxins appeared a few years later (*104*), employing essentially the same analytical methods.

By comparison of the NMR data from the NSTX and JSTX toxins, it seems that the JSTX and NSTX toxins are all (2,4-dihydroxyphenyl)acetyl-asparaginyl-cadaverine derivatives. As shown in the DQF-COSY spectra of JSTX toxins (*106*) in DMSO-d_6, the signals of the 2,4-dihydroxyphenyl moiety, the asparagine, the amide protons of the asparagine, and the cadaverine part are in common with other JSTX toxins and NSTX-3 (for structures see Fig. 23).

During the isolation and purification of JSTX and NSTX toxins Toki *et al.* (*101,107*) discovered that the crude venom had potent activity for mast cell degranulation. They came to the conclusion that the responsible histamine releasers were structurally closely related to the JSTX and the NSTX toxins. The venom extract from about 10 spiders of *Nephila clavata* was separated into 20 fractions by reversed-phase HPLC (Fig. 26). Fraction 7 was rechromatographed, and three active principles, the *Nephila* toxins NPTX-7 to NPTX-9 (see Fig. 30), were isolated. They all showed the same UV absorption at 210, 220, 280, and 290 (shoulder) nm. Each compound was partially hydrolyzed in 50 μl of 5.7 N aqueous HCl under N_2 at 100°C for 20 min, and the hydrolysate was once again separated by reversed-phase HPLC. As an example, the chromatogram of the hydrolysate of NPTX-9 is shown in Fig. 27.

The N-terminal residues of the toxins and those of the partial hydrolysate were determined by the dansylation technique according to the usual method (*107*). To distinguish between aspartic acid and asparagine in each

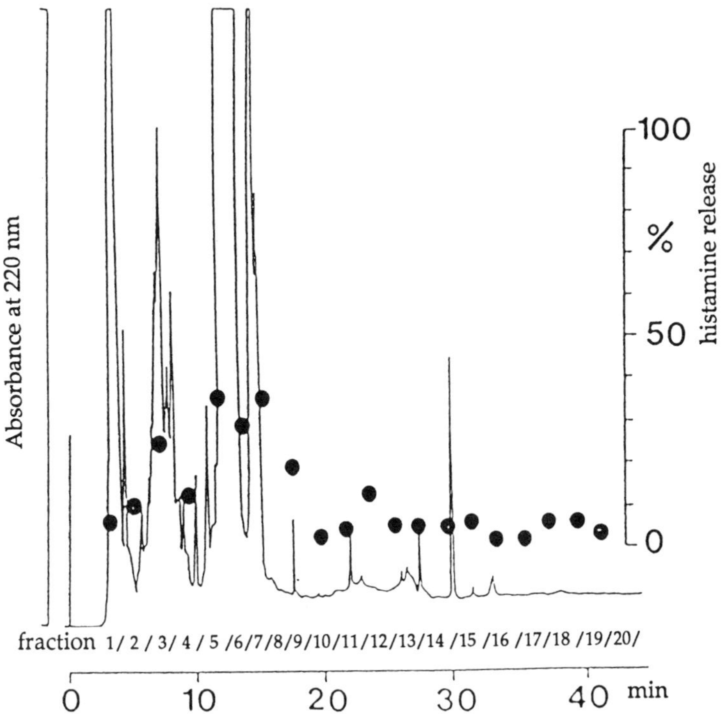

FIG. 26. Separation of components of the venom obtained from *Nephila clavata*. Dots symbolize histamine release in a bioassay. (From Ref. *101*, with permission.)

sequence, the native toxin (100 pmol in 10 μl water) was treated with 10 μl of bis(1,1-trifluoroacetoxy)iodobenzene (at 50°C for 10 min) to oxidize, in the case of asparagine, the carboxyamide of the asparagine residue to α,β-diaminopropionic acid. The reaction mixture was then subjected in the usual manner to total hydrolysis. The disappearance of an aspartic acid residue in the hydrolysate, which was examined by an amino acid analyzer, can thus offer proof of asparagine as the amino acid. NPTX-7 is the only *Nephila* toxin with an aspartic acid unit; all the others contain asparagine. The amino acid and polyamine compositions of NPTX-7 to NPTX-9 were determined (NPTX-7: Asp, Orn, PA5, Pta, Pta; NPTX-8: Asn, PA5, AP-Pta; NPTX-9: Asn, PA5, Orn, Arg).

Peaks 1, 2, and 3 in the chromatogram given in Fig. 27a were composed of amino acids and polyamines. The composition of amino acids and polyamines in each fragment and their sequence based on the fragmentation are summarized in Fig. 27b. Peak 4 did not contain any amino acids or polyamines, but it showed a UV spectrum similar to that of the native

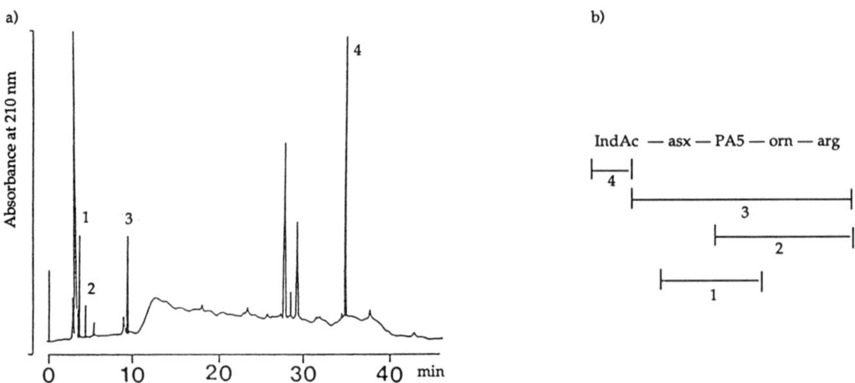

FIG. 27. Separated fragments of NPTX-9 obtained by partial hydrolysis. (From Ref. *101*, with permission.)

toxin [λ_{max} at 220, 280, and 290 (shoulder) nm], suggesting the presence of an indole group in the molecule. The corresponding peak was also detected in the hydrolysate of *Nephila* toxins NPTX-7 and NPTX-8. The component in peak 4 was thus thought to be an indole attached to the N terminus of the asparagine (or, in the case of NPTX-7, ornithine) since no free primary amino group was detected by dansylation of the toxins. The compound corresponding to peak 4 (Fig. 27) appeared to be unstable under the usual hydrolysis conditions of 5.7 *N* aqueous HCl; therefore, a mixture of *Nephila* toxins was hydrolyzed by 25 m*M* oxalic acid for 16 hr. Further data for structural analysis were provided by the ^{1}H-NMR spectrum in D_2O, suggesting that the component represented by peak 4 was an (indol-3-yl)acetic acid [NMR signals at 3.83 (s), 7.12 (t), 7.20 (t), 7.26 (s), 7.46 (d), and 7.58 (d) ppm]. In addition, the elution time of peak 4 in reversed-phase HPLC and its UV spectrum agreed completely with those of (indol-3-yl)acetic acid (*101*), a chromophore which was also found in the *Argiope* toxins (pseudoargiopinines; see Section V,A referring to *Argiope* toxins).

The UV absorption spectra of NPTX-7 to NPTX-9 all showed the same absorption maxima at 210, 220, 280, 290 (shoulder) nm. UV and ^{1}H-NMR data indicated that NPTX-7 to NPTX-9 (Fig. 30) all carry an (indol-3-yl)acetyl moiety as the common substituent at their N-termini, whereas the JSTX and NSTX toxins were reported to have a (2,4-dihydroxyphenyl)-acetyl chromophore (*101,105,106,108,112*). Further investigations enabled Toki *et al.* (*107*) to isolate twelve different *Nephila* toxins (NPTX-1 to NPTX-12, Figs. 29 and 30). The methods employed for the isolation, besides modification of the solvent for the first extraction step, and the characterization were the same as already discussed. Reversed-phase

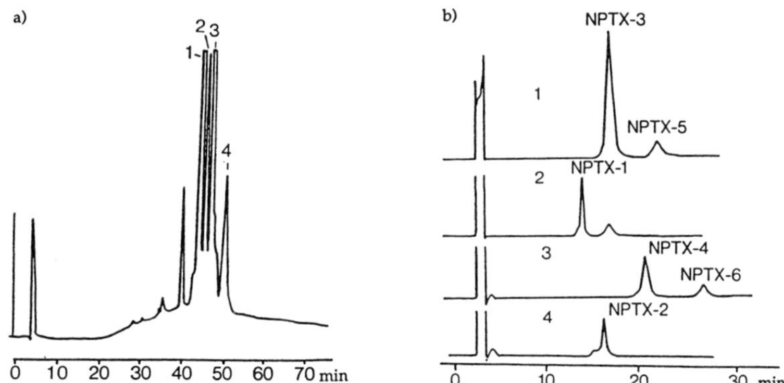

FIG. 28. (a) HPLC chromatogram of the rechromatographed fractions 5 and 6 of Fig. 26. (From Ref. *107*, with permission.) (b) HPLC chromatogram of the purified toxins NPTX-1 to NPTX-6. (From Ref. *101*, with permission.)

HPLC was used for polyamine purification, and the structure elucidation was carried out with data obtained from UV and ^{1}H-NMR spectra, as well as from amino acid and polyamine analysis.

The major toxic compounds extracted by acetonitrile/water (3:2) from the venom glands of Joro spiders were of the JSTX type, namely, (2,4-dihydroxyphenyl)acetyl derivatives. Provided the venom glands were extracted by slightly acidic medium, the extract contained another type of toxic material. Figure 26 shows a chromatogram of the crude venom following acid extraction. Fractions 5 and 6 (Fig. 26) were combined and rechromatographed on an ion-exchange column to yield four active fractions (Fig. 28a). Each fraction was collected and further separated by

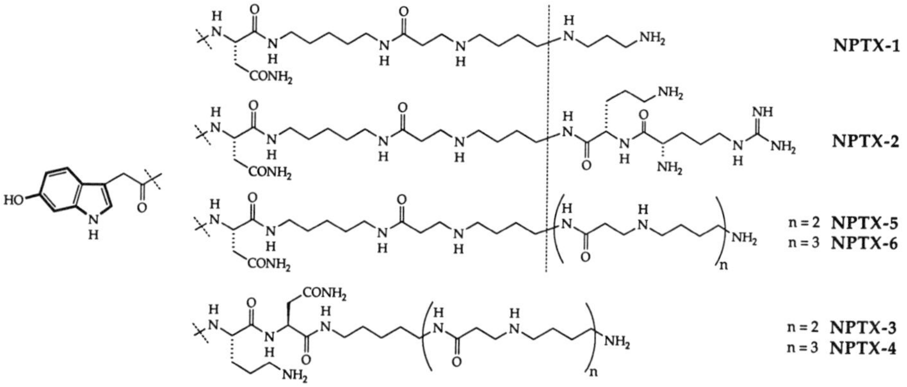

FIG. 29. Structures of NPTX-1 to NPTX-6.

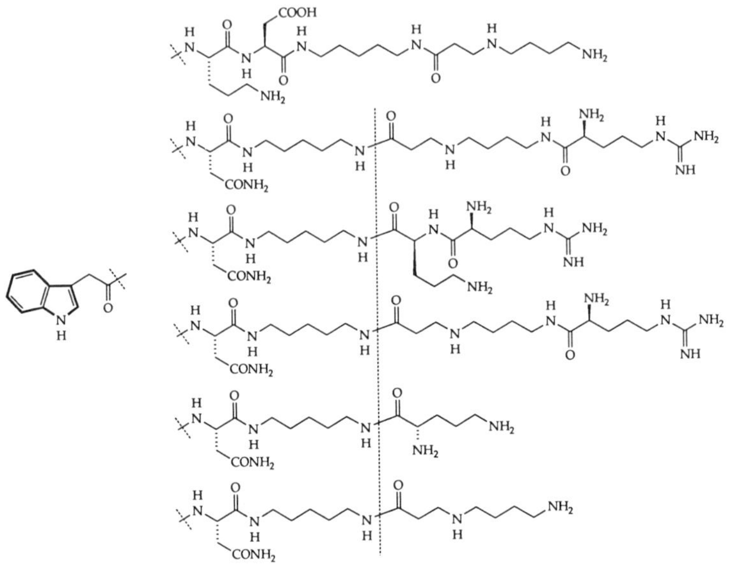

NPTX-7
NPTX-8
NPTX-9
NPTX-10
NPTX-11
NPTX-12

FIG. 30. Structures of NPTX-7 to NPTX-12.

HPLC with an isocratic elution system (Fig. 28b). The purified active materials were named *Nephila* toxins NPTX-1 to NPTX-6 (Fig. 29) according to the order of elution. Fraction 7 in the chromatogram (Fig. 26) was rechromatographed with reversed-phase HPLC, and the *Nephila* toxins NPTX-7 to NPTX-12 (Fig. 30) were obtained.

The UV spectra of NPTX-1 to NPTX-6 (λ_{max} at 220, 265, and 280 nm) were slightly different to those of NPTX-7 to NPTX-12 [λ_{max} at 220, 280, and 290 (shoulder) nm]. Analysis by EI- and field desorption (ionization) (FD)-MS of the chromophoric unit suggested the aromatic component in NPTX-1 to NPTX-6 (Fig. 29) to be a (hydroxyindol-3-yl)acetyl, and that in NPTX-7 to NPTX-12 (Fig. 30) an (indole-3-yl)acetyl moiety. These results were supported by ^{1}H-NMR data after deuterium exchange (*107*).

For NPTX-1 to NPTX-6, EI- and FD-MS (m/z 191, M$^+$·) indicated that these toxins contain a (monohydroxyindol-3-yl)acetic acid moiety; the pattern of the fragments in the EI mass spectrum (signals at m/z 218, 159, 144, 129, and 116) of the dimethylated derivative was identical with that of a (5-hydroxyindol-3-yl)acetic acid moiety. However, the elution time of the corresponding acid on reversed-phase HPLC was different from that known for (5-hydroxyindol-3-yl)acetic acid. ^{1}H-NMR measurements after deuterium exchange gave more information concerning the substitution of the chromophore of toxins NPTX-1 to NPTX-6. The aromatic proton signals appeared at 7.05 and 7.10 ppm (each a singlet, integrating for one proton). Because two neighboring proton signals adjacent to an aromatic hydroxyl group were deuterated in D_2O, the aromatic protons of (4-/7-hydroxyindol-3-yl)acetic acid should appear as three signals (in the case of the 4-hydroxyindol-3-yl type the protons at positions 2, 6, and 7, and in the case of the 7-hydroxyindol-3-yl type the protons at positions 2, 4, and 5). Thus, the ^{1}H-NMR signals of this component differed from those of a (4- or 7-hydroxyindol-3-yl)acetic acid species. The signals were related to those of a similarly deuterated N-acetyl-6-hydroxytryptophan methyl ester [aromatic part: 7.10, 7.15 (2s) ppm]. These data suggested the aromatic part to be a (6-hydroxyindol-3-yl)acetic acid moiety (*107*).

It seems astonishing, considering the high similarity among the spider toxins, to find two different types of hydroxyindole chromophores, namely, (4-hydroxyindolyl-3-yl)acetic acid in spiders of the genera *Argiope*, *Agelenopsis*, and *Hololena* (*1,65,94,102,113*) and (6-hydroxyindol-3-yl)acetic acid in spiders of the genus *Nephila* (*107*). The UV spectra reported for the (4- or 6-hydroxyindol-3-yl)acetic acid moiety are very similar.

The amino acid compositions of toxins NPTX-1 to NPTX-12 are given in Table VIII. The N-terminal amino acids or polyamine in each hydrolysate were analyzed by the dansyl method.

As already mentioned, Aramaki *et al.* (*108*) reported the isolation of four different toxic components (JSTX-1 to JSTX-4) from *Nephila clavata* in

TABLE VIII

AMINO ACID AND POLYAMINE COMPOSITION OF NPTX-1 TO NPTX-12[a,b]

Spider toxin	Molecular weight	Asx	Orn	Pta	Arg	AP-Pta	PA5
NPTX-1	588	1	—	—	—	1	1
NPTX-2	801	1	1	1	1	—	1
NPTX-3	819	1	1	2	—	—	1
NPTX-4	929	1	1	3	—	—	1
NPTX-5	815	1	—	3	—	—	1
NPTX-6	957	1	—	4	—	—	1
NPTX-7	772	1	1	2	—	—	1
NPTX-8	572	1	—	—	—	1	1
NPTX-9	643	1	1	—	1	—	1
NPTX-10	671	1	—	1	1	—	1
NPTX-11	487	1	1	—	—	—	1
NPTX-12	515	1	—	1	—	—	1

[a] From Ref. *107*, with permission.
[b] Asx, Asn or Asp; Pta, putreanine; AP-Pta, ω-(3-aminopropyl)putreanine.

1986. The ^{1}H-NMR spectra suggested that the four toxins were structurally similar. The major component, JSTX-3, was characterized and analyzed in the following 2 years (*106,112*). In 1990, the other compounds of the same group were investigated (*104*). The analytical methods and procedures employed for the structure determination have already been described above. Briefly, amino acids and polyamines were obtained by complete hydrolysis of the toxins and were determined by an amino acid/polyamine analyzer with the OPA method as described by Nakajima *et al.* (*114*). Sequential analysis of the polycationic part of the toxin was carried out by a combination of N-terminal analysis with the dansyl method and fragmentation of the toxins by partial acid hydrolysis (*101,107*). The intact toxin was first dansylated and then hydrolyzed in the usual fashion with 5.7 *N* aqueous hydrochloric acid at 110°C for 20 hr. The dansylated amino acid and the polyamines were identified by reversed-phase HPLC using authentic samples as standards (*104*). A partial hydrolysate (5.7 *N* aqueous hydrochloric acid at 100°C for 20 min) was treated in a similar manner, and the composition of the amino acids and polyamines of each fragment was determined. Confirmation of the presence of the (2,4-dihydroxyphenyl) acetyl group was provided by HPLC analysis of the hydrolyzed component. After the hydrolysis of the JSTX toxins with 25 m*M* oxalic acid at 100°C for 16 hr, the hydrolysate was subjected to reversed-phase HPLC (*106*). The chromphoric unit was identified based on the elution time and the UV spectrum monitored as a (2,4-dihydroxyphenyl)acetyl moiety.

The toxin JSTX-1 elutes at quite a different retention time than the other three JSTX toxins. The toxin JSTX-2 contains, besides the (2,4-dihydroxyphenyl)acetic acid chromophore, one molecule each of asparagine, cadaverine, and ornithine, and two putreanine moieties. This amine component was found also as a part of NPTX-3 (*101*). The amino acid and polyamine components of JSTX-4 are composed of one molecule each of asparagine, cadaverine, ornithine, and arginine, a composition that is identical to that of the known NPTX-9 molecule (*101*). Sequence analysis of JSTX-2 and JSTX-4 gave the same results as that of NPTX-3 and NPTX-9, respectively. According to these data, the structures of JSTX-2 and JSTX-4 were deduced.

The analytical methods employed for the final structure elucidation of the *Argiope* toxins were mainly 2D NMR and MS, whereas for the *Nephila* toxins (JSTX, NSTX, and NPTX toxins), amino acid and sequential analysis of the polyamine by partial acid hydrolysis and dansylation were used as well. The amino acids of the latter unit were determined to be L-configured.

Chemical characterization of the venom that had been called "Joro spider toxin" in 1982 led, in 1990, to the conclusion that the active principle, called at that time JSTX, is a mixture of at least 16 closely related compounds (JSTX-1 to JSTX-4 and NPTX-1 to NPTX-12) (*104*). In the same year, Yoshioka *et al.* (*115*) discovered an additional toxin in the venom glands of the Joro spider *Nephila clavata*, which they called clavamine. This component has insecticidal activity against wrigglers. It did not show irreversible suppression of the excitatory postsynaptic potential (ESP), which is characteristic of the spider toxins NSTX and JSTX (*92,108*). The structure of clavamine (Fig. 31) was deduced by NMR and mass spectral analysis, as well as by Edman degradation and dansylation (*115*), and finally verified by synthesis (*59*) (see Section III,B).

Toki *et al.* (*101,107*) and Aramaki *et al.* (*108*) did not report the presence of clavamine in the venom of *Nephila clavata*. This is most likely due to the fact that they extracted the venom with acetonitrile/water (3:2) instead of pure water at the first isolation step (*115*). Yoshioka *et al.* (*115*) judged from the gel filtration that clavamine is the main active component of the venom. The results were obtained in the following fashion. The venom glands of 4850 spiders were homogenized and extracted with boiling water to yield about 1000 mg lyophilized crude venom. Table IX gives an impression of the amount of toxin present in the venom, and the difficulties involved in purification. On the one hand, care has to be taken not to lose too much of the compound, and on the other hand one has to make sure that the toxin is really pure.

From the gel filtration profile shown in Fig. 32a, the active fractions were found to be the second and third peaks. The molecular weight of the compound in the second peak (Fig. 32a) corresponded to about 800 and was further purified by HPLC (Fig. 32b). The compound in the first peak was almost pure. The Sephadex purification suggested that in the venom clavamine is present in form of a metal complex. This complex is most likely demetallized by the ion-exchange chromatography. It seems that this kind of metal complexation is important for the toxicity of clavamine (*115*).

Twenty-five nanomoles of the toxin were hydrolyzed and derivatized for gas chromatography (GC). From the chromatogram shown in Fig. 33, it was concluded that the toxin consists of glycine, cadaverine,

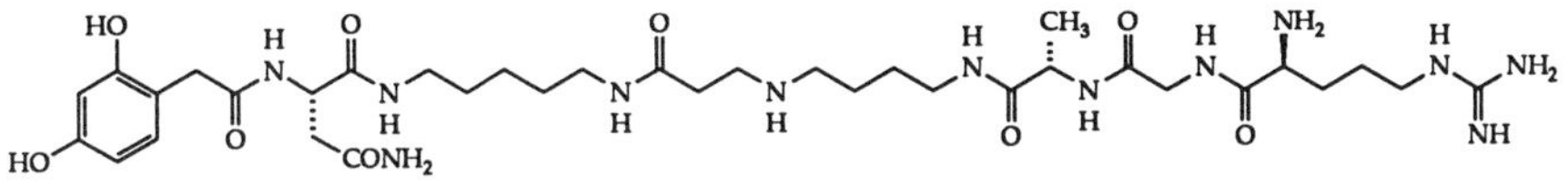

2,4-DiOH-PhAc → asn —→ PA5 ←——————— Pta ←——— ala ←— gly ←——— arg

FIG. 31. Structure of clavamine (*115*).

TABLE IX
RECOVERIES OF CLAVAMINE ON PURIFICATION[a]

Step	Dry weight (mg)	Activity (GU)[b]	Specific activity (GU/mg)	Recovery (%)
Water extraction	1000	9700	10	100
Gel filtration on Sephadex G-10	46	7400	162	77
HPLC	12	6700	570	69
Demetallation through SP-Sephadex	9	6400	690	66

[a] From Ref. *115*, with permission.
[b] GU, Gland unit.

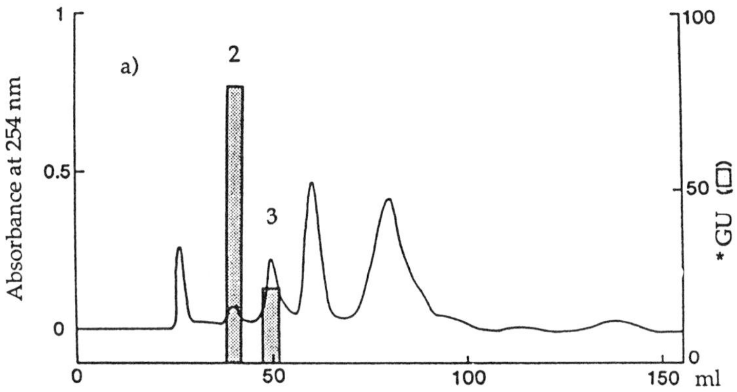

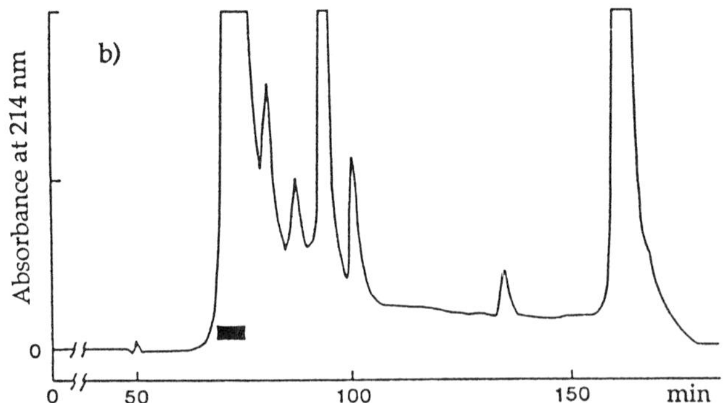

FIG. 32. (a) Chromatogram of the venom extract of *Nephila clavata* (Sephadex G-10; *GU, gland unit. (b) Chromatogram of the partially purified toxin of the second peak of the chromatogram in (a) by preparative HPLC. (From Ref. *115*, with permission.)

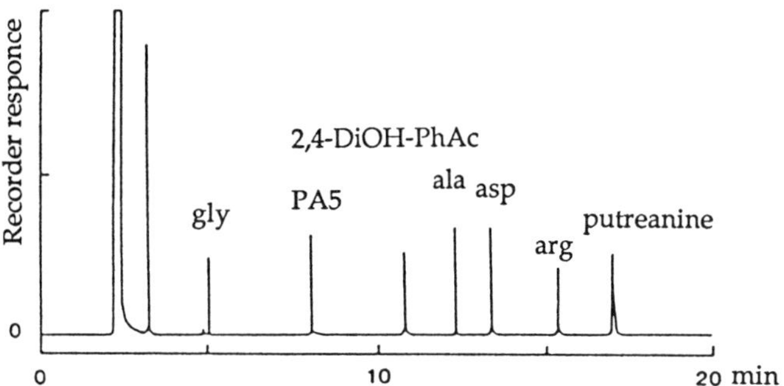

FIG. 33. Gas chromatogram of the derivatized components of hydrolyzed clavamine. The carboxyl groups of the compounds were converted to the butyl esters, and the free amino groups were acylated by heptafluorobutyric acid. (From Ref. *115,* with permission.)

a (2,4-dihydroxyphenyl)acetyl moiety, alanine, arginine, and putreanine. The derivatized compounds corresponding to the peaks in the gas chromatogram were confirmed by GC-MS analysis. Cadaverine, (2,4-dihydroxyphenyl)acetyl, and L-asparagine are in common with the components of JSTX-3. The configuration of all the amino acids of clavamine were defined by HPLC, using a chiral column. Edman degradation and dansylation in each step showed a single spot of Arg, Gly, and Ala in that order. A tripeptide Arg→Gly→Ala was easily postulated, but "after the third step it was difficult to explain the degradation mechanism" (*115*). The other end of the compound was assumed to be (2,4-dihydroxyphenyl)acetyl→Asn→PA5←Pta, deduced from the similarity of the structure to that of JSTX-3. By connecting the ends, the complete structure was proposed to be (2,4-dihydroxyphenyl)acetyl→Asn→PA5←Pta←Ala←Gly←Arg. The structure assignment of clavamine with a molecular weight of 792 was supported by the ^{1}H,^{1}H-COSY NMR spectrum, which is shown in Fig. 34, and by the FAB mass spectrum, given in Fig. 35.

The structure and composition of clavamine differ greatly from those of the other low molecular weight toxins identified so far from *Nephila clavata*. In this case, a tripeptide is connected to the polyamine. This might give rise to the different pharmacological activities. The purification process suggested that clavamine is present in the venom in form of a metal complex (*115*). Owing to the choice of solvent, clavamine had been missed with the initial extraction methods used for the venom of *Nephila clavata* (*101,105–107*).

C. AGELENIDAE TOXINS FROM *Agelenopsis aperta* AND *Hololena curta*

The venom of common funnel-web spiders of the family Agelenidae, found throughout the western United States, contains both peptide and

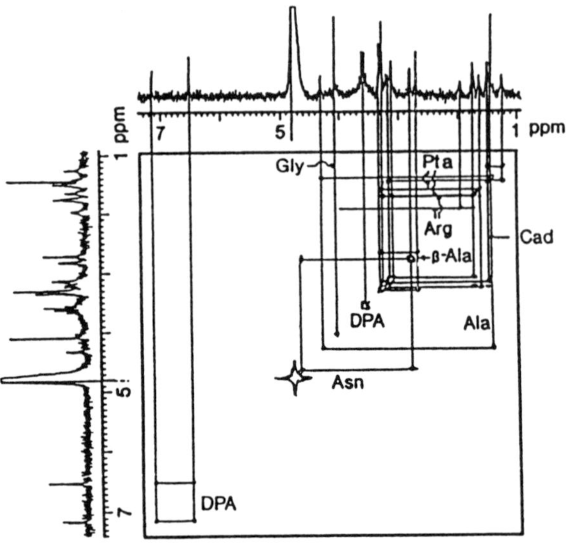

FIG. 34. ¹H,¹H-COSY NMR spectrum of the demetallized toxin clavamine (DPA, 2,4-DiOH-PhAc; Cad, PA5). (From Ref. *115*, with permission.)

polyamine constituents. Intensive investigations on Agelenidae toxins started only a few years ago.

In 1987, the pharmacological effect of the venom of *Hololena curta,* a funnel-web spider of the family Agelenidae, was reported (*116*). In 1988, three types of *Agelenopsis* toxins were found in another funnel-web

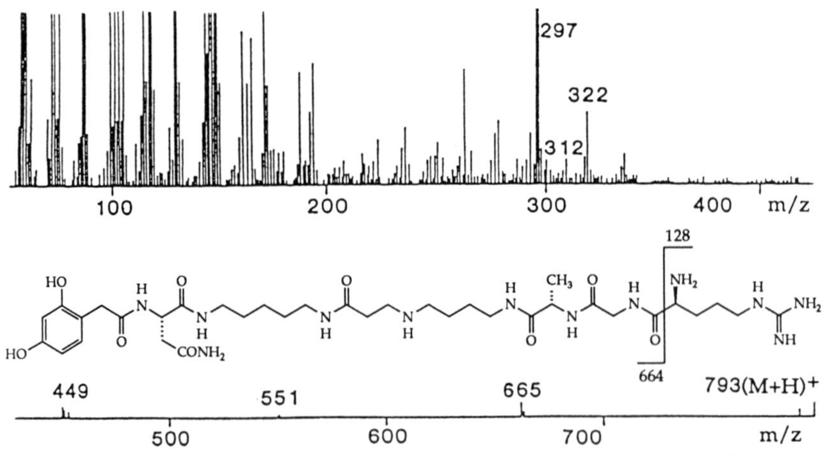

FIG. 35. FAB mass spectrum of clavamine. (From Ref. *115*, with permission.)

spider, *Agelenopsis aperta* (*96*): α-agatoxins, which are low molecular weight toxins (<1000); μ-agatoxins, cysteine-rich peptides with about 36 amino acids; and ω-agatoxins, polypeptides with molecular weights of 7000–15,000 (*96*).

The low molecular weight toxins, the α-agatoxins, compared to the μ- and ω-agatoxins, are quite hydrophilic, which aided in their separation by chromatographic methods (*97*). In 1988, four of the α-agatoxins were isolated and their molecular weights assigned (Agel 488, 489, 505a,* and 505). It could be shown that they contain no amino acids, but rather consist of basic polyamine moieties and a chromphoric unit (*96*). In 1989, an additional α-agatoxin (Agel 452) was isolated and its mass determined (*97*). In 1990, it was possible to isolate 10 low molecular weight *Agelenopsis* toxins (Agel 416, 448, 452, 464, 468, 489, 489a, 505, 505a, and 521). The structure elucidation of 5 of them turned out to be relatively easy owing to their similarity. All five structures could be confirmed by total synthesis (Agel 448, 452, 468, 489, and 505) (*1*).

The isolation and purification were performed similarly to the Araneidae toxins. From *Agelenopsis aperta*, the crude venom was mainly obtained by milking the spiders. An average milking of one spider yielded 0.2 μl venom. By use of synthetic standards, the concentration of individual polyamine constituents in the whole venom was determined (*1*). Thus, 1 μl of whole venom from *A. aperta* contains approximately

3.54 μg of Agel 448	3.80 μg of Agel 521
2.86 μg of Agel 452	0.96 μg of Agel 464
1.32 μg of Agel 468	1.50 μg of Agel 416
4.70 μg of Agel 505a	9.47 μg of Agel 489a
5.59 μg of Agel 505	33.02 μg of Agel 489

The structure of the known polyamine toxins isolated from orb-weaving spiders (Araneidae) include the general sequence arylacetamide-amino acid$_1$-polyamine-(amino acid$_2$). They can be divided according to the nature of the chromophore into three classes (*94,102*) (see Section V,A). Similar results were found for the funnel-web spider *Agelenopsis aperta*. Although it was assumed in the beginning that the *Agelenopsis* toxins all carry "exclusively indole type of chromophores" (*96*), UV analysis of purified constituents obtained with a diode array detector indicated four types of chromophores (cf. Fig. 39).

The structure elucidation of the *Agelenopsis* toxins is mainly based on MS data, whereas for the *Argiope* toxins the main structural information

* In some earlier literature this toxin is called Agel 504, but in more recent literature it is referred to as Agel 505a.

was deduced from NMR data. For the *Nephila* toxins, besides NMR data, polyamine analysis and sequential analysis of the polyamines by partial hydrolysis played the most important rule.

The toxin Agel 489 (Fig. 36) is the most abundant polyamine extracted from *Agelenopsis aperta,* and, as a result, it was analyzed in greatest detail (*1*). A ^{13}C-NMR spectrum of this compound suggested the presence of an (indol-3-yl)acetamide moiety. Subtracting out the elemental composition of the (indol-3-yl)acetamide from the intact molecule leaves a unit with the composition $C_{16}H_{39}N_6O$, the high nitrogen content suggesting the presence of a polyamine substructure. In addition, the presence of an oxygen atom in the substructure implied either a hydroxy or an ether linkage. In the low-resolution FAB mass spectrum, an ion observed 18 amu lower than the molecular ion might be due to the loss of water from the intact molecule, and it could indicate that the oxygen is part of a hydroxyl group. Many low-intensity fragment ions are observed, and on some of them high-resolution FAB-MS was performed (*1*). The fragmentation suggested that the hydroxyl group had to be located somewhere on the first or second propylamine unit. An important clue to the location of the hydroxyl group in Agel 489 was provided by a DEPT ^{13}C-NMR experiment, which suggested that no aliphatic methine group is present in the molecule, implying attachment to one of the nitrogen atoms. However, one has to be aware of the fact that the absence of an appropriate signal in the DEPT ^{13}C-NMR spectrum is not rigorous proof excluding the presence of a methine carbon (*1*). The assignment of the position of the hydroxyl function was a difficult task, since it could still reside on any of the six nitrogen atoms. Based on NMR data, its location in the 1- or 22-aza position could be excluded (*31*). Analysis by FAB-MS/MS (Fig. 38) of the molecular ion gave strong evidence for the 5-aza position, but an uncertainty remained. Final evidence could be given by high-resolution FAB-MS data (*1*), which revealed that the ion at *m/z* 215 coincidentally represents a doublet resulting from two different fragments. Fragment A ($C_{13}H_{15}N_2O$, Fig. 37) contains the chromophore unit and part of the polyamine; fragment B ($C_{11}H_{27}N_4$, Fig. 37) represents the part of the polyamine unit containing no oxygen. From these data, it was possible to

FIG. 36. Structure of Agel 489.

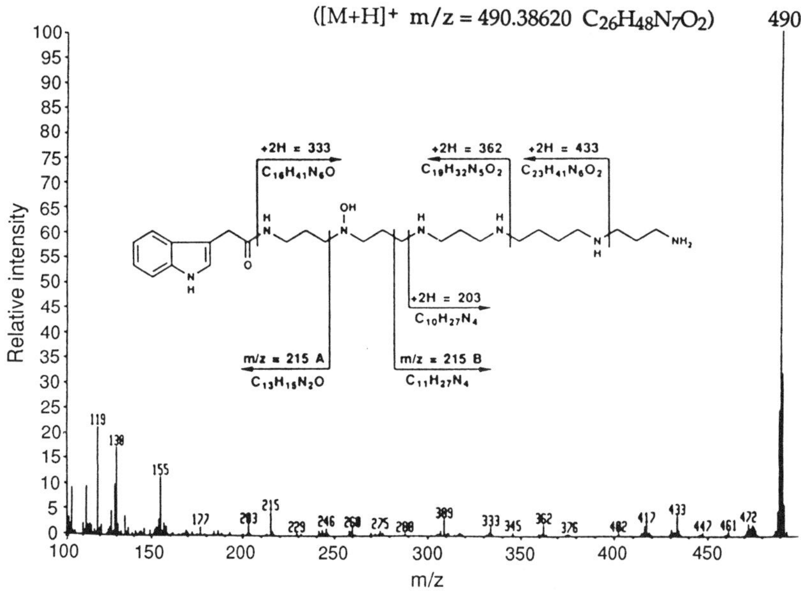

FIG. 37. FAB-MS analysis of Agel 489. (From Ref. *1*, with permission.)

deduce that the hydroxyl function has to be located between the two fragments A and B, that is, in the 5-aza position (see Figs. 37 and 38).

FAB mass spectra of the purified components of polyamine spider toxins yield intense $(M + H)^+$ ions or, in the case of quaternary ammonium salts, M^+ ions, along with a number of fragment ions which are useful for structure elucidation. These fragments provided the key data for establishing the polyamine functionality of the *Agelenopsis* toxins. High-resolution FAB-MS analysis was carried out on four *Agelenopsis* toxins, namely, Agel 489, 468, 452, and 505, and on some fragments of these compounds.

Four of ten toxins isolated from *Agelenopsis aperta* (Agel 489, 505, 452, and 468) exhibit similar FAB-MS fragmentation. As a result, Jasys *et al.* (*1*) concluded that these compounds contain the same polyamine side chain and differ only in the aromatic residue (Fig. 39). The chromophores found in Agel 489 (IndAc) and Agel 505 (4-OH-IndAc) were already familiar from the Araneidae toxins, whereas the 2,5-dihydroxybenzoyl and the 4-hydroxybenzoyl chromophores (Agel 468, 452) had not been known and seem to be typical for the Agelenidae toxins. Subtracting the polyamine side chain from the elemental composition of the molecular ion, which was determined by high-resolution FAB-MS, the elemental composition of the chromophore was obtained. The presence and position of the hydroxyl

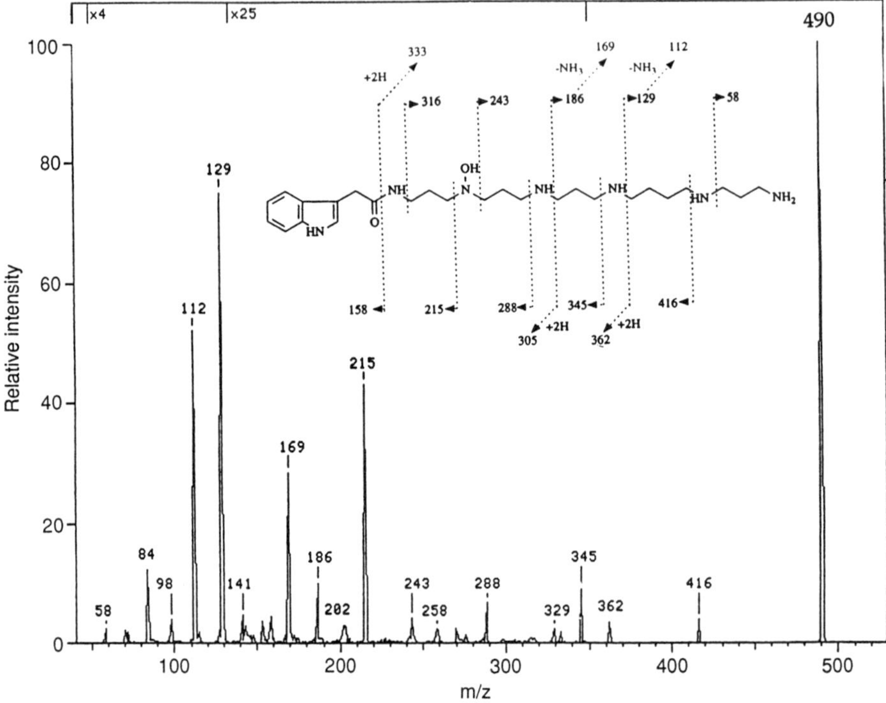

FIG. 38. FAB-MS/MS analysis of Agel 489. (From Ref. *31*, with permission.)

FIG. 39. *Agelenopsis* toxins Agel 468, 452, 489, and 505 show the four types of chromophores found in the Agelenidae toxins and demonstrate the high degree of similarity among them.

groups in the chromophores were determined by UV and NMR analysis (cf. Araneidae toxins).

The crude venom containing acylpolyamine and peptide components was usually fractionated by reversed-phase HPLC, yielding first the smaller and more hydrophilic polyamines (α-agatoxins) followed by the larger, highly disulfide-bridged peptides (μ-agatoxins). The two major acylpolyamine HPLC fractions from the initial separation were subsequently resolved by an additional HPLC run to afford in one instance Agel 489 and a fraction containing Agel 489a [Agel 488 in the former literature (*31,96,97*)] as well as Agel 505 and Agel 505a [Agel 504 (*31,96,97*)] (Fig. 40) (*1*).

The purified toxins were analyzed by Jasys *et al.* (*1,22*) by low- and high-resolution FAB-MS and by ^{1}H-NMR spectroscopy. In the case of Agel 489a the amount of the compound obtained was enough to provide a ^{13}C-NMR spectrum (Fig. 42a). UV spectroscopy and FAB mass spectrometry (Fig. 41) suggested that the two compounds in question (Agel 489a and Agel 505a) differ only in the aromatic chromophore, with one carrying an (indol-3-yl)acetamide (λ_{max} at 218, 279, 287 nm) and the other containing a (4-hydroxyindol-3-yl)acetamide moiety (λ_{max} at 219, 267, 282, 292 nm).

The assignment of the masses m/z 488 and 504 in the earlier literature (*31,96,97*) had been done based on FAB-MS assuming that the signals at m/z 489 and at m/z 505 would represent $(M + H)^+$ ions, without taking into account the possibility of quaternary ammonium ions, which correspond to M^+ ions in FAB-MS. However, the appearance of a intense methyl group singlet at 53.6 ppm in the ^{13}C-NMR spectrum and at 3.1 ppm in the ^{1}H-NMR spectrum suggested the presence of a di- or possibly trimethylammonium salt. This realization necessitated a change in the molecular weight assignment for Agel 488 and Agel 504. Jasys *et al.* (*1,22*) called them Agel 489a and 505a to distinguish them from Agel 489 and Agel 505, the two major polyamine constituents of the venom of the same spider, *Agelenopsis aperta*. In the first purification step, Agel 489/489a and Agel 505/505a, respectively, comigrate, and they happen to have the same molecular weight. High-resolution FAB-MS yielded the elemental compositions $C_{27}H_{49}N_6O_2$ and $C_{27}H_{49}N_6O_3$ for Agel 489a and Agel 505a, respectively. The elemental compositions of the fragment ions were also confirmed by high-resolution mass measurements. The diagnostic fragment ions at m/z 430 (from m/z 489) and m/z 446 (from m/z 505), respectively (Fig. 41), derived by loss of trimethylamine from the chain terminus, is consistent with the FAB mass spectra of other tetraalkylammonium salts (*117*).

Two signals between 57 and 58 ppm in the ^{13}C-NMR spectrum of the Agel 489a affirmed the suspicion that Agel 489a, like other previously

                    ANDREA SCHÄFER *ET AL.*

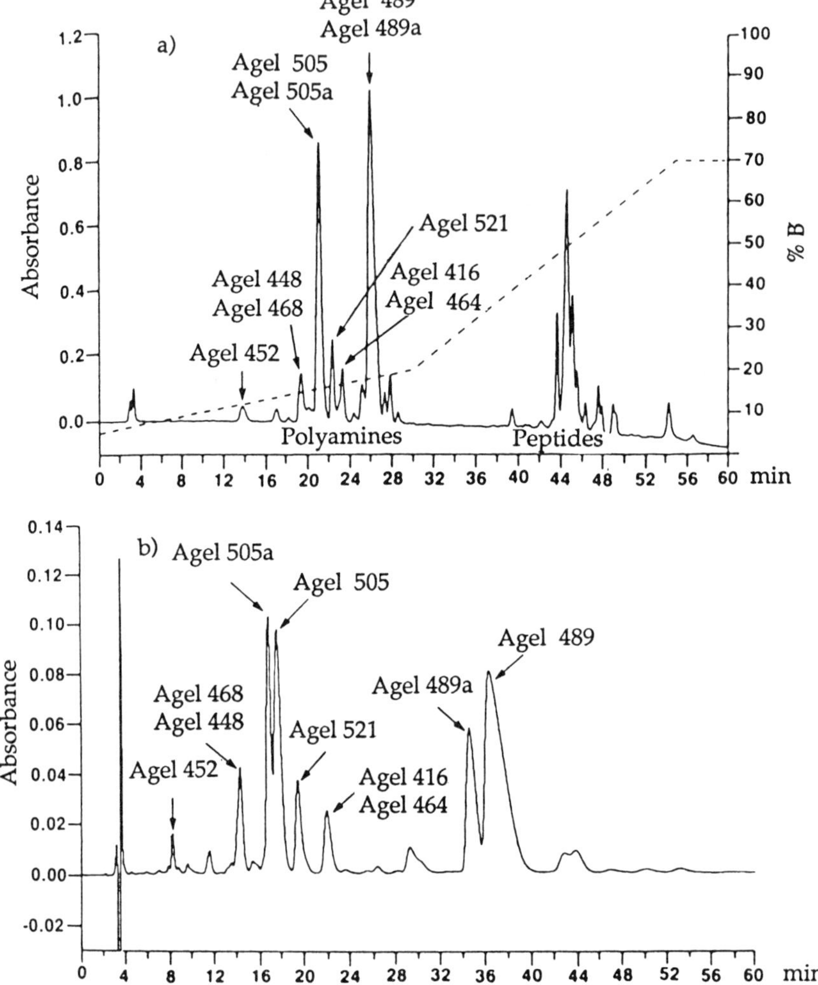

FIG. 40. (a) Reversed-phase HPLC separation of *Agelenopsis* venom using an isocratic 0–20% linear gradient of CH_3CN/H_2O containing 0.1% CF_3COOH for 30 min followed by a 20–70% gradient for 25 min (*1*). (b) Reversed-phase HPLC separation of *Agelenopsis* acylpolyamines using isocratic conditions, 8% CH_3CN/H_2O containing 0.1% CF_3COOH. (From Ref. *1*, with permission.)

identified *Agelenopsis* toxins, contains an internal hydroxylamine functionality. A PA3335 unit was suggested by the ^{1}H,^{1}H-COSY NMR spectrum. Based on these data, new structures for Agel 489a and 505a were proposed bearing a terminal quaternary ammonium group (*22*). Previously, the mass spectrometric data [chemical ionization (CI)- and FAB-MS] for Agel 489a and Agel 505a had been interpreted in a different way,

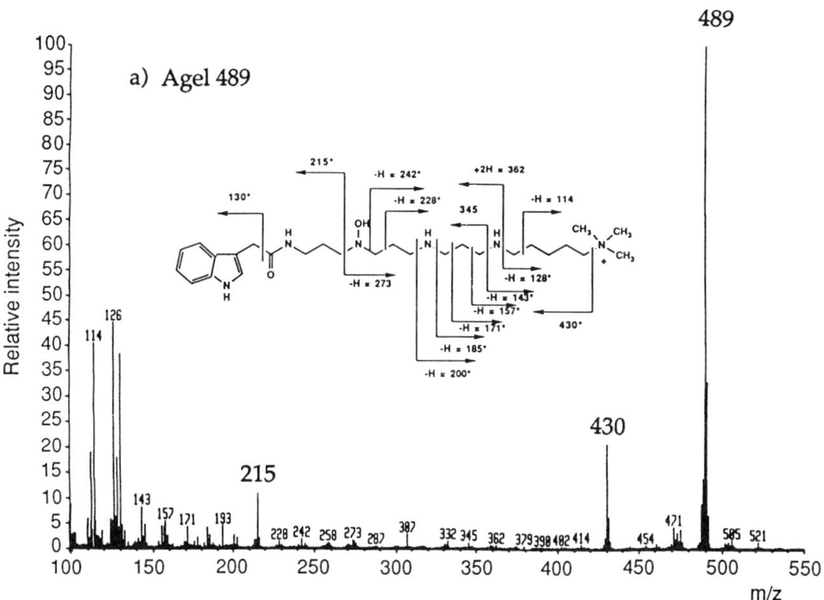

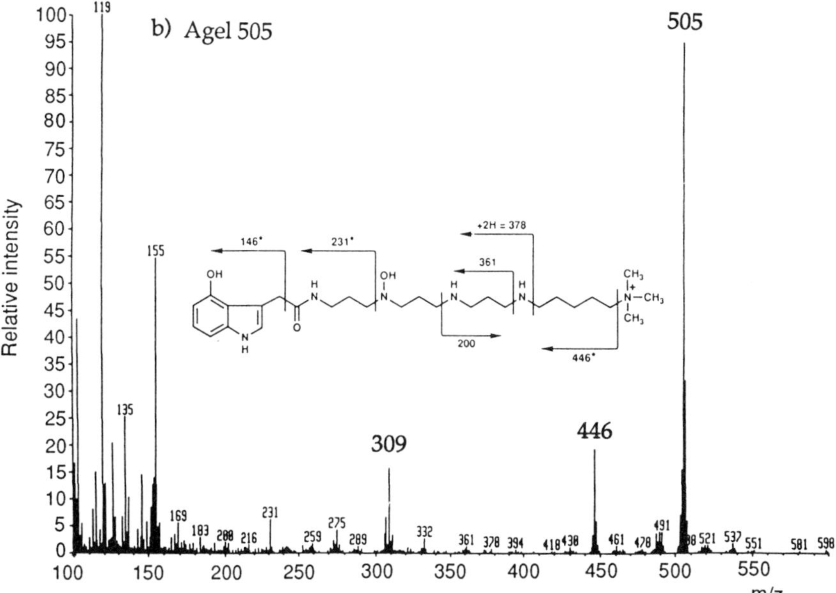

FIG. 41. FAB-MS analysis of the quaternary compounds (a) Agel 489a and (b) Agel 505a. (From Ref. 22, with permission.)

assuming an $(M + H)^+$ ion at m/z 489 and 505, respectively. The most predominant fragment signal in the upper mass range of the FAB mass spectrum of Agel 489a was, as mentioned, registered at m/z 430. The structure of the corresponding ion shown in Fig. 42 is in agreement with the spectra of similar compounds (*118,119*). It was not obvious to interpret the fragment ions at m/z 430 and m/z 414 for Agel 489a as the loss of a terminal guanidine/guanidinoxy moiety (*31*). Important is the participation of the neighboring group for the loss of trimethylamine. This is a good example of how mass spectrometry is a very powerful tool in the structure elucidation of natural products, in particular if only small amounts of the compound are available, but for definitive statements, in practically all cases, proof by several independent methods is required.

In 1991, one peptide and ten acylpolyamines (curtatoxins) were purified and identified from the funnel-web spider *Hololena curta* (*113*). They consist of six different polyamines, which are acylated with three kinds of chromophores, which are found as well in the *Agelenopsis* toxins: (indol-3-yl)acetic, (4-hydroxyindol-3-yl)acetic, and 2,5-dihydroxybenzoic acid (Fig. 43). Fractionation of the crude venom by reversed-phase HPLC gave

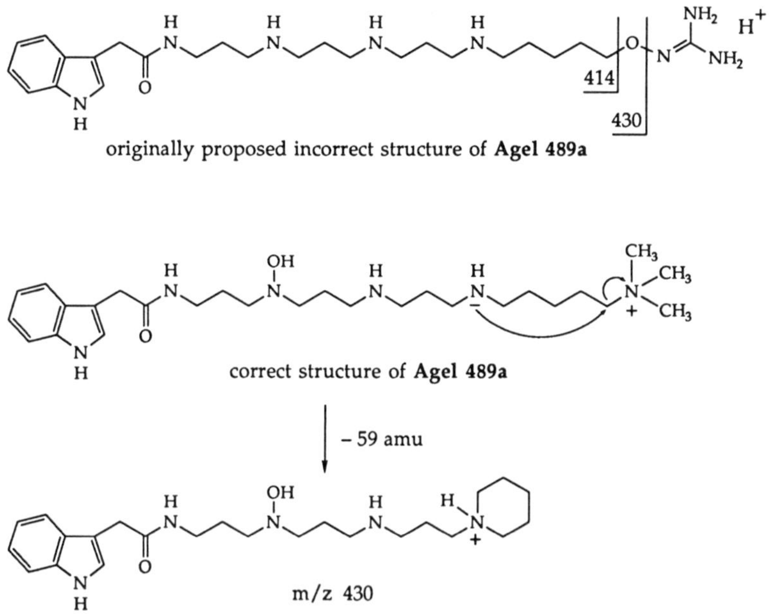

FIG. 42. (Top) Structure of Agel 489a that had been proposed from the first FAB-MS data, assuming a $(M + H)^+$ ion at m/z 489. (Middle) Revised structure of Agel 489a, after it became evident that the signal at m/z 489 represents the quaternary ammonium ion M^+ (*22*) (see Fig. 41).

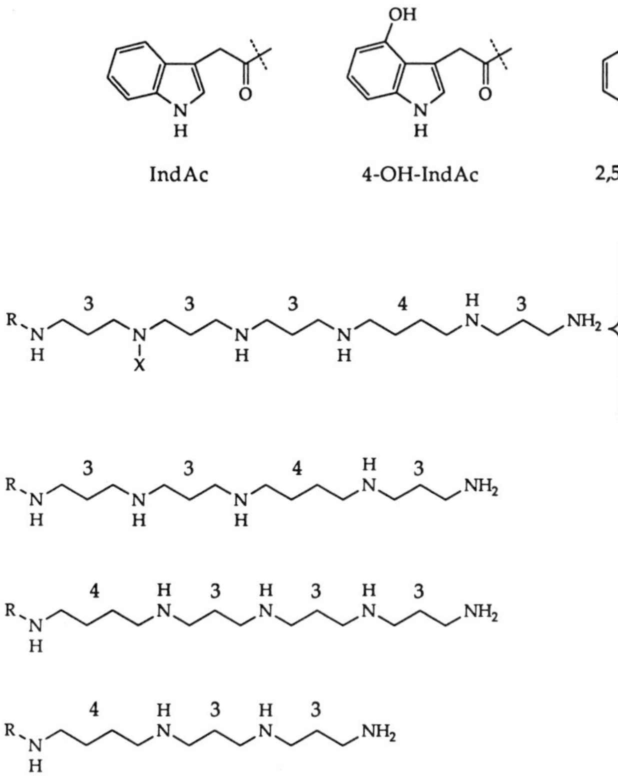

IndAc 4-OH-IndAc 2,5-DiOH-Bz

FIG. 43. Structures of *Hololena* toxins (*113*).

a chromatogram very similar to the one obtained from the funnel-web spider *Agelenopsis aperta* (see Fig. 39). Both groups of spider toxins (*Hololena* and *Agelenopsis*) contain peptide and polyamine constituents. Structural analysis of the polyamine compounds was done in a similar fashion as discussed for the *Agelenopsis* toxins. Size exclusion chromatography demonstrated that over two-thirds of the paralysis, and virtually all of the lethality, of the *Hololena curta* venom is contributed by low molecular weight toxins ($<$10,000) (*113*).

The low molecular weight toxins of funnel-web spiders of the genera *Agelenopsis* and *Hololena* are acylpolyamines. The chromophores are similar, except for 4-hydroxybenzoic acid found only in *Agelenopsis aperta*. Although members of different genera, these spiders belong to the same family (Agelenidae). The composition of the venom and the structural elements present in each of the toxins are very similar. Six of the toxins isolated and characterized from *Agelenopsis aperta* and *Hololena curta,* which have identical molecular weights, might be identical (Agel 416 and HO 416, Agel 448 and HO 448, Agel 452 and HO 452, Agel 468 and HO 468, Agel 489 and HO 489, Agel 505 and HO 505). However, the toxins of the two genera have never been compared directly. Agel 489 and HO 489 predominate in the respective venoms.

Finally, one could summarize that, in contrast to the venom of spiders from the related family Araneidae, the low molecular weight polyamine toxins from Agelenidae contain no amino acids (*113*). There are three types of chromophores found among the *Argiope* toxins, namely, (4-hydroxyindol-3-yl)acetic acid, (indol-3-yl)acetic acid, and (2,4-dihydroxyphenyl)acetic acid. The first two are also found in the Agelenidae toxins, whereas 2,5-dihydroxybenzoic acid is found in *Agelenopsis aperta* and *Hololena curta,* and 4-hydroxybenzoic acid is recognized thus far only in the *Agelenopsis* toxins.

D. Mygalomorphae Toxins from *Hebestatis theveniti* and *Aphonopelma chalcodes*

The fame of tarantulas is mainly due to their widespread lore. Yet, despite all the tales of the extreme potency of their venom, few tarantula venoms have been investigated. In 1957 (*70*), the venoms of nine different spiders of the family Mygalomorphae were analyzed. Free amino acids, mainly glutamic, aspartic, and γ-aminobutyric acid were detected, besides spermine and 1,3-diaminopropane and their aromatic conjugates.

In 1964, Gilbo and Coles (*76*) isolated from the Sydney funnel-web spider *Atrax robustus* the polyamine spermine, associated with phenolic and γ-aminobutyric acid. In 1973, Sutherland suggested (*120*) the presence of (indol-3-yl)acetic acid and spermine as a chemically undefined "complex." Their experimental investigations were insufficient to determine the definite structure. In 1989, Savel-Niemann and Roth (*121*) reported that they had found low molecular weight compounds in the tarantula *Eurypelma californicum,* which were composed of spermine and an aromatic moiety (*4*).

In 1990, Skinner *et al.* (*4*) characterized two acylpolyamines, Het 389 and Het 403 (Fig. 44), from the trap-door spider *Hebestatis theveniti.* The two toxins were isolated by reversed-phase HPLC. Analysis by FAB-MS determined the molecular weights to be 389 and 403, and the compounds were thereafter designated as Het 389 and Het 403, respectively. Amino acid analysis showed that they contain no amino acids like the Agelenidae toxins. Polyamine analysis of the hydrolysates indicated, by HPLC comparison to synthetic standards, a spermine component (PA343) in Het 389 and a PA353 unit in Het 403. Evidence for the complete structural assign-

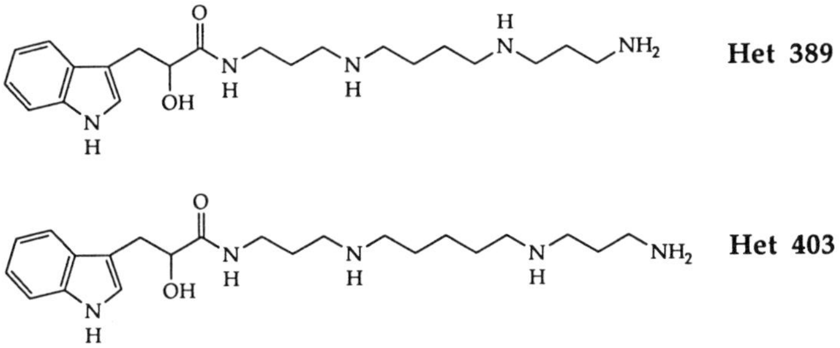

FIG. 44. Structures of toxins Het 389 and Het 403, isolated from *Hebestatis theveniti* (*4*).

ment was given by CI-MS and FAB-MS fragmentation patterns. Het 389 was synthesized, and the synthetic and the natural product showed the same UV and MS data, as well as identical behavior on HPLC.

The concentrations of the toxins Het 389 and Het 403 in the venom are quite high compared to other toxins (e.g., *Argiope* and *Agelenopsis* toxins). The venom of *Hebestatis theveniti* contains 49 and 26 μg/μl of Het 389 and Het 403, respectively. *Harpactriella* venom contains 20 μg/μl Het 389 (*4*). The acylpolyamines Het 389 and Het 403 were the first paralytic toxins of mygalomorph spiders of defined structure (*4*).

In 1990, Skinner *et al.* (*4*) also isolated two toxins, Apc 600 and Apc 728, from the tarantula *Aphonopelma chalcodes*. The molecular weights of the two toxins were determined, and it could be shown that they contain no amino acids. These simple acylpolyamines may serve the same function of immediate immobilization of the prey as do the known low molecular weight toxins of *Argiope*, *Nephila*, and Agelenidae spiders (*4*).

E. Sphecidae Toxins from *Philanthus triangulum*

The paralyzing effect of a sting by a solitary wasp was known by the ancient Chinese. The Erh-Ya (Kuo Po, A.D. 276–324) describes a "green worm" which was paralyzed by the sting of a wasp (*122*). This knowledge was overlooked by Western science, and it was not until 1742 that an observation of a wasp stinging a cockroach, resulting in the loss of the forces of the cockroach was described. Observations during the eighteenth and nineteenth centuries led to conflicting views as to whether the venoms could irreversibly paralyze or kill the prey, with concurrent preservation. It fell to Fabre in 1855 (*123*) to settle the discussion conclusively. His observations led to the conclusion that the prey of solitary wasps were paralyzed and not dead. The ability of the wasps to paralyze other insects or spiders has been the reason to suppose that the investigation of solitary wasp venom could lead to the development of new types of muscle-relaxing compounds. This idea was mainly based on studies by Beard published in 1952 (*124*), and in the 1960s investigation of the paralyzing effects of the venom of the solitary wasp *Philanthus triangulum* started. It was shown that the venom of *P. triangulum* caused paralysis in the prey (honeybee workers) (*125*). The idea was that toxins isolated from solitary wasp venom could lead to new classes of neuroactive drugs (e.g., anti-epileptic drugs), as well as new classes of insecticides (*122,126*).

In 1978 it was possible to isolate one of the active compounds by thin-layer chromatography, and another three compounds were separated using ion-exchange chromatography. A number of fractions were combined to form α-, β-, γ-, and δ-sections. The β-section had no paralyzing activity on its own. The δ-section was biologically active. The compound in the γ-section had no paralyzing activity at all at lower doses. Thin-layer chromatography on cellulose showed that the β-, γ- and δ-sections were nonhomogeneous. A δ-compound could be detected on the thin-layer plate with reagents for phenolic compounds, but apparently the species in the δ-section decomposed on the cellulose layer in the presence of acetic acid, making further structural investigations difficult. Attempts to obtain more information from the FD mass spectrum were without success (*122*).

In 1982 (*86*), two active compounds could be located on the cellulose thin-layer chromatogram, namely, β-PhTX (β-philanthotoxin) and δ-PhTX (δ-philanthotoxin). Although the compounds were isolated, unfortunately the preparations were contaminated by substances originating from the cellulose layers, and so the molecular structure could not be determined. The active components present in the α- and γ-sections were called α-PhTX and γ-PhTX, respectively, although they could not be identified, even on the thin-layer chromatogram, at that time (*86*).

The structure of δ-PhTX (δ-philanthotoxin) was described in 1988 by Piek *et al.* (*89*). Pharmacological studies with synthetic δ-PhTX showed about half the activity of the natural product. The yield of δ-PhTX was about 1 μg per wasp. Injected into ten honeybee workers, this amount caused a 50% recovery time, comparable to that after injection of 2.3 μg synthetic δ-PhTX. This is about twice as much as is needed of the natural toxin. The explanation given is that the natural product represents just one of two optical isomers, whereas the synthetic product is a racemic mixture, and one can assume that only the naturally occurring isomer is active (*47,89*). Identification and separation of the isomers seem not to have been accomplished.

In the same year, 1988, Eldefrawi *et al.* (*37*) also isolated δ-PhTX from *Philanthus triangulum* and assigned the same structure, which was confirmed 2 years later by the same authors (*40*). The commonly accepted data seem to be identical, and in the literature (*48*) are referred to as identical, but so far complete physical and chemical data for δ-PhTX have not been presented. Karst *et al.* (*47*) indicate that "comparison of the synthetic δ-PhTX and the natural product by NMR spectra suggest the correctness of the structure" proposed (*37,89*). "The similarity of the synthetic and the natural toxin was confirmed in both cases by the identical biological activity they possess, examined with several physiological tests" (*47*).

Eldefrawi *et al.* (*37*) extracted 1000 venom glands from *Philanthus triangulum*. The crude venom was fractionated by reversed-phase HPLC, yielding 30 fractions. The pharmacologically most active fraction was collected and rechromatographed by reversed-phase HPLC, separating into four peaks. The pharmacologically most active fraction coincided with the major peak. This fraction yielded 1.1 mg of toxin δ-PhTX (*37*) (Figs. 45 and 46). The UV spectrum of this fraction had a maximum of 274 nm, which shifted to 290 nm at pH 12, in agreement with a tyrosine residue. This conclusion was supported by ^{1}H-NMR data. The NMR spectrum showed the presence of a butyryl group, and FAB-MS gave the $(M + H)^+$ signal at m/z 436. Thus, it was known that the toxin contains one amino acid (tyrosine), butanoic acid, and one polyamine. A ^{1}H-NMR spectrum taken in DMSO-d_6 could clarify the connection of the butyryl, tyrosyl, and polyamine moieties, namely, two protons were observed at 7.82 and 7.86 ppm as a doublet and triplet, respectively. It could be shown that the former was due to a tyrosine and the latter to a polyamine. From the spectroscopic data, Eldefrawi *et al.* (*37*) could not distinguish between three possible structural isomers of the polyamine part, so they synthesized all three isomers. The synthetic compounds with the PA343 moiety were less potent, and those with the PA334 moiety more potent, than the natural toxin δ-PhTX. Hence, they deduced that the natural product contained the PA433 unit (see Fig. 45).

A few years later, a mass spectrometric method, in this case employing secondary ion mass spectrometry (SIMS) in conjunction with metastable ion measurements (linked scanning at constant B/E, where B is the

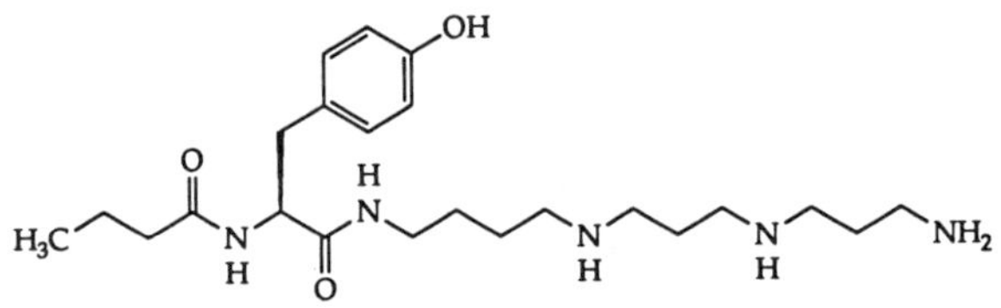

FIG. 45. Structure of the wasp toxin δ-PhTX possessing a PA433 moiety.

ANDREA SCHÄFER *ET AL.*

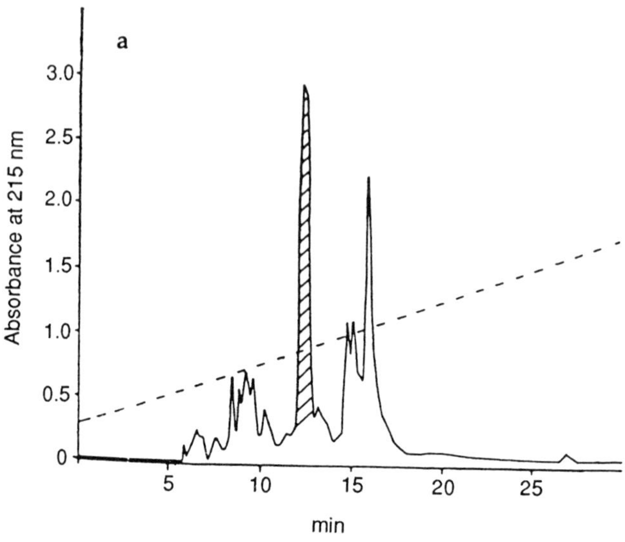

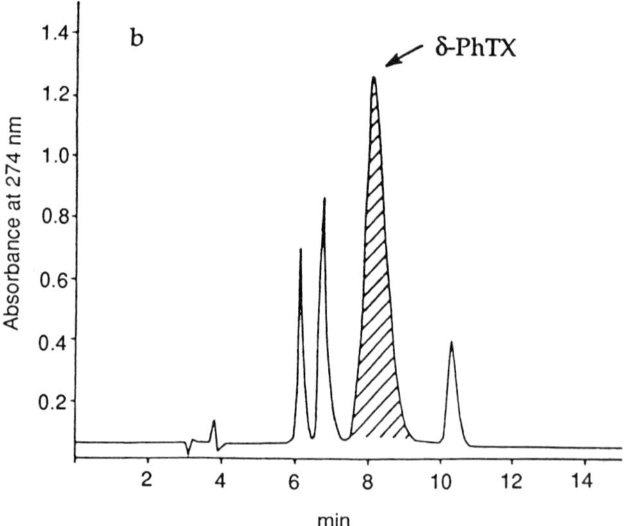

FIG. 46. Reversed-phase HPLC separation of the venom of *Philanthus triangulum*. (From Ref. *37*, with permission.) (a) Fractionation of lyophilized venom glands extracted with CH_3CN/H_2O (1:1) (450 μl crude venom represents the extracts of the glands of 225 wasps) with a linear aqueous gradient of 5–95% $CH_3CN/0.1\%$ aqueous TFA. (b) Separation of the main toxic fraction (hatched peak in a) by an isocratic system of 15% $CH_3CN/0.1\%$ aqueous TFA.

magnetic sector and E the electric sector), led to specific differentiation among the three isomers (40). The different fragmentation patterns seemed to be specific enough to distinguish the three isomers.

The second toxin, β-PhTX (β-philanthotoxin), potentiated the effect of δ-PhTX in intact honeybee workers. The β-PhTX-containing fractions obtained by Spanjer *et al.* (86) with gel ion-exchange chromatography was further purified by Karst *et al.* (68) by HPLC. The yield of pure toxin obtained was low, about 150 ng of β-PhTX from one female wasp. It was mentioned that the overall quantity of the isolated toxin was critical for the acquisition of both NMR and mass spectrometric data, although no figures were given about the total number of wasps used.

From the data known from the investigation of Spanjer *et al.* (86), the two proposed structures for β-PhTX (Fig. 47) could not be distinguished (68). The purified sample was subjected to several mass spectrometric analyses. FD, FAB, FI, and EI mass spectrometry were applied to ionize the sample and to provide both molecular weight and structural information. The spectra showed that the sample was still not absolutely pure, but they left no doubt that the major component has a molecular weight of 243, since the most intense signals in the FD and FAB spectra were detected at m/z 244 [$(M + H)^+$] and at m/z 266 [$(M + Na)^+$], and in the FI at m/z 243 $(M^{+\cdot})$. High-resolution mass measurements of the fragments in the EI mass spectra gave strong support for the elemental composition $C_{13}H_{29}N_3O$. This meant that the molecule contains one double bond or one ring and that the nitrogen atoms are of the aliphatic amine type. A ^{13}C-NMR study gave evidence that a carbonyl group is present. From ^{1}H-NMR spectroscopy, it was likely that the molecular structure is a polyamine.

Both compounds **I** and **II** shown in Fig. 47 were synthesized and compared to the natural product (68). The EI mass spectra of the natural compound resembles most closely that of structure **I**. The FAB mass spectra were less decisive. To obtain further experimental support that the

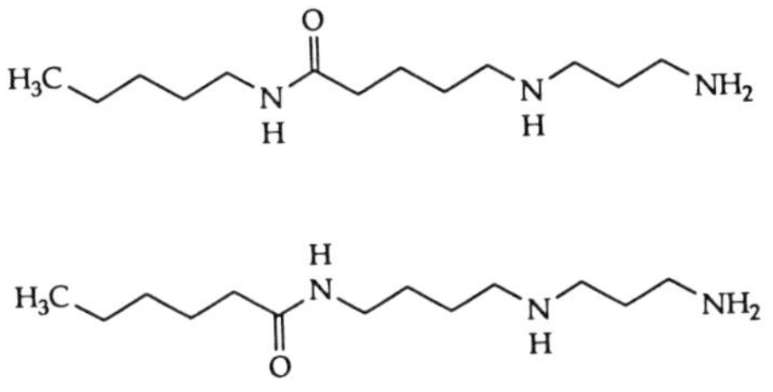

FIG. 47. Proposed structures **I** and **II** for β-philanthotoxin (β-PhTX) (68).

structure of the major component of the natural sample is identical with structure **I,** so-called EI/MIKE (mass analyzed ion kinetic energy) and EI/MIKE/CA (collisional activation) spectra on the molecular ions $M^{+\cdot}$ and FAB/MIKE and FAB/MIKE/CA spectra of the protonated molecules $(M + H)^+$ were taken (68)*. Karst *et al.* (68) draw the conclusion that, "mutual comparison of these spectra shows that the spectra of the major component of the natural sample and compound **I** are very similar, if not identical. This observation, therefore, leaves no doubt that the structure of the major component of the natural sample is identical to that of the synthetic product **I**." The authors admit that the postulated identity of the natural product and the synthetic one is mainly based on the similarity of the MIKE spectra. "Small differences in the NMR spectra are still unexplained. The amount of natural toxin left is too small to perform further experiments" (68). They cite pharmacological tests that indicate the identity of the natural and synthetic toxin and explain small quantitative differences by the fact that the weight of the extremely small amount of the natural toxin left could not be determined with great precision (68).

VI. Final Remarks

The polyamine spider and wasp toxins represent one of the youngest classes of natural products. Spermine (PA343), however, belongs to one of the oldest known groups of organic compounds; it had already been isolated and crystallized as its phosphate in 1678 (*127*). Both groups of compounds can be categorized as alkaloids only with certain reservations. Spermine is classified as a member of the polyamines, which are assigned to the group of biogenic amines. Spermine derivatives such as the spider and wasp toxins represent a borderline case. They are a class of natural products that can be called biogenic amines as well or, with regard to certain derivatives, basic peptides, or even alkaloids, depending on the attitude of the authors. Only since the advent of modern analytical

---

* MIKE spectra can be obtained on a double focusing instrument with *BE* geometry [magnetic sector (*B*) preceding the electric sector (*E*)] by keeping the magnetic sector at a constant m/z value. The fragment ions of a spontaneously decomposing $M^{+\cdot}$ and $(M + H)^+$ ion, respectively, which has been chosen in the magnetic sector are recorded. MIKE/CA spectra register the fragment ions of the respective ion following collision with a target gas such as helium; thus, the intensity of the fragments is increased. To observe MIKE spectra the mass spectrometer is used as a separating device. The data thus obtained can be considered to be structural fingerprints of one specific compound.

methods such as NMR spectroscopy and mass spectrometry [in particular FAB-MS and electrospray ionization (ESI)-MS] has it been possible to elucidate the structures of these new types of compounds.

Section III of this chapter presents syntheses of the natural products and their analogs. The assembly of the basic units of such compounds, namely, the chromophore part, the amino acids, the di- or tripeptides, and the polyamine backbone, takes advantage of methodologies well known from peptide chemistry, especially the well elaborated protective group techniques. The major problem for the construction of polyamine toxins or related compounds is therefore not the final assembling of the subunits to form a particular compound but the preparation of these separate units, in particular the synthesis of suitably protected polyamine portions. Comparing the polyamine parts of the natural toxins, many variations can be found, for example, sequence isomers, constitutional isomers, homologs, and derivatives with N-methyl, N-hydroxyl, or oxo groups. The syntheses of the polyamine units sometimes require quite elaborate and time-consuming procedures.

Two further sections of this chapter concern the isolation, purification, and structure elucidation of the polyamine toxins. As described in detail, the natural toxins were obtained by means of chromatography from venom isolated from the glands of arthropods. It should be emphasized again that for the most part only very small amounts of the compounds were available for structure elucidation. At least in some cases, these small amounts have given rise to unexpected difficulties. Owing to these circumstances some representative compounds were not adequately characterized, and, consequently, in cases of reisolation from different species the specific toxin could not be identified unambiguously as the same substance. It was therefore absolutely necessary to perform in each case a complete structure elucidation with the reisolated compound. A typical example of this is the structure elucidation of Arg 636, one of the principal toxins found in several spider species of the genus *Argiope* (family Araneidae). However, it has to be emphasized that because of this incomplete characterization of this toxin, the different samples of Arg 636 were not identified by comparison with one another, without a degree of uncertainty. Therefore, however paradoxical this may sound, there is no final proof that all of the isolated toxins called Arg 636 are identical. In addition, determination of the absolute configuration of chiral constituents is required for a complete structure elucidation, even if one is only concerned with amino acids, which in general are present in nature in the L configuration. Synthesis of such incompletely determined structures offer no guarantee for the identity of the natural and synthetic products. Furthermore, for some publications, there was no guarantee that the arthropods were classified

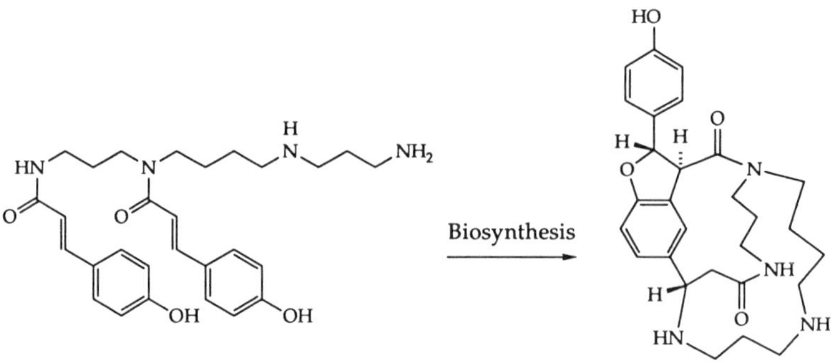

    ANDREA SCHÄFER *ET AL.*

147, 1,5-Bis(cumaroyl) spermine   **146, Aphelandrine**

FIG. 48. 1,5-Bis(coumaroyl)spermine (**147**) can be assumed to be the direct precursor of aphelandrine (**146**).

unambiguously, indicating that difficulties may occur with the definite identification of any biological material (e.g., refer to Ref. *128*).

The discourse on spider and wasp toxins in a treatise reserved for alkaloids can be further justified by their relationship to the polyamine alkaloids. A typical representative is the spermine alkaloid aphelandrine (**146**), which is found in different *Aphelandra* species (family Acanthaceae), for example, *Aphelandra tetragona* (Vahl) Nees (*129*). Biogenetically, 1,5-bis(coumaroyl)spermine (**147**) can be assumed to be the direct precursor of **146** (Fig. 48). This was proved by the incorporation of the ^{14}C-labeled compounds PA4, PA34, PA343, cinnamic acid, and *p*-hydroxycinnamic acid (*130*). Compound **147** was most likely also isolated from pollen of *Aphelandra tetragona* and *A. chamissioniana* Nees (*131*) and identified by comparison to the synthetic product (*132*). The structural similarities of compounds like **147** to almost all spider and wasp toxins is astonishing (Fig. 48). One can assume that similar structures correspond to similar molecular effects, for example, the complexation of certain metal ions. Detailed investigations in this field have, to our knowledge, not yet been carried out. Aside from this, the close structural relationship between compounds originating from such extremely different organisms is particularly interesting (*133*).

Acknowledgments

We are thankful to the Schweizerischer Nationalfonds zur Förderung der wissenschaftlichen Forschung for supporting this work, to Mrs. G. Bienz-Meier for maintaining the literature database, to Ms. A. Latvala for on-line literature search, and to Mr. M. Hefti, Zoologisches Institut der Universität Zürich, for collecting the data summarized in Fig. 2.

References

1. V. J. Jasys, P. R. Kelbaugh, D. M. Nason, D. Phillips, K. J. Rosnack, N. A. Saccomano, J. G. Stroh, and R. A. Volkmann, *J. Am. Chem. Soc.* **112,** 6696 (1990).
2. N. A. Saccomano and R. A. Volkmann, Eur. Pat. Appl. 0395357 (1990); *Chem. Abstr.* **114,** 207832r (1990).
3. N. A. Saccomano and R. A. Volkmann, Eur. Pat. Appl. 436332 (1991); *Chem. Abstr.* **115,** 231996p (1991).
4. W. S. Skinner, P. A. Dennis, A. Lui, R. L. Carney, and G. B. Quistad, *Toxicon* **28,** 541 (1990).
5. W. J. Fiedler, Ph.D. Thesis, Univ. of Zürich, Switzerland (1992); W. J. Fiedler and M. Hesse, *Helv. Chim. Acta.* **76,** 1511 (1993).
6. J. S. Bradshaw, K. E. Krakowiak, and R. M. Izatt, *Tetrahedron* **48,** 4475 (1992).
7. M. Israel, J. S. Rosenfield, and E. J. Modest, *J. Med. Chem.* **7,** 710 (1964).
8. H. Tabor, C. W. Tabor, and L. De Meis, *in* "Methods in Enzymology" (H. Tabor and C. W. Tabor, eds.), Vol. 17B, p. 829. Academic Press, New York, 1971.
9. G. L. Stahl, R. Walter, and C. W. Smith, *J. Org. Chem.* **43,** 2285 (1978).
10. J. B. Hansen, M. C. Nielsen, U. Ehrbar, and O. Buchardt, *Synthesis,* 404 (1982).
11. J. G. Atwell and W. A. Denny, *Synthesis,* 1032 (1984).
12. A. R. Jacobson, A. N. Makris, and L. M. Sayre, *J. Org. Chem.* **52,** 2592 (1986).
13. H. Essien, J. Y. Lai, and K. J. Hwang, *J. Med. Chem.* **31,** 898 (1988).
14. A. P. Krapcho and C. S. Kuell, *Synth. Commun.* **20,** 2559 (1990).
15. R. Geiger, *Liebigs Ann. Chem.* **750,** 165 (1971).
16. E. Wälchli-Schaer and C. H. Eugster, *Chimia* **28,** 728 (1974).
17. G. Kunesch, *Tetrahedron Lett.* **24,** 5211 (1983).
18. P. G. Mattingly, *Synthesis,* 366 (1990).
19. R. Houssin, J.-L. Bernier, and J.-P. Hénichard, *Synthesis,* 259 (1988).
20. U. Kramer, A. Guggisberg, M. Hesse, and H. Schmid, *Angew. Chem., Int. Ed. Engl.* **17,** 200 (1978).
21. H. Benz, Ph.D. Thesis, Univ. of Zürich, Switzerland (1993).
22. V. J. Jasys, P. R. Kelbaugh, D. M. Nason, D. Phillips, K. J. Rosnack, J. T. Forman, N. A. Saccomano, J. G. Stroh, and R. A. Volkmann, *J. Org. Chem.* **57,** 1814 (1992).
23. H. Yamamoto and K. Maruoka, *J. Am. Chem. Soc.* **103,** 6133 (1981).
24. T. Ando and J. Yamawaki, *Chem. Lett.,* 45 (1984).
25. S. A. Davis and A. C. Sheppard, *Tetrahedron* **45,** 5703 (1989).
26. G. Poon, Y.-C. Chui, and F. C. P. Law, *J. Labelled Compd. Radiopharm.* **23,** 167 (1986).
27. B. M. Dunn and T. C. Bruice, *J. Am. Chem. Soc.* **92,** 2410 (1970).
28. V. J. Jasys, P. R. Kelbaugh, D. M. Nason, D. Phillips, N. A. Saccomano, and R. A. Volkmann, *Tetrahedron Lett.* **29,** 6223 (1988).
29. D. M. Nason, V. J. Jasys, P. R. Kelbaugh, D. Phillips, N. A. Saccomano, and R. A. Volkmann, *Tetrahedron Lett.* **30,** 2337 (1989).
30. T. W. Greene, "Protective Groups in Organic Synthesis." Wiley, New York, 1981.

31. G. B. Quistad, S. Suwanrumpha, M. A. Jarema, M. J. Shapiro, W. S. Skinner, G. C. Jamieson, A. Lui, and E. W. Fu, *Biochem. Biophys. Res. Commun.* **169,** 51 (1990).
32. P. N. R. Usherwood, B. W. Bycroft, I. S. Blagbrough, and A. J. Mather, PCT Int. Appl. WO 90/02114 (1990); *Chem. Abstr.* **113,** 184729a (1990).
33. P. N. R. Usherwood, B. W. Bycroft, I. S. Blagbrough, and A. J. Mather, UK Pat. Appl. 2222590 (1990); *Chem. Abstr.* **113,** 226431e (1990).
33a. S. M. Goldin, K. Kobayashi, A. G. Knapp, L. Margolin, S. Katragadda, D. Daly, L.-Y. Hu, and N. Reddy, PCT Internat. Appl. WO 92/14709 (1992); *Chem. Abstr.* **118,** 824b (1992).
34. T. Asami, H. Kagechika, Y. Hashimoto, K. Shudo, A. Miwa, N. Kawai, and T. Nakajima, *Biomed. Res.* **10,** 185 (1989).
35. T. Asami, H. Kagechika, Y. Hashimoto, K. Shudo, A. Miwa, N. Kawai, and T. Nakajima, *in* "Peptide Chemistry 1988" (M. Ueki, ed.), p. 231. Protein Research Foundation, Osaka, 1989.
36. T. Asami, K. Hagiwara, H. Kagechika, Y. Hashimoto, K. Shudo, A. Shirahata, A. Miwa, N. Kawai, and T. Nakajima, *Pept. Chem.*, 233 (1989).
37. A. T. Eldefrawi, M. E. Eldefrawi, K. Konno, N. A. Mansour, K. Nakanishi, E. Oltz, and P. N. R. Usherwood, *Proc. Natl. Acad. Sci. U.S.A.* **85,** 4910 (1988).
38. K. Konno, R. Goodnow, Jr., R. Bukownik, T. Kallimopoulos, K. Nakanishi, A. T. Eldefrawi, M. E. Eldefrawi, P. N. R. Usherwood, and A. Mansour, *Tennen Yuki Kagobutsu Toronkai Koen Yoshishu,* 380 (1989).
39. K. Nakanishi, A. T. Eldefrawi, M. E. Eldefrawi, and P. N. R. Usherwood, PCT Int. Appl. WO 89/07098 (1989); *Chem. Abstr.* **112,** 99246a (1989).
40. K. Nakanishi, R. Goodnow, Jr., K. Konno, M. Niwa, R. Bukownik, T. A. Kallimopoulos, P. N. R. Usherwood, A. T. Eldefrawi, and M. E. Eldefrawi, *Pure Appl. Chem.* **62,** 1223 (1990).
41. R. Goodnow, Jr., K. Konno, M. Niwa, T. Kallimopoulos, R. Bukownik, D. Lenares, and K. Nakanishi, *Tetrahedron* **46,** 3267 (1990).
42. Y. C. Tong, *Pestic. Sci.* **24,** 340 (1988).
43. I. S. Blagbrough, B. W. Bycroft, A. J. Mather, and P. N. R. Usherwood, *J. Pharm. Pharmacol.* **41,** 95 (1989).
44. M. Bruce, R. Bukownik, A. T. Eldefrawi, M. E. Eldefrawi, R. Goodnow, Jr., T. Kallimopoulos, K. Konno, K. Nakanishi, M. Niwa, and P. N. R. Usherwood, *Toxicon* **28,** 1333 (1990).
45. R. A. Goodnow, R. Bukownik, K. Nakanishi, P. N. R. Usherwood, A. T. Eldefrawi, N. A. Anis, and M. E. Eldefrawi, *J. Med. Chem.* **34,** 2389 (1991).
45a. S.-K. Choi, R. A. Goodnow, A. Kalivretenos, G. W. Chiles, S. Fushiya, and K. Nakanishi, *Tetrahedron* **48,** 4793 (1992).
46. T. Piek and B. Hue, *Comp. Biochem. Physiol.* **93C,** 403 (1989).
47. H. Karst, T. Piek, J. van Marle, A. Lind, and J. van Weeren-Kramer, *Comp. Biochem. Physiol.* **98C,** 471 (1991).
48. J. A. Benson, F. Schürmann, L. Kaufmann, L. Gsell, and T. Piek, *Comp. Biochem. Physiol.* **102C,** 267 (1992).
49. N. C. M. Schluter, T. Piek, and F. H. Lopes da Silva, *Comp. Biochem. Physiol.* **101C,** 41 (1992).
50. N. Anis, R. Goodnow, Jr., M. Niwa, K. Konno, T. Kallimopoulos, R. Bukownik, K. Nakanishi, P. N. R. Usherwood, A. Eldefrawi, and M. Eldefrawi, *J. Pharmacol. Exp. Ther.* **254,** 764 (1990).
51. W. J. Fiedler, A. Guggisberg, and M. Hesse, *Helv. Chim. Acta* **76,** 1167 (1993).
52. M. Miyashita, H. Sato, A. Yoshikoshi, T. Toki, M. Matsushita, H. Irie, T. Yanami, Y. Kikuchi, C. Takasaki, and T. Nakajima, *Tetrahedron Lett.* **33,** 2833 (1992).

53. M. Miyashita, H. Sato, M. Matsushita, Y. Kusumegi, T. Toki, A. Yoshikoshi, T. Yanami, Y. Kikuchi, C. Takasaki, T. Nakajima, and H. Irie, *Tetrahedron Lett.* **33,** 2837 (1992).
54. Y. Hashimoto, T. Yasuhara, Y. Endo, K. Shudo, Y. Aramaki, N. Kawai, and T. Nakajima, *Tetrahedron Lett.* **28,** 3511 (1987).
55. Y. Hashimoto, Y. Endo, K. Shudo, Y. Aramaki, N. Kawai, and T. Nakajima, *in* "Peptide Chemistry 1987" (M. Ueki, ed.), p. 363. Protein Research Foundation, Osaka, 1988.
56. T. Teshima, T. Wakamiya, Y. Aramaki, T. Nakajima, N. Kawai, and T. Shiba, *Tetrahedron Lett.* **28,** 3509 (1987).
57. T. Teshima, T. Wakamiya, T. Shiba, Y. Aramaki, and T. Nakajima, *in* "Peptide Chemistry 1987" (T. Shiba and S. Sakakibara, eds.), p. 367. Protein Research Foundation, Osaka, 1988.
58. T. Teshima, T. Matsumoto, T. Wakamiya, T. Shiba, Y. Aramaki, T. Nakajima, and N. Kawai, *Tetrahedron* **47,** 3305 (1991).
59. T. Teshima, T. Matsumoto, M. Miyagawa, T. Wakamiya, T. Shiba, N. Narai, and M. Yoshioka, *Tetrahedron* **46,** 3819 (1990).
60. T. Teshima, T. Matsumoto, T. Wakamiya, T. Shiba, T. Nakajima, and N. Kawai, *Tetrahedron* **46,** 3813 (1990).
61. G. Goto and T. Nakajima, Eur. Pat. Appl. 0339927 (1989); *Chem. Abstr.* **112,** 217537w (1990).
62. L. Kovács and M. Hesse, *Helv. Chim. Acta* **75,** 1909 (1992).
63. T. L. Shih, J. Ruiz-Sanchez, and H. Mrozik, *Tetrahedron Lett.* **28,** 6015 (1987).
64. E. A. Yelin, B. F. de Macedo, V. V. Onoprienko, N. E. Osokina, and O. B. Tikhomirowa, *Bioorg. Khim.* **14,** 704 (1988).
65. M. E. Adams, R. L. Carney, F. E. Enderlin, E. T. Fu, M. A. Jarema, J. P. Li, C. A. Miller, D. A. Schooley, M. J. Shapiro, and V. J. Venema, *Biochem. Biophys. Res. Commun.* **148,** 678 (1987).
66. T. Toki, A. Miwa, N. Kawai, K. Hagiwara, and T. Nakajima, *Biomed. Res.* **13,** 53 (1992).
67. I. Tabushi, Y. Taniguchi, and H. Kato, *Tetrahedron Lett.* **18,** 1049 (1977).
68. H. Karst, R. H. Fokkens, N. de Haan, G. Heuver, B. Hue, C. Kruk, N. M. M. Nibbering, T. Piek, W. Spanjer, Y. C. Tong, and W. van der Vliet, *Comp. Biochem. Physiol.* **97C,** 317 (1990).
69. H. Benz and M. Hesse, *Helv. Chim. Acta* **77,** 80 (1994).
70. F. G. Fischer and H. Bohn, *Liebigs Ann. Chem.* **603,** 232 (1957).
71. W. P. Hurlbut and B. Cecarelli, "Use of Black Widow Spider Venom to Study the Release of Neurotransmitters." Raven, New York, 1979.
72. A. Grasso, *Biochim. Biophys. Acta* **439,** 409 (1976).
73. C. R. Geren and G. V. Odell, *in* "Insect Poisons, Allergens, and Other Invertebrate Venoms" (A. T. Tu, ed.), Vol. 2, p. 441, Dekker, New York, 1984.
74. E. F. D'Amour, *Proc. Soc. Exp. Biol.* **35,** 262 (1936).
75. S. Bettini and M. Maroli, *in* "Arthropod Venoms" (S. Bettini, ed.), p. 149. Springer-Verlag, Berlin, 1978.
76. C. M. Gilbo and N. W. Coles, *Aust. J. Biol. Sci.* **17,** 758 (1964).
77. R. K. Atkinson and P. Walker, *Aust. J. Exp. Biol. Med. Sci.* **63,** 555 (1985).
78. T. Budd, P. Clinton, A. Dell, I. R. Duce, S. J. Johnson, D. L. J. Quicke, G. W. Taylor, P. N. R. Usherwood, and G. Usoh, *Brain Res.* **448,** 30 (1988).
79. C. K. Lee, T. K. Chan, B. C. Ward, D. E. Howell, and G. V. Odell, *Arch. Biochem. Biophys.* **164,** 341 (1974).
80. B. J. Kaston, "How to Know the Spiders." W. C. Brown, Dubuque, Iowa, 1978.

81. K. N. Pinkston, Thesis, Oklahoma State Univ., Stillwater (1973).

82. L. D. Foil, L. B. Coons, and B. R. Norment, *Int. J. Insect Morphol. Embryol.* **8,** 325 (1979).

83. D. L. J. Quicke and P. N. R. Usherwood, *in* "Safer Insecticides: Development and Use" (E. Hodgson and R. J. Kuhr, eds.), Vol. 7, p. 385. Dekker, New York and Basel, 1990.

84. P. E. Meadows and F. E. Russell, *Toxicon* **8,** 311 (1970).

85. A. Bateman, P. Boden, A. Dell, I. R. Duce, D. L. J. Quicke, and P. N. R. Usherwood, *Brain Res.* **339,** 237 (1985).

86. W. Spanjer, T. E. May, T. Piek, and N. de Haan, *Comp. Biochem. Physiol.* **71C,** 149 (1982).

87. P. N. R. Usherwood and I. S. Blagbrough, *Pharmacol. Ther.* **52,** 245 (1991).

88. P. N. R. Usherwood, I. R. Duce, and P. Boden, *J. Physiol. (Paris)* **79,** 241 (1984).

89. T. Piek, R. H. Fokkens, H. Karst, C. Kruk, A. Lind, J. van Marle, T. Nakajima, N. M. M. Nibbering, H. Shinozaki, W. Spanjer, and Y. C. Tong, *in* "Neurotox '88: Molecular Basis of Drugs and Pesticide Action" (G. G. Lunt, ed.), p. 61. Elsevier Science Publ. (Biomedical Division), Amsterdam, 1988.

90. N. Kawai, T. Nakajima, and T. Abe, *Brain Res.* **247,** 169 (1982).

91. T. Abe, N. Kawai, and A. Miwa, *J. Physiol. (London)* **339,** 243 (1983).

92. N. Kawai, A. Miwa, M. Saito, H. S. Pan-Hou, and M. Yoshioka, *J. Physiol. (Paris)* **79,** 228 (1984).

93. E. V. Grishin, T. M. Volkova, A. S. Arseniev, O. S. Reshetova, V. V. Onoprienko, L. G. Magazanik, S. M. Antonov, and I. M. Federova, *Bioorg. Khim.* **12,** 1121 (1986).

94. E. V. Grishin, T. M. Volkova, and A. S. Arseniev, *Bioorg. Khim.* **14,** 883 (1988).

95. P. N. R. Usherwood, I. R. Duce, A. Dell, and G. W. Taylor, Eur. Pat. Appl. 0208523 (1987); *Chem. Abstr.* **107,** 46299k (1987).

96. M. E. Adams, *in* "Neurotox '88: Molecular Basis of Drugs and Pesticide Action" (G. G. Lunt, ed.), p. 49. Elsevier Science Publ. (Biomedical Division), Amsterdam, 1988.

97. W. S. Skinner, M. E. Adams, G. B. Quistad, H. Kataoka, B. J. Cesarin, F. E. Enderlin, and D. A. Schooley, *J. Biol. Chem.* **264,** 2150 (1989).

98. P. N. R. Usherwood, "Neurotoxins and Their Pharmacological Implications." Raven, New York, 1986.

99. L. G. Magazanik, S. M. Antonov, I. M. Federova, T. M. Volkova, and E. V. Grishin, *Biol. Membr.* **3,** 1204 (1986).

100. F. Campagna, A. Carotti, and G. Casini, *Tetrahedron Lett.* **18,** 1813 (1977).

101. T. Toki, T. Yasuhara, Y. Aramaki, N. Kawai, and T. Nakajima, *Biomed. Res.* **9,** 75 (1988).

102. E. V. Grishin, T. M. Volkova, and A. S. Arseniev, *Toxicon* **27,** 541 (1989).

103. M. E. Adams, F. E. Enderlin, R. I. Cone, and D. A. Schooley, *in* "Insect Neurochemistry and Neurophysiology" (A. B. Borkovec and D. E. Gelman, eds.), p. 397. Humana, Clifton, New Jersey, 1986.

104. T. Toki, T. Yasuhara, Y. Aramaki, Y. Hashimoto, K. Shudo, N. Kawai, and T. Nakajima, *Jpn. J. Sanit. Zool. (Eisei Dobutsu)* **11,** 9 (1990).

105. Y. Aramaki, T. Yasuhara, T. Higashijima, A. Miwa, N. Kawai, and T. Nakajima, *Biomed. Res.* **8,** 167 (1987).

106. Y. Aramaki, T. Yasuhara, K. Shimazaki, N. Kawai, and T. Nakajima, *Biomed. Res.* **8,** 241 (1987).

107. T. Toki, T. Yasuhara, Y. Aramaki, K. Osawa, A. Miwa, N. Kawai, and T. Nakajima, *Biomed. Res.* **9,** 421 (1988).

108. Y. Aramaki, T. Yasuhara, T. Higashijima, M. Yoshioka, A. Miwa, N. Kawai, and T. Nakajima, *Proc. Jpn. Acad.* **62B,** 359 (1986).

109. N. Nimura and T. Kinoshita, *J. Chromatogr.* **352,** 169 (1986).
110. Y. Kakimoto, T. Nakajima, A. Kumon, Y. Matsuoka, N. Imaoka, and I. Sano, *J. Biol. Chem.* **244,** 6003 (1969).
111. T. Shiba and T. Kaneko, *J. Biol. Chem.* **244,** 6006 (1969).
112. Y. Aramaki, T. Yasuhara, N. Kawai, and T. Nakajima, *in* "Peptide Chemistry 1987" (T. Shiba and S. Sakakibara, eds.), p. 163. Protein Research Foundation, Osaka, 1988.
113. G. B. Quistad, C. C. Reuter, W. S. Skinner, P. A. Dennis, S. Suwanrumpha, and E. W. Fu, *Toxicon* **29,** 329 (1991).
114. T. Nakajima, T. Yasuhara, N. Yoshida, Y. Takemoto, S. Shinonaga, R. Kano, and H. Yoshida, *Jpn. J. Sanit. Zool. (Eisei Dobutsu)* **34,** 61 (1983).
115. M. Yoshioka, N. Narai, N. Kawai, M. Numata, and T. Nakajima, *Biogenic Amines* **7,** 375 (1990).
116. C. W. Bowers, H. S. Phillips, P. Lee, N. Y. Jan, and L. Y. Jan, *Proc. Natl. Acad. Sci. U.S.A.* **84,** 3506 (1987).
117. H. J. Veith, *Org. Mass Spectrom.* **18,** 154 (1983).
118. L. Bigler, Ph.D. Thesis, Univ. of Zürich, Switzerland, in preparation.
119. E. Schöpp and M. Hesse, *Helv. Chim. Acta* **60,** 2425 (1977).
120. S. K. Sutherland, *Proc. Aust. Soc. Med. Res.* **3,** 172 (1973).
121. A. Savel-Niemann and D. Roth, *Naturwissenschaften* **76,** 212 (1989).
122. T. Piek and W. Spanjer, *in* "Pesticide and Venom Neurotoxicity" (D. L. Shankland, ed.), p. 211. Plenum, New York, 1978.
123. J. H. Fabre, *Ann. Sci. Nat.* **4,** 129 (1855).
124. R. L. Beard, *Conn. Agric. Exp. Stn. New Haven Bull.* **562,** 1 (1952).
125. W. Rathmayer, *Z. Vgl. Physiol.* **45,** 413 (1962).
126. T. Piek, *Comp. Biochem. Physiol.* **96C,** 223 (1990).
127. D. A. Lewenhoeck, *R. Soc. London, Philos. Trans.* **12,** 1040 (1678).
128. M. E. Endress, M. Hesse, S. Nilsson, A. Guggisberg, and J. Zhu, *Plant Syst. Evol.* **171,** 157 (1990).
129. H. Bosshardt, A. Guggisberg, S. Johne, and M. Hesse, *Pharm. Acta Helv.* **53,** 355 (1978).
130. G. Papazoglou, J. Sierra, K. Homberger, A. Guggisberg, W.-D. Woggon, and M. Hesse, *Helv. Chim. Acta* **74,** 565 (1991).
131. C. Werner and M. Hesse, unpublished results.
132. F. Veznik, A. Guggisberg, and M. Hesse, *Helv. Chim. Acta* **74,** 654 (1991).
133. H. Jackson, M. Urnes, W. R. Gray, and T. N. Parks, *Soc. Neurosci. Abstr.* **11,** 107 (1985).
134. P. B. Usmanov, J. Kalikulov, K. E. Nasirov, K. J. Akhmedov, and H. S. Nurtaev, *Biol. Membr.* (Engl. transl.) **6**(11), 1687 (1992).

CHAPTER 2

THE MORPHINE ALKALOIDS

Csaba Szántay and Gábor Dörnyei

Central Research Institute for Chemistry
Hungarian Academy of Sciences
H-1525 Budapest, Hungary

AND

Gábor Blaskó

EGIS Pharmaceutical Ltd.
Budapest, Hungary

I. Introduction

The morphinans represent a large group of alkaloids, related to the isoquinoline alkaloids. However, they are not considered to be a part of that large family of alkaloids because of the saturated character of the isoquinoline C ring. The first representative of the morphinan group of alkaloids, morphine (**1**), was discovered about 200 years ago by Serturner. In 1971, at the time of the last review in this treatise (*1*), more than 30 alkaloids were known which possessed the morphinan nucleus. In the period since that time, many significant advances have been made in the isolation, structure elucidation, and chemistry of the morphinan alkaloids. Many books and chapters of monographs (*2–11*) have been devoted to the chemistry of morphinan alkaloids. The morphinans have also been reviewed regularly in the *Specialist Periodical Reports* of the Chemical Society (*12*) and later in *Natural Products Reports* (*13*). This chapter is a sequel to the previous review published in 1971 this treatise and covers the literature on the occurrence, isolation, structure elucidation, synthesis, transformations, reactions, biogenesis, and pharmacological activity of morphinan alkaloids to the middle of 1992.

II. Occurrence and Structure Elucidation

A. 3-METHOXY-4,6-DIHYDROXYMORPHINANDIEN-7-ONE

The isolation of 3-methoxy-4,6-dihydroxymorphinandien-7-one (mocrispatine, **2,** mp 143°C), a compound isomeric with norsinoacutine (**3**), was

	R^1	R^2
2	CH$_3$	H
3	H	CH$_3$

reported independently from two different sources, namely, *Croton bonplandianus* Baill. (*14*) and *Monodora crispata* Engl. (*15*). The structure was determined by UV, IR, and low-resolution ^{1}H-NMR methods, and the absolute configuration of **2** was established by comparison of its *O,O*-dimethylmethiodide derivative with *N,O*-dimethylnorsinoacutine methiodide (*14*).

B. *N*-NORSALUTARIDINE

The first two *N*-normorphinandienone alkaloids were found in natural sources. *N*-Norsalutaridine (**4**) was isolated from *Croton salutaris* Casar (*16*) as an amorphous material, and its structure was established by ^{1}H-NMR spectroscopy. The two methoxy singlets appear in the spectrum at 3.78 and 3.92 ppm, H-8 and H-5 absorb at 6.32 and 7.58 ppm, respectively, and the AB pattern of H-1 and H-2 appears at 6.72 ppm.

C. *N*-NORPALLIDINE

N-Norpallidine (**5,** mp 102°C) was found in *Fumaria vaillanti* Loisel var. *Schrammii* (*17*). The ^{1}H-NMR spectrum of *N*-norpallidine (**5**) contains two methoxy signals at 3.73 and 3.82 ppm and four singlets at 6.22, 6.30, 6.62, and 6.73 ppm corresponding to H-8, H-1, H-4, and H-5, respectively. The absolute configuration of *N*-norpallidine was established by CD measurements (*17*).

D. Isosinoacutine

Isosinoacutine (**6,** mp 120°C) was isolated from *Stephania elegans* Hook. f. *et* Thoms. (*18*). Unambiguous structure elucidation was completed by UV, IR, ¹H-NMR, and mass spectrometry. The *N*-methyl resonance appears at 2.33 ppm in the ¹H-NMR spectrum, the two methoxy singlets are at 3.74 and 3.86 ppm, the signals of H-1 and H-2 overlap at 6.73 ppm, and the two olefinic protons appear at 6.26 and 7.25 ppm. Diazomethane *O*-methylation of isosinoacutine (**6**) resulted in *O*-methylsinoacutine, identical with that obtained from the *O*-methylation of sinoacutine.

E. *O*-Methylpallidine

O-Methylpallidine (**7,** mp 118–120°C) was found in *Ocotea acutangula* Mez. (*19*). In the ¹H-NMR spectrum the *N*-methyl signal could be found at 2.45 ppm, the three methyl singlets at 3.75, 3.81, and 8.86 ppm, and the H-8, H-1, H-4, and H-5 signals at 6.25, 6.33, 6.59, and 6.78 ppm, respectively. Diazomethane methylation of the previously known pallidine (**8**) gave *O*-methylpallidine (**7**). The absolute configuration of **7** was established by CD measurements (*19*).

F. Pallidinine

Pallidinine (**9,** mp. 234–236°C) was isolated from *Ocotea acutangula* (*19*). It was determined by combined UV, IR, ¹H-NMR, and mass spectrometry that **9** is an 8,14-dihydro derivative of pallidine (**8**). Comparative ¹H-NMR data of pallidine (**8**) and pallidinine (**9**) are shown in Table I.

	R
7	CH_3
8	H

	R
9	H
10	CH_3

G. O-METHYLPALLIDININE

O-Methylpallidinine (**10**) was isolated from the same source as pallidinine (**9**), *Ocotea acutangula* (*19*), as an amorphous material. The hydrochloride salt of **10** has a melting point of 195–200°C. The structure of O-methylpallidinine (**10**) was established by comparison of the ^{1}H-NMR data with those of pallidinine (**9**) (see Table I).

H. AMURININE

Amurinine (**11**) was found, together with epi-amurinine (**12**), in *Papaver pilosum* Sibth. *et* Smith (*20*), as well as in *Papaver apokrinomenon* (*21*) as an amorphous material. A careful ^{1}H-NMR and mass spectrometry study was performed in order to establish the structure and relative stereochemistry of **11**. The ^{1}H-NMR data of amurinine (**11**) are shown in Table II.

TABLE I

SELECTED ^{1}H-NMR SIGNALS OF PALLIDINE (**8**),[a]
PALLIDININE (**9**), AND O-METHYLPALLIDININE (**10**)

	8	9	10
1-H	6.33	6.38	6.32
4-H	6.68	6.71	6.62
5-H	6.77	6.86	6.85
2-OCH$_3$	—	—	3.85
3-OCH$_3$	3.79	3.70	3.69
6-OCH$_3$	3.89	3.90	3.89
N-CH$_3$	2.37	2.36	2.37
8-H	6.29	Not available	Not available

[a] Data from Ref. *98*.

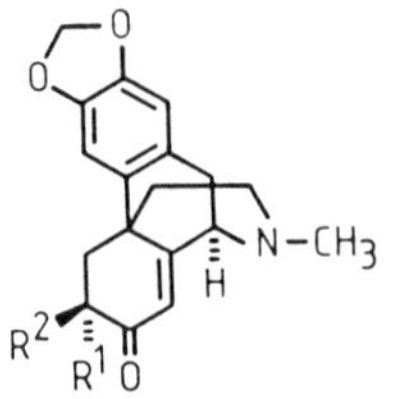

	R¹	R²		R
11	CH₃O	H	**13**	H
12	H	CH₃O	**14**	CH₃

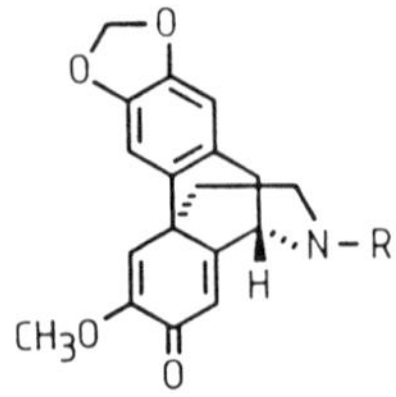

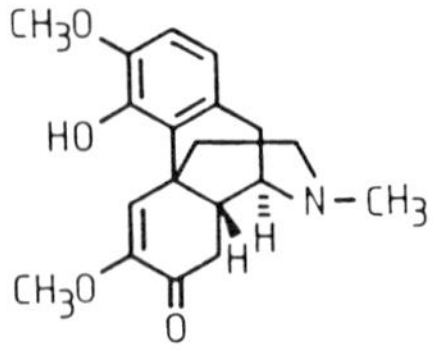

15

I. Epi-amurinine

Epi-amurinine (**12**) was isolated from *Papaver pilosum* Sibth. *et* Smith (*20*), as well as from *Papaver apokrinomenon* (*21*), as an amorphous minor alkaloid. The structure of **12** was determined by high-resolution ¹H NMR spectroscopy. Relevant NMR data and the comparison with amurinine (**11**) are presented in Table II.

J. Noramurine

(−)-Noramurine (**13**) was isolated as an amorphous substance from *Roemeria refracta* DC. (*22*), together with several isopavine alkaloids and the known amurine (**14**) and flavinantine. In the ¹H-NMR spectrum of **13**, singlets at 6.26, 6.33, 6.61, and 6.82 ppm revealed the aromatic and vinyl

TABLE II
¹H-NMR COMPARISON OF AMURININE (**11**) AND EPI-AMURININE (**12**) (*20*)

	11	12
5-H$_{ax}$	2.27 (1*H*, dd, *J* = 13 and 14 Hz)	2.12 (1*H*, dd, *J* = 12.5 and 14 Hz)
N-CH₃	2.46 (3*H*, s)	2.48 (3*H*, s)
5-H$_{eq}$	2.67 (1*H*, dd, *J* = 13 and 6 Hz)	2.79 (1*H*, dd, *J* = 4.5 and 14 Hz)
10α-H	2.96 (1*H*, dd, *J* = 18 and 6 Hz)	2.91 (1*H*, dd, *J* = 18 and 6 Hz)
10β-H	3.24 (1*H*, d, *J* = 18 Hz)	3.35 (1*H*, d, *J* = 18 Hz)
9-H	3.56 (1*H*, d, *J* = 6 Hz)	3.60 (1*H*, d, *J* = 6 Hz)
6-OCH₃	3.61 (3*H*, s)	3.55 (3*H*, s)
6-H	4.07 (1*H*, dd, *J* = 14 and 6 Hz)	3.67 (1*H*, dd, *J* = 12.5 and 4.5 Hz)
OCH₂O	5.93 (2*H*, s)	5.94 (2*H*, s)
8-H	5.95 (1*H*, s)	5.77 (1*H*, s)
1-H	6.57 (1*H*, s)	6.60 (1*H*, s)
4-H	6.65 (1*H*, s)	6.85 (1*H*, s)

protons of the morphinandienone structure. Besides an *O*-methyl singlet (3.78 ppm) and a set of methylenedioxy doublets at δ 5.91 and 5.94, there was no *N*-methyl signal, reflecting the presence of *N*-nor structure. The down-field chemical shift of H-9 (4.09 ppm), as well as MS measurements, supported this conclusion. The absolute configuration (9*S*) of **13** was determined by comparison of the CD spectrum with that of known natural morphinandienones.

K. OCOBOTRINE

Ocobotrine (**15**, mp 97–99°C) was found in *Ocotea brachybotra* (Meiss.) Mez. (*23,24*). The 8,14-dihydrosalutaridine structure of ocobotrine (**15**) was established by UV, IR, ^{1}H-NMR, and mass spectral means. The ^{1}H-NMR spectrum of **15** displays an *N*-methyl signal at 2.35 ppm, two methoxy signals at 3.65 and 3.90 ppm, H-1 and H-2 signals at 6.73 and 6.75 ppm, and an H-5 singlet at 7.76 ppm.

L. 14-EPISINOMENINE

14-Episinomenine (**16**, mp 118–120°C) was isolated from *Ocotea brachybotra* (Meiss.) Mez. (*23,24*). The structure of 14-episinomenine (**16**) was established by UV and IR comparison with sinomenine.

M. CAROCOCCULINE

Carococculine (**17**, mp 219–220°C) was found in *Cocculus carolinus* DC. (*25*). The UV spectrum of **17** proved to be similar to that of sinomenine or 14-episinomenine (**16**) owing to the 7,8-double bond present in the molecule. The ^{1}H-NMR spectrum of carococculine (**17**) displays an *N*-methyl group at 2.46 ppm, two methoxy signals at 3.81 and 4.09 ppm, H-5 as a doublet of doublets at 2.36 and 4.46 ppm (*J* = 14 Hz), and H-1 and H-2 overlapping at 6.72 ppm as a two-proton singlet.

N. Stephodeline and Isostephodeline

Stephodeline (**18**) and isostephodeline (**19**) were first isolated by Russian authors (*26,27*) from the roots of *Stephania delavay*. Detailed characterization was offered later, when both morphinenone alkaloids were found in *Stephania zippeliana* Miq. (*28*). On the basis of careful spectroscopic analysis, the structure of isostephodeline (**19**) was revised. It is worth mentioning that an alkaloid given the name stephaphylline (mp 197°C) was isolated from *Stephania suberosa* Forman (*29*) and showed identical ^{1}H- and ^{13}C-NMR data with those of isostephodeline. As the optical properties are also similar, it is not surprising that the identical structure **19** was deduced for stephaphylline.

18

19

O. Erromangine and Tannagine

Erromangine (**20**) and tannagine (**21**) were found in *Stephania zippeliana* Miq. as minor alkaloids (*28*). These two new alkaloids were found to be C-14 epimers of the major alkaloids of the plant, isostephodeline (**19**) and stephodeline (**18**), respectively. The structures were elucidated by comparing the spectral data with those of **19** and **18**.

20

21

P. ZIPPELINE AND ZIPPELIANINE

Zippeline (**22**) and zippelianine (**23**) are also minor alkaloids of *Stephania zippeliana* Miq. (*28*), and the structures were determined by ^{1}H-NMR spectroscopy [nuclear Overhauser effect (NOE) measurements]. Methylation of **23** with diazomethane gave a material identical with stephodeline (**18**).

Q. TRIDICTYOPHYLLINE

Tridictyophylline (**24**) was isolated from the whole plant of *Triclisia dictyophylla* Diels as colorless prisms (mp 204°C, after softening at 190°C) (*30*). The structure of **24** was established by consideration of the spectral data (IR, UV, ^{1}H and ^{13}C NMR, and MS) and was confirmed by X-ray crystallographic analysis.

R. OREOBEILINE

Oreobeiline (**25**) was obtained as colorless prisms (mp 116–118°C) from *Beilschmiedia oreophila* Schlechter (*31*). The general features of the mass spectrum suggest a morphinan alkaloid with a saturated C ring and a

	R^1	R^2
25	OCH$_3$	H
26	H	OCH$_3$

B/C-trans junction. The ^{1}H-NMR spectrum exhibits two 1*H* aromatic singlets at 6.67 and 6.60 ppm, typical for 2,3-disubstituted morphinans. The remaining spectral data, as well as chemical transformations (oxidation–reduction), established the structure and absolute stereochemistry of oreobeiline (**25**).

S. 6-Epi-oreobeiline

6-Epi-oreobeiline (**26**) was obtained as colorless needles (mp 207–209°C) from *Beilschmiedia oreophila* (*31*). The mass and ^{1}H-NMR spectra were almost indistinguishable from those of oreobeiline (**25**). The ^{1}H-NMR spectrum differs significantly from that of oreobeiline only in the multiplicity of the H-6 signal, which is a ddd with a large trans-diaxial coupling constant ($J = 14.5$ Hz; $J' = 5$ Hz; $J'' = 3$ Hz) at 3.50 ppm. This feature is in favor of structure **26** for 6-epi-oreobeiline. The structure was verified and the absolute stereochemistry determined by chemical correlations.

T. *O*-Methylpallidine *N*-Oxide

O-Methylpallidine *N*-oxide (**27**) was isolated in amorphous form from *Sacrocapnos enneaphylla* (*32,33*) as the first example of a morphinandienone *N*-oxide. Structure **27** was determined by comparing the ^{1}H and ^{13}C-NMR spectra with those of *O*-methylpallidine (**7**). The *N*-oxide functionality was supported by the down-field chemical shift of the *N*-methyl singlet (3.40 ppm), and of the carbon atoms bound to the quaternary nitrogen (CH$_3$, C-9, C-16) ($\sim$15 ppm lower field). Mass spectral measurements supported this assumption. *N*-Oxide **27** might be an artifact formed during the isolation process. *O*-Methylpallidine (**7**) is, however, a very minor alkaloid in the plant, from which 30 other alkaloids have been isolated (*33*), none of which was obtained as the *N*-oxide.

U. Salutadimerine

Salutadimerine (**28**), the first reported dimeric morphinandienone alkaloid, was isolated from Turkish *Papaver pseudo-orientale* Fedde (Medw.) and *Papaver lasiothrix* Fedde (Papaveraceae) (*34*) as an amorphous material. Salutadimerine is most probably formed biogenetically by oxidative coupling of two (+)-salutaridine entities. The structure **28** was supported by ^{1}H- and ^{13}C-NMR data, as well as by NOE investigations. A molecular peak at *m/z* 652 (100%) unambiguously established the dimeric structure. It should be noted, however, that although (+)-salutaridine is a major alkaloid in the two plants, the stereochemistry indicated in formula **28** requires further verification.

28

29

V. Dihydronudaurine

Dihydronudaurine (**29**) was originally isolated from *Papaver pannosum* O. Schwarz (*35*), and later from *Papaver pilosum* Sibth. *et* Smith (*20,36*), *Papaver apokrinomenone* (*21*), and *Papaver strictum* Sibth. *et* Smith (*36*). When **29** was initially isolated (*35*), the position of the C = C double bond was not established. After reisolation the structure of dihydronudaurine (**29**) was determined by ^{1}H- and ^{13}C-NMR spectroscopy.

W. Sinococuline

Sinococuline (**30**) was isolated from *Cocculus trilobus* (Menispermaceae) (*37*), and its structure was determined by careful spectroscopic investigation (IR, UV, ^{1}H and ^{13}C NMR, MS, CD). The alkaloid was found to possess significant antitumor activity.

X. Alkaloids FK 2000* and FK 3000*

Alkaloids FK 2000* (**31**) and FK 3000* (**32**) were isolated from *Sinomenium acutum* (*38*) by a Japanese group. The structures were determined by spectroscopic analysis.

30

31

32

Y. *O*-METHYLFLAVINANTINE

O-Methylflavinantine [(9*R*)-**7**], the antipode of *O*-methlypallidine (**7**), was reported (*39*) to have been isolated first from *Croton flavens* L. (*40*), in the cited publication. However, only the isolation of flavinantine (**92b**) is described; *O*-methylflavinantine was synthesized from the latter by diazomethane methylation (*41*). *O*-Methylflavinantine was later found as a natural product in a number of plants, for example, *Nemuaron vicillardii* Baill. (*42*), *Rhigiocarya racemifera* Miers (*43,44*), *Papaver bracteatum* Lindl. (*45*), *Cocculus laurifolius* DC. (*46*), *Fissistigma oldhamii* (*47*), and *Glaucium corniculatum* (*48*), as a glassy solid material without defined melting point [the methiodide has mp 223–225°C (*41*)]. The structure was unambiguously elucidated both by spectroscopic and chemical methodologies.

III. Synthesis

Because of the great medicinal importance of the morphinan alkaloids, as well as of codeine and semisynthetic derivatives, intensive research has been directed toward finding an effective and commercially feasible total synthesis for these compounds. During the more than two decades since the last review published in this treatise (*1*), many publications dealing with attempted or successful syntheses of morphines have appeared.

The majority of the approaches follow two main strategies in the key step of the procedure. Some utilize the original idea of Grewe *et al.* (*49*), who described an acidic intramolecular cyclization of properly substituted 1-benzylhexahydroisoquinolines into morphinan-type compounds. At the same time, many research groups intended to realize a so-called biomimetic synthesis with more or less success. The biomimetic approach involves intramolecular oxidative cyclization of 1-benzyltetrahydroisoquinolines into morphinandienones as a key step. The remainder of the syntheses will be discussed separately Section III,C under the heading of miscellaneous methods.

A. GREWE CYCLIZATION

The Grewe cyclization, already studied in the 1940s, was applied, independently to the synthesis of morphines as early as 1967 by two research groups: Grewe and Friedrichsen (*50,51*) and Morrison *et al.* (*52*), who found that the hexahydroisoquinoline derivative **34** or the octahydroisoquinolone **35,** both obtained via Birch reduction of the corresponding 1-benzyltetrahydroisoquinoline (**33**), yields the desired (±)-dihydrothebainone derivative (**36a**), but only as a minor product (<3%), besides the isomeric 2-hydroxy-3-methoxy-6-oxomorphinan (**37**) as the main product (30–90%).

Since the transformation of dihydrothebainone (**36a**) into thebaine, codeine, and morphine had already been carried out by that time (*53*), significant effort was devoted to modifying the disadvantageous regioselectivity of the cyclization. Beyerman and co-workers (*54*) achieved the desired regioselectivity by blocking position 1 in the envisaged morphinan ring (position 6′ in the 1-benzylisoquinolines), thus synthesizing (±)-*N*-formyl-1-methyldihydrothebainone (**39a**) in high yield. Some years later, a full paper (*55*) completed the preliminary publication (*54*), in which the authors reported on the synthesis of (+)-**38** in detail, including the resolution of the corresponding 1-benzyltetrahydroisoquinoline intermediate. Furthermore, they carried out the transformation of the formyl into a methyl group, and thus the total synthesis of "natural" (−)-1-methyl-dihydrothebainone (**39b**) was completed.

	R
a	CHO
b	CH_3

(**39**)

The above-mentioned Dutch group (*56,57*) used a symmetrically substituted 1-benzyl moiety in the starting material (**40a**), thus avoiding the regioselectivity problem. As the two expected sites of the cyclization in compounds of type **40** are equivalent, only *N*-formyl-2-hydroxy-dihydrothebainone (**36b**) was obtained in 75% yield. The 2-hydroxy substituent could be converted selectively with 5-chloro-1-phenyltetrazole to the 2-phenyltetrazolyl ether **36c,** hydrogenolysis of which on Pd/C gave (±)-*N*-formylnorthebainone (**36d**). Selective reduction of the formyl group on Pd/C yielded (±)-dihydrothebainone (**36a**). Transformation of the formyl into a methyl group was subsequently carried out by the same authors (*58*) in two different ways.

40	R
a	CHO
b	CH_3

36	R^1	R^2
b	CHO	OH
c	CHO	$OC_7H_5N_4$
d	CHO	H
a	CH_3	H

Some years later the synthesis of (−)-dihydrothebainone [(−)-**36a**] was similarly achieved; the resolution of the starting material (**40a**) was accomplished during its synthesis via the Bischler–Napieralski method (*59*). A minor modification of this approach in which cyclization of both the racemic and the chiral *N*-methyl derivative **40b** was performed, in excellent yield, resulted in shortening the reaction sequence by two steps (*60*). The same principle was adopted by Brossi and co-workers for the synthesis of

41

42

	R^1	R^2
a	CHO	OH
b	CHO	$OC_7H_5N_4$
c	CHO	H
d	CH_3	H

43

	R
a	CHO
b	CH_3

44

	R
a	OH
b	$OC_7H_5N_4$
c	H

45

($\pm$)-3-demethoxy-dihydrothebainone (**42d**) (*61*) and ($\pm$)-3-deoxy-7,8-dihydromorphine (**45**) (*62–66*), both of which were found to possess strong antinociceptive activity.

Several groups have utilized this approach to synthesize codeine and morphine by introducting bromine at the 6 position of the benzyl moiety in order to direct the cyclization to the desired dihydrothebainone oxygenation pattern. The application of bromine seems to be ideal since the oxide ring closure of 4-hydroxymorphinans such as dihydrothebainone (**36a**) involves bromination at C-1 of the morphinan system and subsequent removal of the C-1 bromine atom by hydrogenolysis (*53,67*). The first successful cyclization of *N*-acyl derivatives of α,β-unsaturated bromoketones **46** to *N*-acyl-1-bromo-dihydrothebainone **36e** and **36f** using an 80% H_2SO_4–Et_2O system has been claimed by Merck and Co., Inc. (*68*). No data, however, to support this claim have been presented, and two subsequent attempts (*55,69*) to reproduce these results were unsuccessful. According to the supposition of the authors the failure was due to a significant deactivation of the aromatic ring caused by the bromine.

46	R
a	CHO
b	CO_2CH_3

36	R
e	CHO
f	CO_2CH_3

Some years later, Rice's communication supported the conclusion of the previous papers, namely, questioning the possibility of the **46** → **36** cyclization. His group similarly failed while using $NH_4F \cdot HF$ in dry trifluoromethanesulfonic acid. However, the less stable isomeric β,γ-ketone (**52**) could be converted to the desired *N*-formyl-1-bromo-dihydrothebainone (**36e**) in 60% isolated yield applying the above acid mixture as the cyclizing agent (*65,66,70,71*). Since this is the first method that, starting from 3-methoxyphenylethylamine (**47**) and homoisovanillic acid (**48**), yields (±)-dihydrothebainone (**36a**) in a relatively reasonable yield (overall 37%) via Grewe cyclization, it is worth discussing the whole pathway in more detail.

The starting hexahydroisoquinoline derivative (**50b**) was prepared from **47** and **48** by a classic Bischler–Napieralski cyclization followed by reduction with sodium cyanoborohydride and subsequent Birch reduction. Modification of the original procedure of Grewe and Friedrichsen (*50,51*) using unprotected phenolic intermediates afforded the enol ether **50a** in 74% yield. After formylation with phenyl formate, the enol ether function of **50b** was converted to the ethylene ketal **51a,** bromination of which was performed with *N*-bromoacetamide. Mild acid hydrolysis (formic acid–water 7:1, 25°C, 1 hr) gave the β,γ-unsaturated ketone **52**. This important ketone is not stable toward the acid normally used in Grewe type cyclizations and isomerizes to the α,β-unsaturated ketone **46a,** previously prepared by DeGraw and co-workers (*69*). The desired ring closure, however, could be accomplished, as Rice found, starting from **52** in trifluoromethanesulfonic acid, but not with **46a,** to give (±)-1-bromo-*N*-formyl-nordihydrothebainone (**36e**), in relatively good yield.

Conversion of the *N*-formyl derivative **36e** to (±)-dihydrothebainone (**36a**) was performed without isolation of the intermediate **36g** by hydrolysis and subsequent hydrogenation over Pd/C in the presence of HCOOH in almost quantitative yield. If the hydrolysis is followed by bromination and hydrogenation over Pd/C, *N*-nordihydrocodeinone (**53a**) is obtained in 80% yield. The latter step represents the first example of oxide ring closure in the basic N-nor series and is of potential interest in the synthesis of different N-substituted morphinan derivatives (70,71). If the last step of this reaction series, namely, the hydrogenation, was performed in the presence of formic acid, (±)-dihydrocodeinone (**53b**) was readily isolated as the tosylate in 79% yield. In view of these results, the high-yielding

144   CSABA SZÁNTAY *ET AL.*

conversion of (−)-**53b** to (−)-thebaine and (−)-codeine (*72*), and the facile
O-demethylation (*73*) of the latter to (−)-morphine to be discussed in detail
in Section IV,A, a practical total synthesis of these alkaloids* via Grewe
cyclization is in hand.

B. Biomimetic Approach by Phenolic Oxidative Coupling

Since the original conception (*74,75*) and refinement (*76,77*) of the idea
that morphine alkaloids arise in the plant via intramolecular phenolic
oxidative coupling of an appropriately substituted 1-benzyltetrahydroi-
soquinoline derivative, a substantial amount of effort has been devoted
both to elucidate the actual biosynthetic pathway and to apply the biogenetic postulate to a practical synthesis of the alkaloids. The regioselective *para–ortho* oxidative coupling of (−)-reticuline (**54b**) into (+)-salutari-
dine (**57b**) is regarded as the most important step in the proposed pathway,
delineated by Barton and others (*76–80*).

Although syntheses of (±)-*N*-norreticuline (**54a**) and (±)-reticuline (**54b**)
have been reported by many investigators (*6*), it is worth noting that two
novel procedures provide optically active *N*-norreticuline derivatives.
Brossi and co-workers (*66,81*) succeeded in performing the classic
Bischler–Napieralski method with unprotected phenolic intermediates. At
the end of the procedure, (±)-**54a** was resolved with the aid of tartranilic
acid. N-Methylation of **54a** to give **54b** was carried out by *N*-formylation
and reduction of the *N*-formamido group with diborane.

Different *N*-norreticuline derivatives (**54a,c,e–h**) obtained by two different pathways, both starting from intermediates of an industrial papaverine
synthesis, have been prepared by Szántay and co-workers. One is based
on the monodemethylation of homoveratronitrile (**61**) and subsequent separation of the isomers. Hydrogenolysis or hydrolysis of the two isomers **62**
and **63,** respectively, produces the two building blocks of *N*-norreticuline
(*82*). An enantioselective procedure for (+)-*N*-norreticuline [(+)-**54a**], the
appropriate antipode for the synthesis of natural morphines, has also been
published by the same authors. For the reduction of the prochiral iminium
function of 1,2-dehydro-*N*-norreticuline (**64a**), chiral sodium borohydride
has been used. Asymmetric induction occurs if the reducing agent contains
the auxiliary D-(+)-*N*-carbobenzyloxyproline, affording (*R*)-(+)-*N*-
norreticuline [(+)-**54a**] in 80% optical purity (*83,84*). The

---

* In fact, Rice's synthesis requires a resolution at some stage of the reaction sequence.
The authors succeeded in resolving (±)-**49** into the (*R*) and (*S*) enantiomers in high optical
purity. The conversion of (*R*)-**49** to the natural alkaloids, as well conversion of the (*S*)
enantiomer to the unnatural morphine series, has not been reported up to now.

54–58	R
a	H
b	CH_3
c	CHO
d	CO_2CF_3
e	$CO_2C_2H_5$
f	$CO_2C(CH_3)_3$
g	$CO_2CH_2CCl_3$

other method (*85*) made use of the regioselective ether cleavage of
3,4-dihydropapaveraldine (**66**), which could be obtained from
3,4-dihydropapaverine (**65**) in good yield (*85–87*). Elimination of the
oxygen from the methylene bridge of **67**, and thus the synthesis of *N*-
formylnorreticuline (**54c**), could be performed in three steps. Both research groups synthesized not only *N*-norreticuline and reticuline, but also
a series of *N*-acyl- and 6-bromo-*N*-acylnorreticuline derivatives (*81,88*) as
potential starting materials for the phenolic oxidative coupling reaction.

CH3O
HO
59
COOH
CH3O
OH
60a
1. Δ
2. POCl3
3. NaBH4
CH3O
HO
N—R
CH3O
OH
54
R
a   H
b   CH3

CH3O
CH3O
CN
61
AlCl3/toluene
CH3O
HO
CN
62
+
HO
CH3O
CN
X
63   (X=H or Hlg)
H2/cat
CH3O
HO
NH2
59
HO
CH3O
COOH
X
60   X
a   H
b   Br
CH3O
HO
X
N
CH3O
OH
64   X
a   H
b   Br
CH3O
HO
X
NH
CH3O
OH
54   X
a   H
h   Br

It is worth summarizing the great number of unsuccessful efforts directed at simulating nature in performing the phenolic oxidative coupling (*89*) with the desired *para–ortho* regioselectivity. Many attempts to achieve this goal started with (±)-reticuline (**54b**) itself, using $K_3Fe(CN)_6$ (*79,90–94*), MnO_2 on silica gel (*95,96*), Ag_2CO_3–Celite (*93,94*), and $VOCl_3$ (*93,94,97*). However, with one exception, these approaches afforded only the *para–para* coupled product, (±)-isosalutaridine (**58b**),* in 0.3–4% yield, and/or, *ortho–para* coupled (±)-isoboldine (**56b**) in 0.4–53% yield. The exception was the detection by Barton and co-workers via isotope dilution techniques of 0.024% (±)-salutaridine (**57b**) from the ferricyanide oxidation of **54b** (*79*). Similarly, negative results have been obtained with *N*-acylnorreticuline derivatives using MnO_2–silica gel (*95,96*), $K_3Fe(CN)_6$ (*91,99,100*), or VOF_3 in the presence of trifluoroacetic acid (TFA) (*101*). These failures are partly a consequence of the four theoretically possible couplings in the process, of which the transition state for the *para–ortho* coupling is the most disadvantageous, and partly arise from the fact that salutaridine (**57b**) is highly sensitive to further oxidation (*79*).

The first breakthrough with a considerably improved yield was attained by Schwartz and Mami (*102–104*). By using 1 equivalent of thallium tristrifluoroacetate (TTFA) as the oxidizing agent, and starting with

---

* All oxidations resulting in (±)-isosalutaridine from (±)-reticuline can be considered as a total synthesis of (±)-pallidine (**58b**) (or its *O*-methyl ether, *O*-methylflavinantine), which was isolated by Kametani and co-workers (*98*). Accordingly, after the isolation the latter authors claimed the total synthesis of these alkaloids (*93,94*).

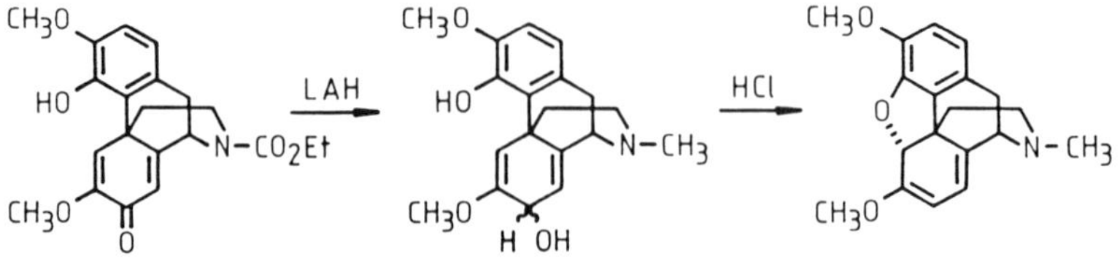

($\pm$)-*N*-trifluoroacetylnorreticuline (**54d**), 11% ($\pm$)-*N*-trifluoroacetylnor-salutaridine (**57d**) was isolated together with 26% unreacted starting material and 7% of the isoboldine derivative **56d**. An even higher yield of the desired morphinandienone (23% of **57e**) was obtained when the same oxidation was performed on ($\pm$)-*N*-ethoxycarbonylnorreticuline (**54e**). ($\pm$)-*N*-Ethoxycarbonyl-*N*-norsalutaridine (**57e**), on the other hand, could be transformed into an approximately 1:1 mixture of the epimeric salutaridinols (**68a**) in 82% yield. Treatment of the mixture with 1 N HCl at room temperature afforded ($\pm$)-thebaine (**69**), transformation of which into morphine and codeine via codeinone has already been achieved (see Section IV,B). Thus, a biogenetically patterned practical synthesis of morphine alkaloids has been presented for the first time. The remarkable regioselectivity of the oxidation is attributed to a possible coordination effect of TTFA. By interconnecting the two phenol groups via chelation, the reagent forces the starting norreticuline derivatives into a quasi-axial conformation, which facilitates the desired *para–ortho* coupling.

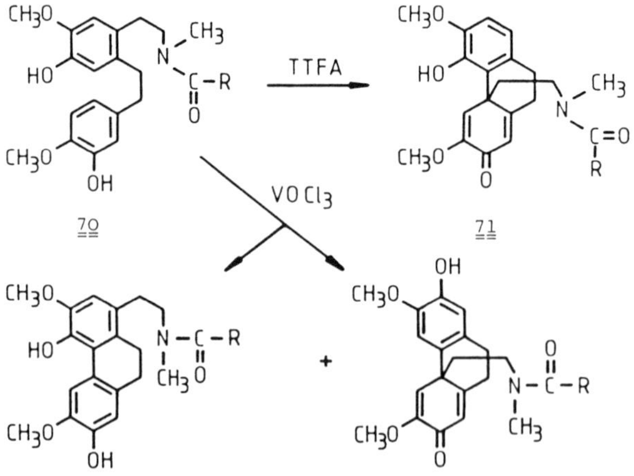

Although not directly related to the theme of this chapter, it is noteworthy that Schwartz could cyclize not only N-acyl-N-norreticuline derivatives, but N-acylsecoreticulines (**70**) with TTFA to produce N-acylsecosalutaridines (**71**) in 15–24% yield. The latter compounds were converted to alkaloids of the hasubanan and 9,17-secomorphine classes. Using VOCl₃, only isosalutaridine (**73**) and isoboldine (**72**) type seco compounds could be obtained (**105**).

Following this basic principle, Szántay and co-workers thoroughly investigated this reaction with a great number of oxidants and substrates. Successful phenolic oxidative coupling of N-acylnorreticulines (**54c,e**) into the corresponding salutaridine-type compounds (**57c,e**) in 14–30% yield with lead tetraacetate or with *I,I*-bisacetoxyiodobenzene derivative **74** in the presence of strong organic acids (e.g., trichloroacetic, trifluoroacetic, or picric acid) has been reported (*83,84,106*). In the coupling reaction, a small amount of N-acylnorisoboldine (**56c,e**) could also be isolated. To improve the regioselectivity toward the exclusive formation of the target morphinandienone structure, the role of the 6′-halogen substituent in the coupling reaction was reinvestigated. According to the findings of Jackson and Martin (*91*) and Kametani *et al.* (*100*), bromine actually did not prove to be useful as a protecting group; moreover, under the conditions applied by these authors, the new linkage arose only between the 6′ and 8 carbon atoms to supply aporphinoid (**56**-type) derivatives.

As a result of systematic experimentation, Szántay and co-workers have found that N-acylnorreticulines containing a bromine or chlorine substituent at the 6′ position (**54i–l**) furnished the N-acylnorsalutaridine derivative (**57i–l**) in significantly higher yields than those without halogen. The most pronounced protecting effect of the 6′-halogen was observed when tetraethyl ammonium bis[acyloxyiodate (I)]-type oxidizing agents (**75**) were used in the oxidative cyclization. Without protection at the 6′ position (**54c,e**), the aporphinoid isoboldine derivative (**56c,e**) was the main product. In the presence of the 6′-halogen group, however, phenolic coupling supplied the 1-halogen-N-acylnorsalutaridine derivative (**57i–k**) as the only isolable product in 35–58% yield (*83,84,106*).

$$\text{Et}_4\text{N}^{\oplus}\ [(\text{CX}_3\text{CO}_2)_2\ \text{I}]^{\ominus}$$

<u>74</u>   <u>75</u>

X = H, F or Cl

54, 57	R	X
i	CHO	Br
j	CHO	Cl
k	$CO_2C_2H_5$	Br
l	$CO_2C_2H_5$	Cl

Evaluation of a great number of these coupling reactions has revealed that it is not the chelating effect of a metal ion that is responsible for the success of a regioselective phenolic oxidative coupling, as was suggested earlier (*102–104*), but rather it is the oxidant and the applied anion that play a crucial role in the reaction in question. Only oxidizing agents transferring two electrons ($Tl^{3+} \rightarrow Tl^+$, $Pb^{4+} \rightarrow Pb^{2+}$, $I^{3+} \rightarrow I^+$, $I^+ \rightarrow I^-$) seem suitable to perform the desired coupling reaction in reasonable yield. Szántay and co-workers stressed, however, that successful accomplishment of the reaction with the desired regioselectivity requires the presence of an anion derived from strong organic acids, either as a ligand of the oxidant or as an additive in the form of the free acid or a soluble salt (*83,84,106*). It has also been shown that besides the presence or absence of the 6'-halogen group, the regioselective formation of *N*-acylnorsalutaridines (**57**) greatly depends on the bulkiness of the *N*-acyl substituent. Under identical reaction conditions, *N*-norreticulines containing a bulky *N-tert*-butoxycarbonyl or *N*-2,2,2-trichloroethoxycarbonyl protecting group (**54f,g**) supplied a considerably higher yield of the corresponding salutaridine derivative (**57f,g**) than *N*-formyl- or *N*-ethoxycarbonylnorreticuline.

Finally, Szántay's findings on phenolic oxidative coupling were successfully adopted for the *in vitro* realization of the (±)-reticuline (**54b**) to (±)-salutaridine (**57b**) transformation on a preparative scale. Using lead tetraacetate in the presence of trichloroacetic acid, crystalline (±)-salutaridine (**57b**) has been isolated in 2.7% yield, although in the reaction (±)-isoboldine (**56b**) was the main product (14%) (*107*). (±)-Salutaridine could also be prepared by Eschweiler–Clarke methylation of *N*-norsalutaridine (**57a**), obtained either from *N*-formylnorsalutaridine (**57c**) by deformylation or from *N-tert*-butoxycarbonylnorsalutaridine (**57f**) by removing the protecting group with *p*-toluenesulfonic acid. Sodium borohydride reduction of (±)-salutaridine (**57b**) results in a 1:1 epimeric mixture of (±)-salutaridinol (**68a**), which can also be prepared from (±)-1-bromo-*N*-ethoxycarbonylnorsalutaridine (**57k**) in one step by LiAlH₄ reduction (*83,84,106*).

In the course of a comprehensive, systematic investigation, Szántay and co-workers found manganese or vanadyl acetylacetonate to be excellent oxidizing agents for the regioselective *para–para* phenolic oxidative coupling. Treatment of *N*-ethoxycarbonylnorreticuline (**54e**) with 5 equivalents of manganese tris(acetylacetonate) in boiling acetonitrile afforded *N*-ethoxycarbonylnorisosalutaridine (**58e**) in 32% yield along with a small amount of the isoboldine derivative **56e**. Even if the 6′ position is protected by a halogen atom, the oxidation results in the formation of the new bond between the C-4a and C-6′ carbon atoms, thus providing isosalutaridine derivatives. The same high regioselectivity was also observed with *N*-formylnorreticulines (**54c,i,j**) and vanadyl bis(acetylacetonate) as an oxidizing agent. (±)-Pallidine (**58b**) could be obtained from *N*-formylnorisosalutaridine (**58c**) via deformylation and subsequent Eschweiler–Clarke methylation (*108*).

54	R	X
c	CHO	H
e	CO₂C₂H₅	H
i	CHO	Br
j	CHO	Cl
k	CO₂C₂H₅	Br
l	CO₂C₂H₅	Cl

58	R
b	CH₃
c	CHO
e	CO₂C₂H₅

As a result of parallel investigations, White, Brossi, and co-workers presented a full paper (*109*) in which the biomimetic total synthesis of (−)-codeine (**77**) from (±)-*N*-norreticuline (**54a**) in eight steps was reported. Resolution of the starting material was performed as previously described (*81*), and the key step of the procedure, the phenolic oxidative cyclization, was thoroughly investigated with hypervalent iodine derivatives of type **74** as oxidants (*110*). The best yield (21%) was achieved when (*R*)-(−)-6′-bromo-*N*-trifluoroacetyl-*N*-norreticuline [(−)-**54m**] was treated with phenyliodoso bistrifluoroacetate; the (*R*)-(−)-1-bromo-*N*-trifluoroacetylnorsalutaridine [(−)-**57m**] obtained could be converted to a 1:1 mixture of epimeric (−)-1-bromosalutaridinols (**68b**). Both of the epimers, separately or together, furnished (−)-1-bromothebaine [(−)-**76**] by dehydration with dimethylformamide dineopentylacetal in good yield. Hydrolysis and subsequent reduction by LiAlH₄ completed the total synthesis of (−)-codeine [(−)-**77**].

An alternative approach to limit the number of possible directions of the phenolic coupling reaction has been published by Schwartz and Zoda (*111*). Treatment of triphenolic *N*-ethoxycarbonyl-5′-hydroxy-*N*-norreticuline (**79**), instead of the norreticuline derivative, can afford only two cyclized products because of the symmetry of the benzyl-aromatic ring (see Beyerman's principle (*54*)]. The substrate **79** was prepared in six steps, and 50% overall yield, via the Bischler–Napieralski route from 4-hydroxy-3-methoxyphenylethylamine (**59**) and 3,5-bisbenzyloxy-4-methoxyphenylacetic acid (**78**). Treatment of **79** with $VOCl_3$ in diethylether afforded the morphinandienone **80** as the sole product in 64% yield. The same procedure in CH_2Cl_2 showed much lower regioselectivity; besides 13% unreacted starting material, morphinandienone **80** (37%) and an aporphinoid derivative (24%) were also isolated. Morphinandienone **80** was converted to (±)-2-hydroxycodeine (**81**) or, via oxidization with singlet oxygen and elimination of the 2-hydroxy function by phenyltetrazolyl ether (*54*), to (±)-noroxycodone (**82**), a very important intermediate of synthetic 14-hydroxymorphinan analgesics and/or narcotic antagonists (*112,113*).

Protection of the basic nitrogen in reticuline derivatives usually requires N-acylation of a norreticuline derivative. As a clever solution, Vanderlaan and Schwartz (*114*) presented a paper in which the synthesis, as well as the oxidative coupling, of (±)-3-oxoreticuline (**86**) is described. Synthesis of the starting material (**86**) was begun with from isovanillin derivatives **83** and **84** via bisbenzyl **85,** which was cyclized to **86** with α-chloro-α-(methylthio)acetyl chloride in three steps. Oxidative coupling of 3-oxoreticuline (**86**) with iodosobenzene diacetate (**74**), in the presence of TFA (*83,84,106*), gave racemic 16-oxosalutaridine (**87**) in 27% yield, besides 16-oxopallidine (**88**), 5-oxoisoboldine, and 13-oxothalidine in yields of 8, 6, and 10%, respectively. Reduction of 16-oxosalutaridine (**87**) with excess $LiAlH_4$ in tetrahydrofuran (THF) led to an epimeric mixture of salutaridinols (**68a**), conversion of which to thebaine, codeine, and morphine has already been elaborated (see Section IV,B).

A recently published enzyme-catalyzed biotransformation of (*R*)-reticuline [(+)-**54b**] to the morphine precursor (*R*)-salutaridine [(−)-**57b**] concludes this section. Although attempts to mimic this biosynthetic step had already been made previously using a rat liver enzyme preparation (*115,116*), a product with *para–ortho* coupling could not be detected from the reaction mixture. Zenk and co-workers found, however, that (*R*)-reticuline [but not the (*S*) isomer], supplies (*R*)-salutaridine regio- and stereoselectively when incubated in a buffer system containing cytochrome *P*-450-linked NADPH and O_2-dependent microsomal fractions of *Papaver somniferum* in the flowering stage (*117*), or a microsomal-bound cytochrome *P*-450 NADPH-dependent enzyme discovered in the liver of pigs and other mammals (*118*).

C. MISCELLANEOUS METHODS

Among the different approaches listed in this section, those starting with reticuline derivatives are mentioned first. Although a number of investigators have dealt with the electrochemical oxidation of 1-benzylisoquinolines, none succeeded in performing the desired *para–ortho* coupling. Morphinandienones with the isosalutaridine-type substitution pattern (**58** type), suitable for the synthesis of pallidine or flavinanthine alkaloids have been isolated in 10–94% yield by anodic oxidation of laudanosine derivatives (**89**) (*119*), of *N*-acylnorreticulines (**54**) (*120*), or of other tetraalkoxy-1-benzylisoquinolines (**90**) (*121*). Even when 6'-bromolaudanosine derivatives were oxidized anodically, no salutaridine derivative with a morphine substitution pattern was isolated. Instead, cleavage of the bromo substituent took place, and the isosalutaridine derivative **58** was formed (*119*). The mechanism of nonphenolic electrochemical oxidations has also been published (*122*).

The structures and substituent table for the anodic oxidation scheme (leading to the **58** type product) are:

	R^1	R^2	R^3	X
54	H	H	CO_2R	H
89	CH_3	CH_3	CH_3	H or Br
90	CH_3	CH_2Ph	CH_3	H

At the end of the 1960s, numerous Pschorr cyclizations were carried out to obtain different isoquinoline alkaloids. This reaction, performed via diazotization of an amino group, properly attached to a 1-benzylisoquinoline derivative, and subsequent thermal decomposition of the diazonium salt in the presence or absence of some metal catalyst, seemingly possesses the advantage of limiting the directions of the cyclization. As one coupling site is given by the position of the amino group, only one aporphine and one morphinandienone product should be expected from the reactions. It was Kametani and co-workers who first thoroughly investigated the possibilities in synthesizing different alkloids via Pschorr cyclization (89). The method, unfortunately, produced in all cases a number of side reactions (reductive deamination, cleavage of the 1-benzyl group, etc.), and therefore extremely low (1–8%) yields resulted. Many publications (89) dealt with the Pschorr reaction of 1-(2-amino-4,5-dialkoxybenzyl)-tetrahydroisoquinolines (**91**) and reported on the total synthesis of (±)-pallidine (**58b**) (123), (±)-O-methylflavinantine (**92a**) (41), (±)-flavinantine (**92b**) (124), (±)-amurine (**92c**) (125), and (±)-demethoxy-O-methylandrocymbine (**92d**) (126).

Perhaps because of the rather poor yields of the required morphinandienones, there are only two publications which accomplish the Pschorr cyclization with 1-(2-amino-3,4-disubstituted) benzyltetrahydroisoquinolines (e.g., **91f,g**), which are regarded as suitable starting materials for the salutaridine skeleton. The reaction sequence, starting from **91f,** was already cited in the last review in this treatise (1) and furnishes salutaridine **57b** either in racemic or in both enantiomeric forms in 1% yield (127).

Neumeyer and co-workers, while synthesizing corytuberine-type aporphines (**93**) starting from tetrahydroisoquinoline (**91g**), obtained (±)-N-propyl-O-methylnorsalutaridine (**92e**) as a minor product (2%) (128). If the starting material contains a 6,7-methylenedioxy substituent,

91	R^1	R^2	R^3	n
a	BzO	CH_3	H	1
b	CH_3O	CH_3	H	1
c	CH_3O	Bz	H	1
d	$-OCH_2-$		H	1
e	CH_3O	CH_3	H	2
f	H	CH_3	BzO	1

	R^1	R^2	R^3	n
58b	OH	CH_3	H	1
92a	CH_3O	CH_3	H	1
92b	CH_3O	H	H	1
92c	$-OCH_2-$		H	1
92d	CH_3O	CH_3	H	2
57b	H	CH_3	OH	1

the cyclization leading to a morphinandienone by-product is accompanied by cleavage of the dihydrofuran ring, and thus the 6-demethylsalutaridine derivative **92f** is obtained. This result is in good accord with early findings by Kametani (*129*).

91	R^1	R^2	93
g	CH_3	CH_3	a
h	$- CH_2 -$		b

92	R
e	CH_3
f	H

A series of alkaloids has been synthesized by photolysis of appropriately substituted and brominated 1-benzylisoquinolines. Irradiating, for example, a solution of **94a** with a 500-W UV lamp in the presence of sodium hydroxide and sodium iodide afforded (±)-salutaridine (**57b**) in 1% yield. Similarly, (±)-pallidine (**58b**) could be isolated in low yield from the reaction mixture when the 2,8-dibromo derivative **94b** was irradiated (*130*).

Several research groups have pursued benzomorphan and morphine syntheses by strategies which involve late formation of the C-10–C-11 bond. Stella *et al.* (*131*) described the $AlCl_3$-promoted cyclization of a 2-(chloromethyl)-4-phenylpiperidine to a benzomorphan in 60% yield. Gless and Rapoport (*132*) made use of Friedel–Crafts cyclization in the synthesis of 9-methyl-6,7-benzomorphan from 4-aryl-2-piperidones.

$\underline{94}$	X	Y	Z
a	H	Br	H
b	Br	H	Br

	R^1	R^2
$\underline{57b}$	H	OH
$\underline{58b}$	OH	H

However, this methodology could not be extended to cyclizations leading to dihydrothebainone derivatives *(133)*.

Evans and *et al.* *(134)* have developed a reasonable approach to the *trans*-morphinan skeleton (**100**) from bicyclic enamine **95** via *cis*-immonium (**97**) and *cis*-aziridinium perchlorates (**98**). An analoguous transformation of the trans-fused immonium perchlorate **96** has not been reported. A total synthesis of ($\pm$)-morphine, based on this approach, however, has been accomplished by the same authors *(135)* and is presented here in brief. Tetrahydropyridine **101,** prepared by the addition of 2,3-dimethoxyphenyllithium to *N*-methyl-4-piperidone, was cyclized with dibromide **102** in two steps. The resulting bicyclic enamine **103,** in accord

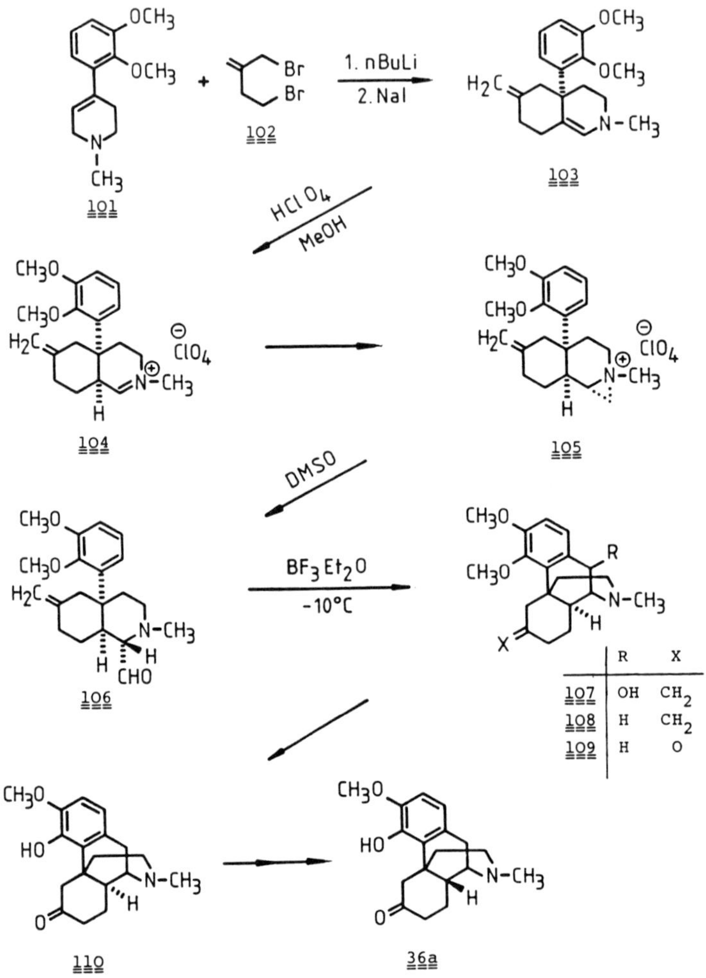

with earlier studies (*134*), afforded the "kinetic" *trans*-immonium salt, which in methanol solution, was equilibrated to the thermodynamically preferred cis isomer (**104**). Conversion of the latter to the morphinan skeleton was performed in 72% overall yield by the following three-step sequence. Reaction of **104** with diazomethane resulted in diastereomerically pure aziridinium salt **105,** which was readily transformed into α-aminoaldehyde **106** by Kornblum oxidation (with dimethyl sulfoxide). Conversion of **106** to the morphinan carbinol **107** was accomplished by Lewis acid catalysis, and subsequent reductive elimination of the benzylic hydroxy function afforded morphinan **108** in excellent yield. Completion of a formal total synthesis of (±)-morphine was accomplished by Lemieux–Johnson oxidation (OsO$_4$,NaIO$_4$) of **108** to ketone **109,** which had previously been employed by Gates in the pioneering total synthesis of morphine (*53*).

The above authors noted, furthermore, that direct cyclization could be observed when immonium salt **97** was treated with diazomethane, so that in addition to the main product, aziridinium **98,** some *trans*-morphinan **100** was also obtained. This direct cyclization did not take place when the immonium perchlorate **104** was reacted with diazomethane, and the closure of ring B could only be accomplished via α-aminoaldehyde **106.** McMurray *et al.* (*136*) supported these findings, and while investigating the cyclization step, found that the *O*-alkyl substituent adjacent to the future C-4, was responsible for the lack of ring closure in the above case. To prove this hypothesis, they synthesized *O*-methylpallidinine (**10**) from the appropriately substituted 4a-aryloctahydroisoquinoline derivative **111.** Ring closure of the "C-4"-unsubstituted starting material to morphinan **112** proceeded in fairly good yield. A three-step oxidation sequence afforded diketone **113,** "kinetic" enol methylation of which with *p*-toluenesulfonic acid in methanol, led to *O*-methylpallidinine (**10**) as the main product.

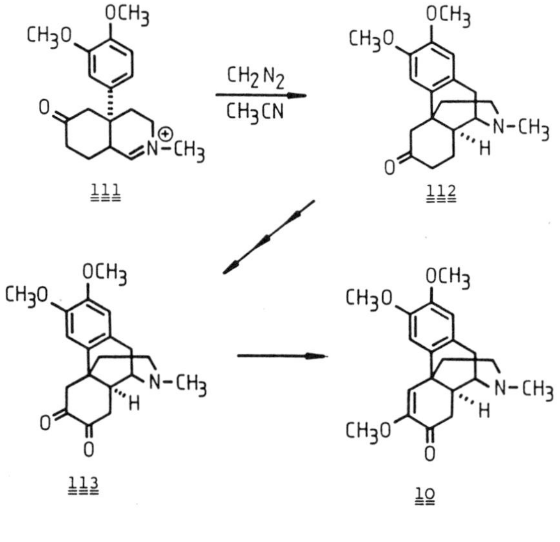

Having investigated the stereochemistry and methodology in detail (*137–140*), Rapoport and co-workers reported on a successful formal total synthesis of (±)-morphine via a basically identical route (*133*) as had been published by Evans and Mitch (*135*). They, however, suggested utilization

160 CSABA SZÁNTAY *ET AL.*

of ($\pm$)-dihydrothebainone (**36a**) instead of the 14-epi-ether **109.** Selective ether cleavage of **109** to **110,** epimerization to the "natural" morphinan dihydrothebainone (**36a**), as well as further manipulations to codeine and morphine are discussed in Section IV,A in connection with transformations.

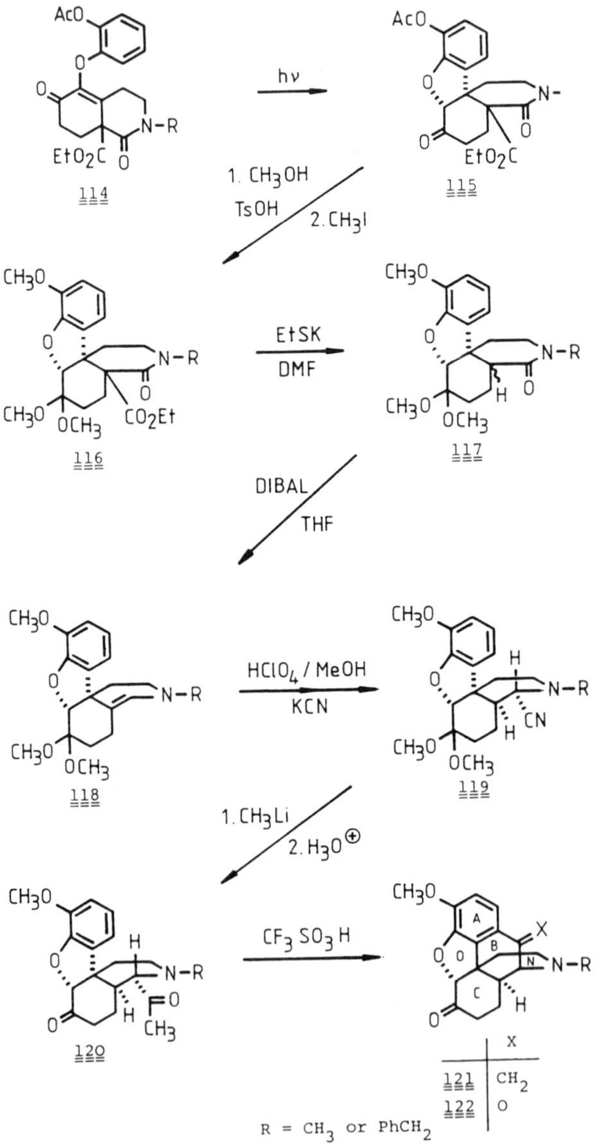

R = CH$_3$ or PhCH$_2$

	X
121	CH$_2$
122	O

The same strategy was applied by Schultz and co-workers (*141*), and the closure of ring B has been accomplished from tetracyclic intermediate **120,** already containing the future oxide ring (Ring O). Formation of the tetrahydrofuran ring was realized in an elegant photocyclization process of aryl vinyl ether **114.** The final cyclization step, affording the *trans*-morphine derivative **121,** was accomplished via stereochemically pure cyano (**119**) and methyl ketone (**120**) compounds. Cyclodehydration of the latter with trifluoromethanesulfonic acid resulted in *trans*-morphinan **121** in 75% yield. Transformation of the vinyl function to an oxo group has been performed, and conversion of the *trans*-morphinan intermediate to the natural alkaloids was stated to be in progress. Similar ACNO ring fragments of morphine, which are potential intermediates for synthetic purposes, have been synthesized by Ciganek (*142*) via intramolecular Diels–Alder reaction (**125**) and by Weller and co-workers (*143,144*) by Dieckmann condensation (**128**).

A fundamentally different strategy of morphine construction, based on the development of a vinyl sulfone, has been published by Fuchs and

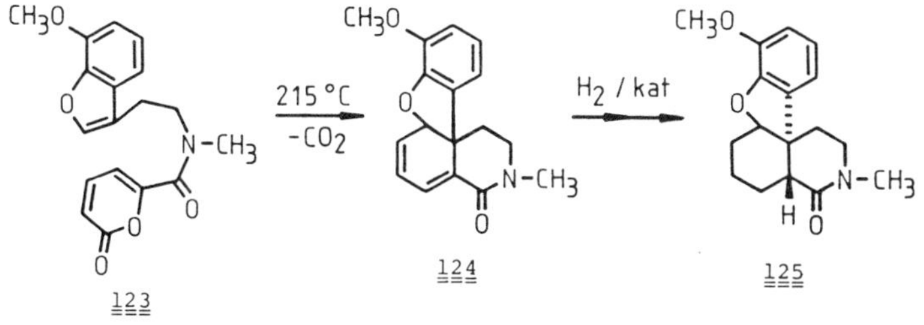

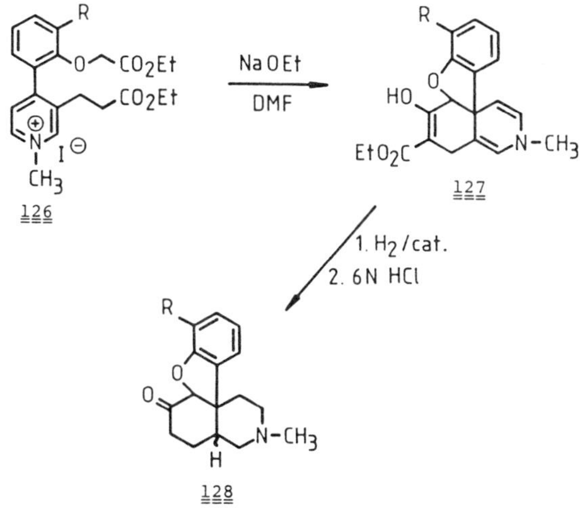

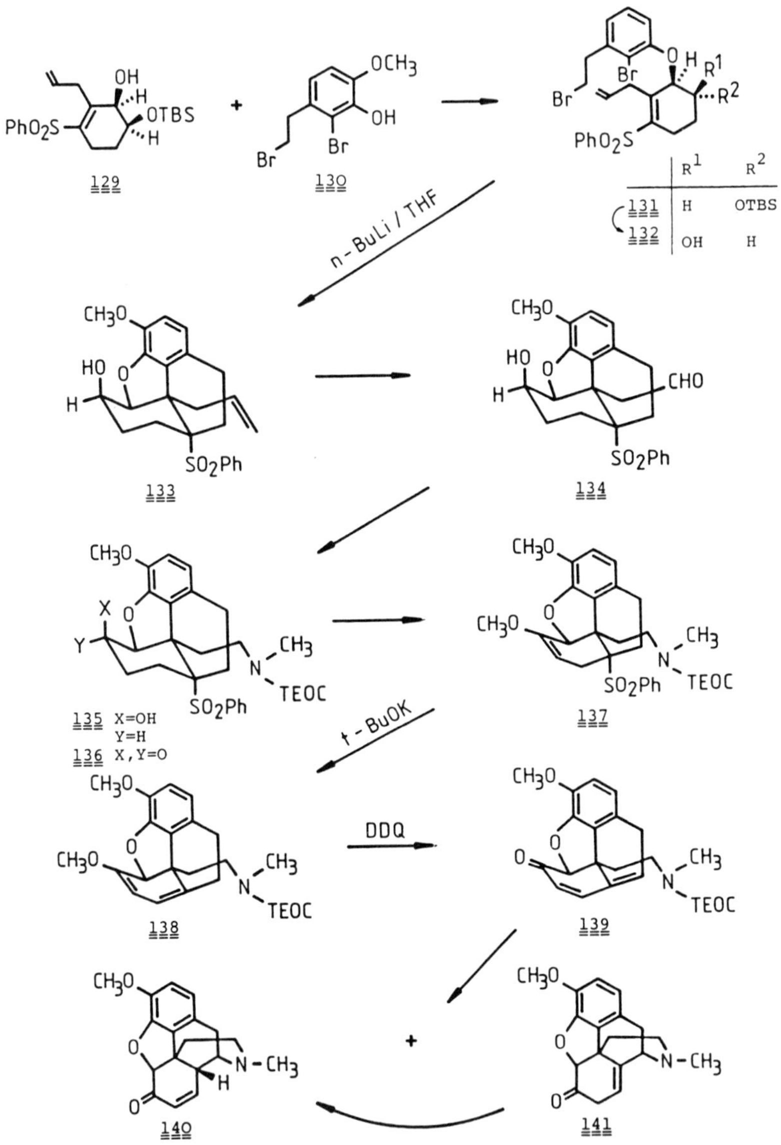

co-workers (*145*). γ-Aryl oxyvinyl sulfone **131** was synthesized by modified Mitsunobu coupling of sulfone **129** and phenol **130.** The trans relative
configuration of the two oxygenated functions in **131** has been altered to cis
in three steps and in 85% overall yield. Double cyclization of *cis*-hydroxy
ether **132** under the conditions of deprotonation and metal–halogen ex-

change afforded a 63% yield of the tetracyclic sulfone **133.** Aldehyde **134** was produced via biphasic osmium tetroxide/sodium periodate cleavage of the allylic double bond of **133.** Reductive amination of **134** with excess methylamine and sodium cyanoborohydride, and subsequent Schotten–Baumann acylation with (trimethylsilyl)ethoxycarbonyl chloride (TEOC-Cl), led to the protected amine **135,** Swern oxidation of which gave ketone **136** in good overall yield. Treatment of ketone **136** with trimethylortho-formate and *p*-toluenesulfonic acid in methanol afforded the methyl enol ether **137.** Potassium *tert*-butoxide-mediated elimination of benzenesulfi-nate from **137** furnished the methyl dienyl ether **138** in excellent yield. 2,3-Dichloro-5,6-dicyano-1,4-benzoquinone (DDQ) oxidation of the latter then led to dienone **139,** trifluoroacetic acid deprotection and biphasic neutralization of which afforded a mixture of (±)-codeinone (**140**) and (±)-neopinone (**141**). This mixture was isomerized into (±)-codeinone (**140**) exclusively by the method of Rapoport (*146*). This step, as well as further transformations to codeine and morphine, is discussed in Section IV.

IV. Transformations

This part is divided into four sections. The first section is devoted to the conversion of the dihydrothebainone product of Grewe cyclization to codeine and morphine. The second section covers the transformation of the biomimetic intermediate salutaridine and its derivatives. The third deals with semisynthetic approaches resulting in 14-hydroxy morphinan derivatives with high medicinal importance, and the last section comprises miscellaneous transformations.

A. TRANSFORMATION OF DIHYDROTHEBAINONE INTO MORPHINE

In pioneering work, Gates and Tschudi (*53*) published a multistep synthesis of β-dihydrothebainone (**110**), the C-14 epimer of the natural dihydrothebainone (**36a**). All investigators having arrived at this *trans*-morphinan intermediate during synthesis, have had to solve the C-14 epimerization problem in order to reach the natural morphine series. Both Evans and Mitch (*135*) and Rapoport and co-workers (*137–140*) utilized the same principle. This conversion, originated by Small and Browning (*147*), and later applied by Gates and Tschudi (*53*), involves the acid- or base-catalyzed epimerization of 1-bromo-β-thebainone **143** to the thermodynamically more stable 1-bromothebainone (**144**). For the selective demethylation at C-4 of *O*-methyldihydro-β-thebainone (**109**), Rapoport and co-workers published a high-yielding and simple method using sodium thioethylate (*133*).

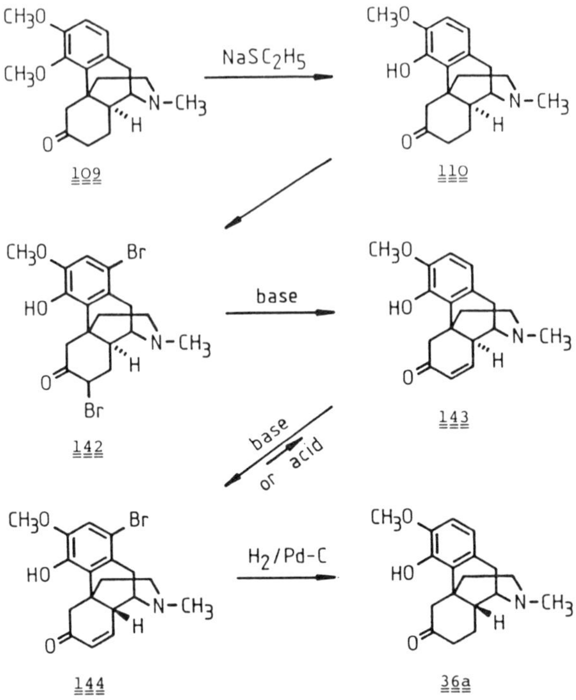

The original method of Gates and Tschudi, starting from β-dihydrothebainone (**110**) and ending in the total synthesis of morphine (**1**), could be adopted to dihydrothebainone (**36a**) (*148,149*). However, the procedure, already summarized in the previous review on this subject (*1*), involves many steps and gives a rather poor (5%) overall yield.

Rapoport and Weller (*72*) succeeded in developing the conversion of (−)-dihydrothebainone [(−)-**36a**] to (−)-thebaine or (−)-codeine in much higher yield. There are two goals to be achieved in such a process, namely, closure of the 4,5-oxide ring and introduction of the 7,8-double bond. The first step has been accomplished via a well-studied (*148–154*) bromination–dehydrobromination method. Optimum results were achieved when 2.5 equivalents of bromine was applied in acetic acid at room temperature, the mixture was evaporated after 2 hr, and biphasic NaOH treatment (1 N NaOH/CHCl₃) afforded a mixture of brominated dihydrocodeinones

53	X	Y
b	H	H
c	Br	H
d	Br	Br

(**53c,d**). Hydrogenation of the crude mixture in acetic acid/sodium acetate buffer over Pd/C resulted in a 100% crude yield of dihydrocodeinone (**53b**).

The second step, that is, introduction of the $\Delta^{7,8}$ double bond into the skeleton, has been solved in a multistep, but high-yield sequence. Dihydrocodeinone (**53b**) was first converted to enol ether **146** in two steps via dimethyl ketal **145** in practically quantitative yield. Addition of CH_3OBr to **146** furnished the crude bromoketal **147**. Alkoxide-catalyzed elimination in dimethyl sulfoxide at 60°C afforded codeinone dimethyl ketal (**148**) in excellent yield; at higher temperatures (120°C) conversion of **148** to thebaine (**69**) was complete in 1 hr, and the latter was thus obtained in 95% yield. Hydrolysis of ketal **148** in 3 N acetic acid at room temperature yielded 82% pure codeinone (**140**).

Without exception, all of the reactions give extremely pure products, and the entire sequence is thus applicable to working with crude intermediates throughout the whole synthesis. Starting from **36a**, codeinone (**140**) could be obtained in 85.5% yield using this approach. For the final step, codeinone was reduced with sodium borohydride, as described earlier (*53*). The overall yield in the transformation of **36a** into pure codeine (**77**) is 68%, a remarkable improvement over the 5% reported previously (*53*).

The single steps of the 4,5-oxide ring closure as well as the ^{1}H- and ^{13}C-NMR spectroscopy and stereochemistry of the brominated intermediates have been investigated in detail by Beyerman and co-workers (*155*). Basically, they followed the method of Rapoport, and the following pathway was delineated for the conversion of (−)-dihydrothebainone [(−)-**36a**] to (−)-dihydrocodeinone [(−)-**53b**]. Bromination of (−)-**36a** yielded successively 1-bromo-, 1,7α-dibromo-, and 1,5β,7α-tribromodihydrothebainone (**149, 150,** and **151**). Closure of the oxide ring in boiling ethanol afforded an equilibrium mixture of 1,7α- and 1,7β-dihydrocodeinone (**53d**) in the ratio 35:65. Starting from 1-methyldihydrothebainone (**39b**), and following the same route, 1-methyldihydrocodeinone (**53e**) was obtained.

(±)-1-Bromo-*N*-formylnordihydrothebainone (**36e**), as described in Section III,A, could be similarly transformed either into (±)-*N*-nordihydrocodeinone (**53a**) or into (±)-dihydrocodeinone (**53b**) (*70*), via Rapoport's bromination–dehydrobromination method (*72*) combined with other steps. This work of Rice represents the first example of an oxide ring closure in the *N*-nor series.

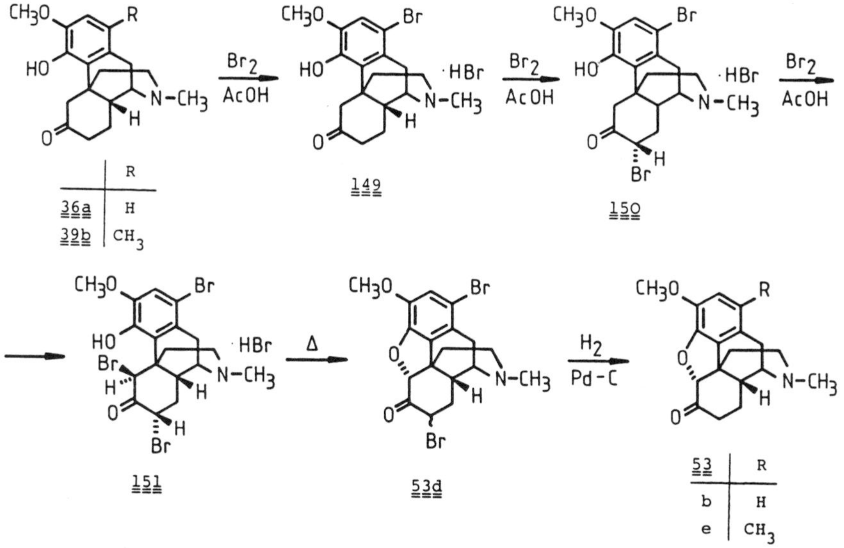

In another publication, Iijima and Rice described a four step-transformation of (+)-dihydrocodeinone [(+)-**53b**] into unnatural (+)-codeine [(+)-**77**] via *N*-carbethoxy-7-phenylselenonordihydrocodeinone (**152**) in 40% overall yield (*156*). In accordance with earlier results of Rapoport (*72*), the sequence could not be realized if a basic amine function was present in the molecule. The structure and stereoarrangement of intermediate **152** were supported by NMR and single-crystal X-ray analysis.

In this section, a reverse process, namely, the opening of the 4,5-oxide ring, should also be mentioned. Using formamidinesulfinic acid under either heterogeneous or homogeneous conditions, Brine *et al.*, as well as

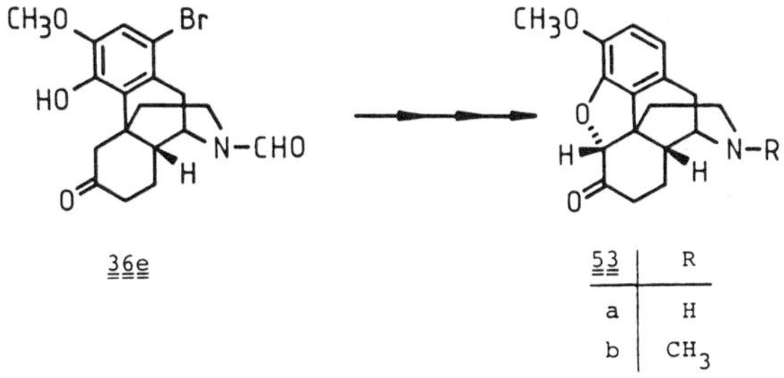

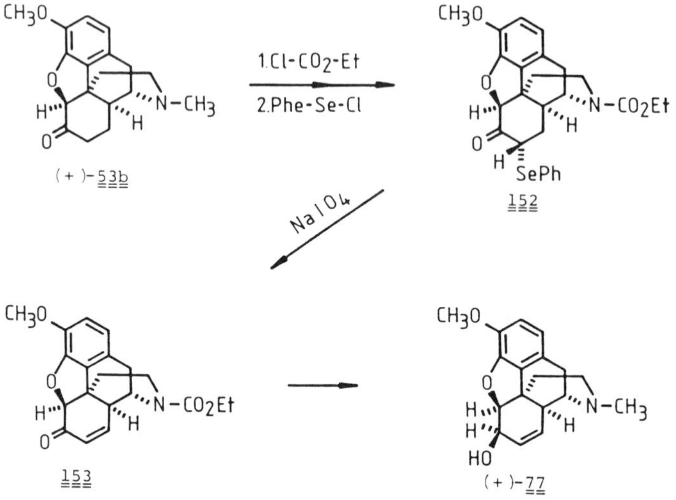

Chatterjie *et al.* (*157,158*), succeeded in performing a reductive cleavage of the oxide ring at C-5, thus transforming dihydrocodeinone (**53b**) into dihydrothebainone (**36a**). Depending on the conditions applied, various quantities of dihydroisothebainol (**154**) were also formed during the reduction. As the stereoselectivity of the process supplying **154** is opposite (6β-OH) to that of hydride reduction, the preparative potential is obvious.

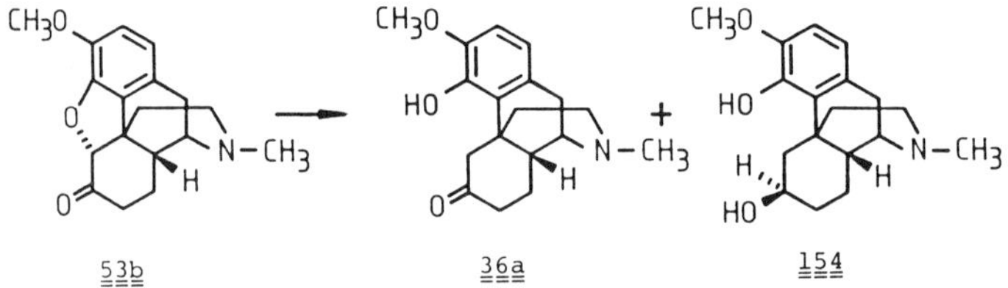

Demethylation of codeine to give morphine has been carried out by Rice (*73*) and Lawson and DeGraw (*159*) in much higher yield than earlier. Rice treated codeine with boron tribromide in chloroform at room temperature to obtain a 90–91% yield of morphine. In parallel work, DeGraw achieved an 80% yield of morphine when sodium propylmercaptide in dimethylformamide (125°C) was used for demethylation. The reverse process, namely, selective 3-*O*-methylation of morphine to codeine, has also been mentioned in the literature. Russian authors published an improved technique for this transformation, which was accomplished with trimethylanilinium hydroxide in xylene to produce codeine in relatively high yield (*160*).

B. TRANSFORMATION OF SALUTARIDINE DERIVATIVES INTO MORPHINE

The biomimetic synthesis of morphines was thoroughly investigated during the 1970s and 1980s. Achievement of the key step involving the regioselective *para–ortho* coupling of reticuline derivatives into morphinandienones was described in Section III,B. Transformation of salutaridine and its derivatives into thebaine, codeine, and morphine, as well as some retro procedures, are presented in this section.

It was Barton and co-workers who first investigated the conversion of salutaridine (**57b**) into morphine. In early communications (*77,78*), it was shown that reduction of **57b** with sodium borohydride gave two epimeric alcohols, salutaridinol-I and -II (**68a**). Salutaridinol-I, but not its epimer, was found to be a biological precursor of thebaine, codeine, and morphine (*79*).

7β-OH "salutaridinol-I"
7α-OH "salutaridinol-II"

	R	X	Y
57e	CO$_2$Et	H	H$_2$
57k	CO$_2$Et	Br	H$_2$
103	CH$_3$	H	O

More recently, Zenk and co-workers (*161*) revised these findings. They reduced salutaridine (**57b**) enzymatically, and only one epimer, identical with salutaridinol-I previously described by Barton and co-workers, was obtained. Although it had been proved that only this epimer is to be regarded as the biological precursor of morphines, X-ray crystallography unambiguously showed that salutaridinol-I has the opposite configuration at C-7 to that hitherto supposed. Accordingly, in the future, the

(7*S*)-configured biologically active alkaloid will be called salutaridinol
(**68a**, 7α-OH) and its inactive epimer, 7-episalutaridinol (**68a**, 7β-OH).
The same, approximately 1:1, epimeric mixture has been obtained
from *N*-ethoxycarbonylnorsalutaridine (**57e**) (*102–104*), from 1-
bromo-*N*-ethoxycarbonylnorsalutaridine (**57k**) (*83,84,106*), or from
16-oxo-salutaridine (**87**) (*114*) in a one-step reduction with LiAlH₄.

In contrast to biological experimentation, both epimeric salutaridinols,
as well as their mixture, could be converted to thebaine (**69**) chemically.

$$CH_3O \cdots \longrightarrow \cdots CH_3O$$

68a   69

Treatment of either epimer or the epimeric mixture (**68a**) with 1 N HCl at
room temperature for 1 hr afforded (±)-thebaine (**69**) in 31–37% yield (*78*).
This method was used by different groups (*102–104,127*) in total synthesis
until 1977, when a better procedure, patented in many countries, succeeded in performing this step in 80% yield using thionyl chloride in
pyridine (*162,163*).

155	R¹	R²
a	H	H
b	Ac	Ac
c	Ac	H
d	Ac	CH₃

69

156a   57b

As a reverse transformation, Rapoport and Bjeldanes (*164*) reported on an oxidative hydrolysis of thebaine with aqueous sodium bisulfite (pH 4) in the presence of oxygen to give 6-*O*-demethylsalutaridine (**155a**), which could also be obtained by the action of alkali on 14β-bromocodeinone* (**156a**) (*165,166*). Treatment of diketone **155a** with diazomethane afforded salutaridine (**57b**) as the predominant product, but only in 27% yield. An improved yield for the methylation of **155a** to give salutaridine (**57b**) was published by Horváth and Makleit (*168*). The parallel 4-hydroxy methylation was avoided by diacetylation followed by selective deacetylation of **155b** with diethylamine–ethanol; methylation of **155c** was performed with diazomethane, and finally alkaline hydrolysis gave salutaridine (**57b**) in 65.5% overall yield.

As codeine is among the most effective and widely used analgesic and antitussive agents, conversion of thebaine to codeine is of commercial importance. A highly effective thebaine to codeine conversion became even more significant when it transpired that, in *Papaver bracteatum*, thebaine is the major alkaloid (*169–171*). The conversion involves two fundamental steps: (1) transformation of the dienol ether of thebaine (**69**) into the α,β-unsaturated ketone codeinone (**140**), and (2) reduction of the α,β-unsaturated ketone to the allylic alcohol codeine (**77**). Step 2 has been achieved (*172,173*), and it proceeded stereoselectively in almost quantitative yield with sodium borohydride in methanol. The loss in the overall yield therefore lies in step 1. Among the earlier procedures, the best reported method (*174,175*), involved the addition of 2 mol of anhydrous hydrogen bromide or chloride to thebaine to form 6,8-dibromotetrahydrothebaine (**157**) followed by alkaline hydrolysis and dehydrohalogenation, which is claimed to proceed with 74% yield.

Basically in the same way, codeinone (**140**) was claimed to have been prepared in 98% yield by treating thebaine with HBr under anhydrous conditions in the presence of iodine above 10°C, followed by hydrolysis (*176–178*). A more complex process has been reported by Barber and

* 14β-Bromocodeinone (**156a**) is easily available from thebaine (**69**) in single step (*167*).

Rapoport (*146*). Oxymercuration of thebaine (**69**) with mercuric acetate in refluxing methanol gave a mercurated neopinone dimethyl ketal (**158**), hydrolysis of which with 3 N acetic acid, or, alternatively, reduction of the organomercury compound with sodium borohydride and mild acid hydrolysis of the resulting neopinone dimethyl ketal (**159**), afforded neopinone (**141**) in 95–100% yield. Either acid- or alkali-catalyzed isomerization led to the equilibrium mixture consisting of codeinone (**140**) and neopinone

(**141**) in a ratio of 3:1. Complete conversion to codeinone in 85–90% yield followed by treatment of neopinone with anhydrous hydrogen chloride or hydrogen bromide in ether–methylene chloride, followed by elimination of hydrogen halide from the intermediate 8-halodihydrocodeinone (**160**). The known borohydride reduction of codeinone then gave codeine in 85% overall yield from thebaine.

Conversion of thebaine to codeinone has also been accomplished photochemically. Thebaine (**69**) was irradiated in methanol using a 450-W Hanovia lamp and a Corex filter to give neopinone dimethyl ketal (**159**) in 78% yield. If the irradiation was performed in slightly acidic aqueous solution under the same conditions, a 9:1 mixture of neopinone (**141**) and codeinone (**140**) was obtained. As conversion of neopinone to codeinone and codeine had been previously realized (*146*), a new photochemical transformation of thebaine to codeine was thus described (*179,180*).

In the same study, Dauben and co-workers (*179,180*) investigated the hydrolysis of neopinone dimethyl ketal (**159**) to form neopinone (**141**) and found that starting from pure ketal the process was very slow and the yield low. Traces of mercuric ions, remaining from the previous step, were supposed to catalyze the hydrolysis performed by Barber and Rapoport (*146*). As a proof of this suggestion, addition of a catalytic amount of mercuric acetate (0.7 mol%) to the solution of the photochemically produced ketal resulted in smooth hydrolysis and an 89% yield of the enone mixture (**141** plus **140**). This finding was extended to the hydrolysis of thebaine itself. As a result of systematic investigation, utilization of 6.7 mol% of mercuric acetate in 3 N formic acid (6.5 hr, room temperature) was found to optimize the reaction conditions for the direct conversion of thebaine to the enone mixture (**141** plus **140**). This mixture, without purification, was treated with hydrogen chloride, followed by dehydrochlorination and sodium borohydride reduction of the crude codeinone (**140**), to produce codeine (**77**) in 71% overall yield (*179,180*).

Having investigated enzyme-catalyzed processes in the biosynthesis of morphine, Hodges and Rapoport found that cell-free enzyme extracts, prepared from *Papaver somniferum* or *Papaver bracteatum*, reduce codeinone to codeine (*181*). As a reverse transformation, a publication by Barber and Rapoport (*182*) is worth mentioning, in which the semisynthesis of thebaine and oripavine from codeine and morphine, respectively,

	R^1	R^2
77	CH_3	H
161	CH_3	CH_3
1	H	H
162	H	CH_3
163	Ac	CH_3

	R
69	CH_3
164	Ac
165	H

was described. Methylation of the potassium salt of codeine with methyl iodide furnished codeine methyl ether (**161**) selectively in 83% yield (no quaternization of the nitrogen occurred). Codeine methyl ether (**161**) was then oxidized to thebaine with manganese dioxide in 67% overall yield from codeine. Similarly, the potassium salt of the di-*O*-anion of morphine was selectively alkylated to give morphine 6-methyl ether (heterocodeine) (**162**) in over 90% yield. Heterocodeine was then acylated to give **163**, oxidation of which with MnO_2 gave oripavine 3-acetate (**164**). Finally, hydrolysis of the phenol ester group furnished oripavine (**165**) in 73% yield from morphine.

Although this last transformation was not performed on a natural product, it is the only case when a salutaridine derivative, namely, dihydro-thebaine-ϕ 4-phenyl ether (**166**), was converted to a 1-benzylisoquinoline compound (**167**). The reaction provides a unique example of carbon–carbon cleavage among morphinandienones and can be regarded as a retro process of the oxidative ring closure, the key step of biomimetic-type

166

	R
167	OPh
168	H

syntheses. The cleavage of **166** was actually performed with potassium amide in liquid ammonia, and the structure of the 1-benzylisoquinoline **167** was elucidated spectroscopically, and verified by its conversion to the known benzylisoquinoline **168** (*183*).

C. Transformation of Morphine Alkaloids into Biologically Active Semisynthetic Derivatives

14-Hydroxymorphinans, such as naloxone (**170**), naltrexone (**171**), and nalbuphine (**172b**), behave as potent analgesics and narcotic antagonists (*112,113*). Synthesis of these medicinally significant compounds, performed and patented by the end of the 1960s (*184–187*), requires transformation of a natural alkaloid into a common intermediate, noroxymorphone **169**. Modifications and improvements of these procedures are presented in this section.

Transformation of thebaine (**69**) into **169** centers on its oxidation to 14-hydroxycodeinone (**173**) with hydrogen peroxide–formic acid (*188,189*) or *m*-chloroperbenzoic acid in acetic acid–trifluoroacetic acid (*190,191*). Subsequent hydrogenation of **173** to oxycodone (**174**), followed by *O*- and *N*-demethylation procedures, leads to noroxymorphone (**169**). The sequence was performed both in the "natural" (*192*) and in the enantiomeric, unnatural series (*189*). *O*-Demethylation of **174** could be accomplished by the method of Rice (*73*) with boron tribromide; *N*-demethlyation required diacetylation of oxymorphone (**175**) to give **176,** which was *N*-demethylated with cyanogen bromide in chloroform. Refluxing the *N*-cyano compound **177** in 25% sulfuric acid effected deacetylation, hydrolysis, and decarboxylation and led to the desired intermediate noroxymorphone (**169**). The latter was converted to naloxone (**170**) on treatment with allyl bromide. The published overall yield of (−)-naloxone from oxymorphone (**175**) was only 20% (*184*). Refinement of the procedure by Brossi and co-workers resulted in a significant increase of the yield; unnatural (+)-naloxone [(+)-**170**] has been prepared basically by the same pathway in 59% yield from **175** (*189*).

N-Demethylation could preferably be accomplished from **176** with vinyl chloroformate (VOC-Cl) in dichloroethane at reflux. The *N*-VOC compound (**178**), obtained in essentially quantitative yield, was then deacylated in a three-step process in which anhydrous HCl was first bubbled through a CH_2Cl_2 solution of **178** to produce the normal adduct. Next, solvent evaporation, followed by warming in methanol, gave the *N*-nor salt **179,** which was refluxed in 25% sulfuric acid for 5 hr. Neutralization of the reaction mixture precipitated noroxymorphone (**169**). When the reaction sequence was performed without isolation of the intermediates, crude

	R	X
169	H	O
170	$-CH_2-CH=CH_2$	O
171	$-CH_2-\triangleleft$	O
172a	$-CH_2-\diamondsuit$	O
172b	$-CH_2-\diamondsuit$	α-OH β-H

	R^1	R^2	R^3
174	CH_3	H	CH_3
175	H	H	CH_3
176	Ac	Ac	CH_3
177	Ac	Ac	CN
178	Ac	Ac	$CO_2CH=CH_2$
179	Ac	Ac	H

69 →(ox.)→ **173**

* Olofson and Pepe consistently call the 6-oxo compound **172a** nalbuphine.

noroxymorphone (**169**) was obtained in 98% overall yield from oxymorphone (**175**). Allylation of the crude product with 1.1 equivalents of allyl bromide in ethanol gave naloxone (**170**) in 71% recrystallized yield (70% from **175**) (*192*).

Nalbuphine (**172b**) had earlier been prepared in a very similar way from thebaine via oxycodone (**174**), oxymorphone (**175**), and noroxymorphone (**169**) (see above). The last step, however, deviates from the commercial route to naloxone because cyclobutylcarbinyl bromide is a very poor alkylating agent. Instead, the *N*-cyclobutyl methyl moiety is attached by *N*-acylation followed by $LiAlH_4$ reduction of the *N*-cyclobutylcarbonyl derivative. To protect and reconstitute the phenol and keto functions, extra steps had to be incorporated into the patented procedure (*185*). For the transformation of thebaine into nalbuphine, a 18–20% yield would be a most optimistic estimate (*193*).

Olofson and Pepe (*194*) found a more efficient route from oxycodone (**174**) to nalbuphine* (**172a**). First, oxycodone (**174**) was converted in quantitative yield to its 14-cyclobutylcarbonyl derivative **180** by acylation with cyclobutanecarboxylic acid anhydride in dioxane at 100°C. Next, the

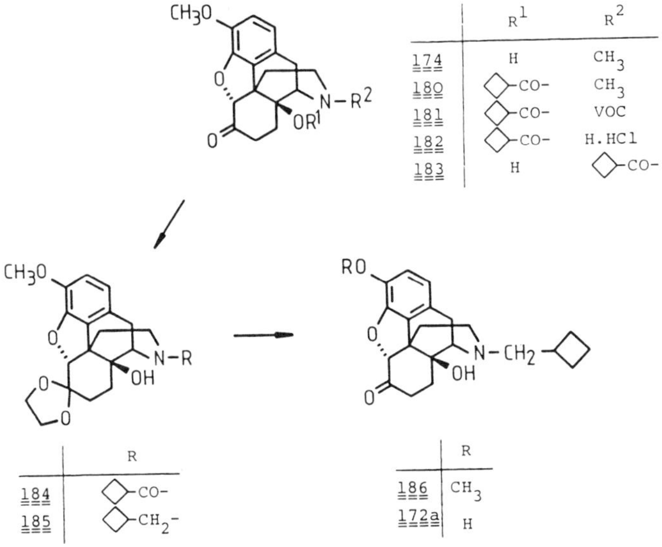

	R^1	R^2
174	H	CH_3
180	⬦-CO-	CH_3
181	⬦-CO-	VOC
182	⬦-CO-	H·HCl
183	H	⬦-CO-

	R
184	▱-CO-
185	▱-CH₂-

	R
186	CH_3
172a	H

crude **180** was *N*-demethylated with VOC-Cl to give the *N*-VOC compound **181.** Removal of the VOC group was accomplished by bubbling anhydrous HCl through a CH_2Cl_2 solution of **181,** followed by evaporation and subsequent dissolution of the intermediate adduct in methanol. The CH_2Cl_2 solution of 14-*O*-acylnoroxycodone hydrochloride (**182**) thus obtained was treated with aqueous $NaHCO_3$, whereupon an immediate quantitative rearrangement to *N*-acylnoroxycodone (**183**) occurred. For further manipulation the 6-keto function had to be masked, and the reduction of the amide **184** to a cyclobutylmethylamine (**185**) was achieved with excess $LiAlH_4$ in THF. Aminoketal **185** was not isolated, but rather hydrolyzed during the work-up to the amino ketone **186.** Finally, *O*-demethylation with BBr_3 afforded nalbuphine* (**172a**). The overall yield of pure, recrystallized **172a** from oxycodone (**174**) was 58.4%, and from thebaine, 49%. Adaptation of this synthetic method for the improved preparation of other 14-hydroxymorphinan pharmaceuticals (e.g., naltrexone, **171**) is claimed to be easy.

In view of the relative scarcity of natural thebaine, either a practical conversion of codeine to thebaine or an alternate synthetic route to noroxymorphone from morphine or codeine seemed to be attractive. The first task has been accomplished by Barber and Rapoport (*182*) in relatively

---

* In fact, a final stereoselective reduction to **172b** completes the synthetic approach to nalbuphine.

178   CSABA SZÁNTAY *ET AL.*

high yield (67%) (see Section IV,B). Direct conversion of codeine or codeinone to 14-hydroxymorphinans was solved by Seki (*195*), and later by Schwartz and Wallace (*196*). The former reported the preparation of the pyrrolidine dienamine **187a** and various dienol ether derivatives (e.g., **187b**) of codeinone. However, subsequent oxidation of the thebaine analogs with hydrogen peroxide–formic acid gave only poor yields (15–30% overall from codeinone) of 14-hydroxycodeinone (**173**) (*195*).

A superior procedure was published by Schwartz and Wallace (*196*). Codeine (**77**) was first *N*-demethylated according to the method of Rice and May (*197*) with ethyl chloroformate, and then it was oxidized with manganese dioxide to give *N*-ethoxycarbonylnorcodeinone (**188**) in 90% overall yield (*198*). Derivative **188** could be readily converted to the dienol acetate **189** by treatment with sodium acetate in refluxing acetic anhydride. Oxidation with photochemically generated (Rose Bengal sensitization) singlet oxygen afforded the 14-hydroxy derivative **190** as the major product. The entire four-step sequence for the conversion of codeine (**77**) to **190** could be accomplished, without purifying any of the intermediates, in 66% overall yield. Completion of the synthesis of noroxycodone (**82**) and noroxymorphone (**169**) from **190** is routine. Hydrogenation of **190** over Pd/C gave *N*-ethoxycarbonylnoroxycodone (**191**), and the *N*-acyl group was removed by acid hydrolysis (*184,189*) to afford noroxycodone (**82**) in 80% overall yield (from **190**). Alternatively, **191** was demethylated with BBr$_3$ and then subjected to acid hydrolysis to afford noroxymorphone (**169**) in 73% overall yield.

Brossi and co-workers published a series of papers (*199–206*) dealing with the structure–activity relationships of morphinan derivatives synthesized from natural compounds. Some transformations are worth mentioning in this section. Starting from oxymorphone **175,** the synthesis and biological properties of several 14-hydroxymorphinans were described (*199*). 3-Dehydroxy-oxymorphone (**193**) was prepared via tetrazolyloxy compound **192.** Cleavage of the oxide bridge of **193** afforded the 4,14-dihydroxy ketone **194,** from which the methoxy ketone **195** could be easily

CH_3O / HO / $N-CH_3$ / **77**

1. $ClCO_2Et$
2. MnO_2

CH_3O / $N-CO_2Et$ / **188**

CH_3O / AcO / $N-CO_2Et$ / **189**

ox. / $h\nu$

CH_3O / OH / $N-CO_2Et$ / **190**

R^1O / OH / $N-R^2$

	R^1	R^2
191	CH_3	CO_2Et
82	CH_3	H
169	H	H

R / OH / $N-CH_3$

	R
175	OH
192	$OC_7H_5N_4$
193	H

RO / OH / $N-CH_3$

	R
194	H
195	CH_3

RO / OH / $N-CH_3$

	R
201	CH_3
202	H

CH_3O / OR / $N-R$

	R
196	Cl_3CCH_2CO
197	H

CH_3O / OH / $N-R$

	R
198	$-CH_2CH=CH_2$
199	$-CO-\triangleleft$
200	$-CH_2-\triangleleft$

obtained by *O*-methylation. By appropriate manipulations, further 14-hydroxymorphinans (**196–202**) were prepared, and the role of the various functional groups in antinociceptive properties was studied.

Several 3-deoxy opioids and 3,6-dideoxydihydromorphine were synthesized from dihydromorphine (**203**) to ascertain the effect of the phenolic hydroxy group on antinociceptive potency and receptor binding activity (*200*). Similarly to the previous scheme, catalytic hydrogenation of the 3-tetrazolyl ether derivative **204** provided entry into the 3-deoxydihydro series. Through 3-deoxydihydromorphine (**205**), which had previously been obtained directly by hydrogenolysis of the 3-tetrazolyl ether of morphine (*207*), the *N*-nor (**206**) and *N*-allyl derivatives (**207**), as well as 3,6-dideoxydihydromorphine (**208**) were synthesized. The semisynthesis of 3-deoxymorphine (**213**) itself could be accomplished from 3-deoxydihydromorphine (**205**) (*200*) by the sequence outlined above (i.e., **205** → **209** → **210** → **211** → **212** → **213**), as a parallel procedure to the earlier publication of Rice and co-workers (*201*).

	R^1	R^2	R^3
203	OH	OH	CH_3
204	$OC_7H_5N_4$	OH	CH_3
205	H	OH	CH_3
206	H	OH	H
207	H	OH	allyl
208	H	H	CH_3

	R
209	CH_3
210	CO_2Et

(−)-4-Hydroxy-*N*-formylmorphinan-6-one (**214**), prepared from ketone **209** in three steps and 70% overall yield, was described as a versatile intermediate for the synthesis of 4-hydroxymorphinans and 3-deoxy opioids (*202–204*). Other publications (*205,206*) described the synthesis of 4-methoxy- (**215a**) and some *N*-substituted 4-methoxymorphinans (**215b–e**) from (−)-4-hydroxy-*N*-formylmorphinan-6-one and summarized their biological properties.

Finally, the synthesis of some 3,14-dihydroxymorphinans is briefly mentioned in this section, since a number of these derivatives have shown

215	R
a	H
b	CH_3
c	$CH_2=CH-CH_2-$
d	▷-CH_2-
e	◇-CH_2-

interesting biological properties. Butorphanol (**216a**), a clinically effective analgesic, and oxilorphan (**216b**), a strong narcotic antagonist, can both be obtained either by modification of thebaine (*208*) or by total synthesis (*209,210*).

216	R
a	◇-CH_2-
b	▷-CH_2-

Azidoethylmorphine (**219a**) was found to be 60 times more effective as a cough-relieving agent than codeine. Synthesis of **219a** was accomplished from ethylmorphine (**217**) by mesylation or tosylation and subsequent azidolysis (*211,212*). Demethylation with BrCN, followed by alkylation afforded *N*-allyl- and *N*-cyclopropylmethyl-*N*-demethyl-azidoethylmorphine (**219b,c**) (*213*).

	R
217	H
218	Ms or Ts

219	R
a	CH_3
b	$CH_2=CH-CH_2-$
c	▷-CH_2-

D. Miscellaneous Transformations

The absolute configuration of the morphinan skeleton of (−)-sinomenine [(−)-**217**] is enantiomeric to that of natural (−)-morphine; in this way, "unnatural" (+)-morphine was synthesized from [(−)-**217**] by Goto and co-workers (*214–217*). A markedly improved transformation, based in principle on the original scheme by Goto, was published by Brossi and co-workers (*218*). Catalytic reduction of sinomenine [(−)-**217**] afforded a mixture of two diastereomers **218a** and **218b.** Because the acid-catalyzed S_N2' cyclization of the epimeric mixture to dihydrocodeinone [(+)-**53b**], with loss of methanol, proceeds under conditions equilibrating the two epimers, this mixture was treated, without separation, with polyphosphoric acid. (+)-Dihydrocodeinone [(+)-**53b**] was then converted to (+)-codeinone [(+)-**140**] by the method of Weller and Rapoport (*72*) (see Section IV,A) in 51% overall yield. Reduction of (+)-codeinone with sodium borohydride in methanol (*172,173*) afforded (+)-codeine [(+)-**77**], which was converted to (+)-morphine [(+)-**1**] by *O*-demethylation with boron tribromide (*73*). (+)-Heroin [(+)-**219**], previously unknown, was obtained from (+)-morphine by treatment with acetic anhydride. (+)-Codeine, (+)-morphine, and (+)-heroin showed no analgesic activity (*218*).

	R^1	R^2
a	H	OCH_3
b	OCH_3	H

	R
(+)-**77**	CH_3
(+)-**1**	H
(+)-**219**	Ac

The first synthesis of neopine (**221**) from thebaine via 14β-bromocodeinone (**156a**) and neopinone (**141**) or 14β-bromocodeine (**220a**) was accomplished by Conroy (*167*). The yield of the two-step thebaine to neopinone conversion was 69%; reduction of neopinone with sodium borohydride in aqueous alcohol was reported to afford neopine (**221**) as the only product (no yield was mentioned). Some years later, Okuda and co-workers (*219,220*) showed that the reduction is not stereospecific; indeed, a mixture of neopine (**221**) and isoneopine (**222**) is obtained in approximately equal amounts.

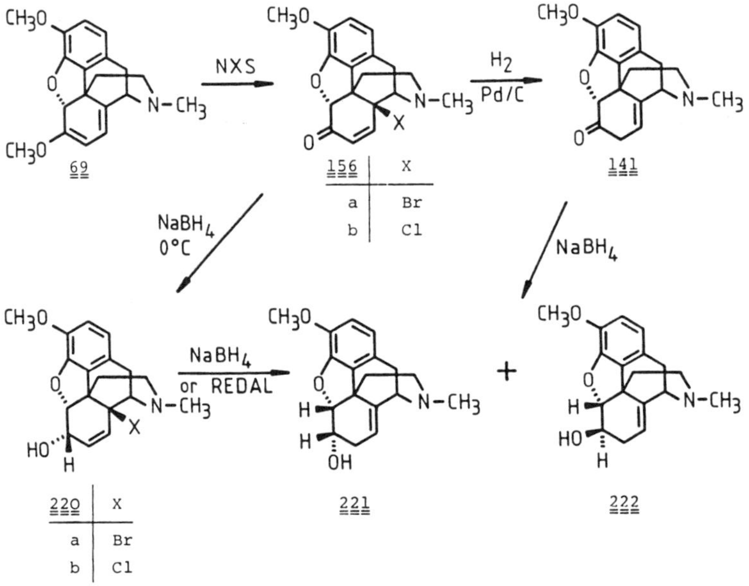

The intermediate neopinone (**141**) was prepared subsequently by Barber and Rapoport (*146*) via the mercurated neopinone dimethyl ketal (**158**) (see Section IV,A) in almost quantitative yield. The problem of stereoselectivity in the hydride reduction has been solved by two groups. Studying the conformational characteristics of neopinone, Wunderly and Brochmann-Hanssen (*221*) found stereoselective conditions for the reduction to produce either neopine or isoneopine. When neopinone was reduced with a bulky hydride reducing agent (e.g., lithium triethyl- or tri-*sec*-butylborohydride) in THF, neopine (**221**) was the sole reaction product (95% yield). Sodium borohydride reduction in aqueous solution, however, gave isoneopine (**222**) as the major product (85%), along with a small amount of neopine (11%).

As an improvement of the previously mentioned procedure by Conroy, Makleit and co-workers (*222*) elaborated a synthesis of neopine via 14β-chlorocodeinone (**156b**), reduction of which with sodium borohydride gave 14β-chlorocodeine (**220b**) stereoselectively in 90% yield. The second reductive step to neopine was performed with sodium dihydrobis(2-methoxyethoxy)aluminate (REDAL) in practically quantitative yield (overall yield from thebaine was 72%). However, if 14β-bromocodeinone dimethyl ketal is reduced with REDAL in benzene solution, thebaine is obtained in high yield (*223*). The same result was obtained by the use of chromous sulfate in aqueous dimethylformamide.

Berényi and Makleit also succeeded in synthesizing pure isoneopine (**222**) from neopine (**221**) by simple, unambiguous, and totally stereocontrolled steps (*224*). Neopine was first transformed into 6-*O*-mesylneopine (**223**), which on treatment with sodium azide in dimethylformamide gave the azido derivative **224** in 75% yield. Compound **224** was then reduced with zinc in dimethylformamide to obtain the amino derivative **225**, which gave the known 6-*O*-acetylisoneopine (**226**) (*225*) in 90% yield on treatment with sodium nitrite in acetic acid. Saponification of the ester group of **226** with alcoholic KOH afforded isoneopine (**222**) quantitatively. Thus isoneopine (**222**) was prepared in a 44% overall yield from neopine (**221**) (*224*).

	R
221	H
223	Mesyl

	R
224	N_3
225	NH_2
226	OAc
222	OH

In the late 1960s, a series of publications was presented by Kugita and co-workers (*225–228*) in which hydroboration of isoneopine derivatives afforded an entry into the B/C-*trans* morphines. As the majority of this work has already been discussed in the previous review in this treatise (*1*), only a recent development is presented here. Acetylation of the organoborane intermediate obtained from isoneopine (**222**), and subsequent hydrogen peroxide oxidation, afforded the B/C-*trans* fused 6,8-diol acetate **227** in 75% yield. Elimination of the elements of water was accomplished via

222 → 1. B_2H_6 / 2. Ac_2O / 3. H_2O_2 → **227, 228**

	R
227	H
228	Tosyl

	R^1	R^2
229	CH_3	Ac
230	CH_3	H
231	H	H

tosylate derivative **228** with collidine. One of the two isomeric olefins was identified as possessing the Δ^7 structure (**229**), hydrolysis of which gave *trans*-isocodeine (**230**). Demethylation of *trans*-isocodeine with diphenylphosphine anion afforded *trans*-isomorphine (**231**) in 61% yield (*225*).

In a series of publications, Makleit and co-workers reported on the synthesis of 6-demethoxythebaine (**235**) and 6-demethoxyoripavine (**236**), as well as their transformation into the aporphinoids apocodeine (**237**) and apomorphine (**238**), respectively. Conversion of neopine mesylate (**223**) to 6-demethoxythebaine (**235**) was accomplished by the same group on elimination of the elements of methanesulfonic acid with tetrabutylammonium fluoride, or preferably with potassium *tert*-butylate (*229,230*). Synthesis of 6-demethoxythebaine was simultaneously reported by Beyerman and co-workers (*231*) in a basically analogous way, and by Rapoport and co-workers (*232*) via isocodeine and its sulfenate ester starting from codeine. Both groups performed the methanesulfonic acid-catalyzed rearrangement of 6-demethoxythebaine (**235**) into apocodeine (**237**), thus supplying the first experimental proof for the mechanism of the acid-catalyzed codeine to apocodeine transformation assumed by Bentley (*233*).

	R^1	R^2
221	CH_3	H
223	CH_3	Mesyl
232	H	H
233	Ac	H
234	Ac	Mesyl

	R
235	CH_3
236	H

	R
237	CH_3
238	H

The first synthesis of 6-demethoxyoripavine (**236**) was published by Makleit and co-workers (*234*). Direct demethylation of 6-demethoxythebaine failed to afford the desired product; in every case considerable decomposition and rearrangement to give aporphine derivatives were observed. Neopine (**221**) was therefore demethylated to neomorphine (**232**) with 48% hydrogen bromide. Partial acetylation and subsequent mesylation of neomorphine gave 3-*O*-acetyl-6-mesylneomorphine (**234**),treatment of which with potassium *tert*-butylate afforded 6-demethoxyoripavine (**236**) in 60% overall yield (from neopine). Methanesulfonic acid treatment, finally, resulted in apomorphine (**238**) in 88% yield. It is worth noting that by making use of these transformations, N-substituted *N*-norapocodeine (*229,230*) and apomorphine derivatives (*234*) have also been prepared.

The rearrangement of morphine into apomorphine (**238**), and of thebaine into morphothebaine (**239**), in strong mineral acids has been known for about 120 years. A possible mechanism for the rearrangement was suggested by Stork (*235*) and by Bentley (*233*). The procedure, starting from morphine and codeine derivatives and effected by 85% H_3PO_4 at 125–140°C under reduced pressure, was later patented (*236*).

Owing to the significant dopaminergic activity of aporphine derivatives, Neumeyer and co-workers (*237,238*) reinvestigated the procedure. They found that the rearrangement proceeds with methanesulfonic acid at 90–100°C, and, under these mild conditions, 10-*O*-methylmorphothebaine (**240**) could be obtained from thebaine. Furthermore, *N*-nor as well as *N*-substituted *N*-northebaine and -codeine derivatives could be transformed into the corresponding *N*-nor- and N-substituted *N*-normorphothebaine or -apocodeine derivatives, respectively, with some improvement in the yields over those reported previously (*239*). A suggestion for the detailed mechanism in the case of thebaine derivatives was presented (*237*). By utilizing the mild method outlined by Neumeyer for

	R^1	R^2
238	H	H
239	OH	CH_3
240	OCH_3	CH_3

the morphinan to aporphine rearrangement, some 6- and 7-halogeno-6-demethoxythebaines were transformed into 2- and 3-haloapocodeines and -apomorphines (*240–242*).

V. Reactions

In this section, reactions and modifications of morphine alkaloids are presented, with the exception of the Diels–Alder type cyclization of thebaine and other morphinandienes which were discussed in detail in the previous review (*1*). Reactions supplying natural compounds or biologically active derivatives (e.g., 14β-hydroxymorphinans) have already been described in the section on transformations (Section IV) and are mentioned here only occasionally.

A. *N*-DEMETHYLATION AND RELATED REACTIONS

Efforts to modify the original framework of morphine alkaloids to produce semisynthetic *N*-substituted derivatives with preferable biological behavior initiated widespread interest in selective *N*-demethylation processes. The first and classic method to accomplish *N*-demethylation of tertiary amines, such as morphine and codeine, had already been elaborated by von Braun (*243*) at the beginning of the twentieth century. The general method, later summarized by Hageman (*244*), involves cyanogen bromide treatment of the alkaloid. In the first step, the *N*-cyanodemethyl derivative (**243** or **244**) was formed, which could be converted to normorphine (**247**) or norcodeine (**248**), respectively, by acidic (*243*) or alkaline

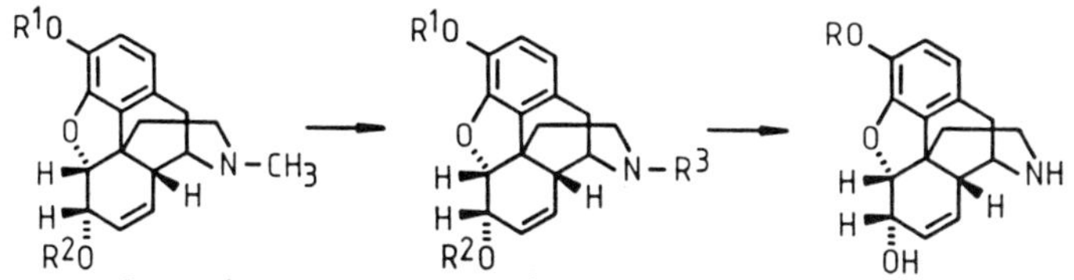

	R^1	R^2
1	H	H
241	Ac	Ac
77	CH_3	H
242	CH_3	Ac

	R^1	R^2	R^3
243	Ac	Ac	CN
244	CH_3	Ac	CN
245	CO_2R	CO_2R	CO_2R
246	CH_3	CO_2R	CO_2R

	R
247	H
248	CH_3

(*245*) hydrolysis or by LiAlH$_4$ reduction (*246,247*). In the procedure, the phenolic and alcoholic hydroxy functions must be protected by acetylation.

The application of chloroformates for demethylation subsequently became preferred and widely utilized in alkaloid chemistry. *N*-Demethylation of tertiary methylamines with phenyl and ethyl chloroformate was discovered by Hobson and McCluskey (*248*) and was first applied to morphine and codeine by Abdel-Monem and Portoghese (*249*). The intermediate urethane (**245** or **246,** R = Ph or Et) could be hydrolyzed with KOH in ethanol. Later, Rice (*197,250*) modified this method by using 80% hydrazine for the second step, thus improving the yield significantly. For demethylation of morphine and codeine, methyl chloroformate was used by Brine and co-workers (*251*), and in this way *N*-demethylmorphine (**247**) and codeine (**248**), obtained by hydrazinolysis of the urethane in 70% yield, were not contaminated with phenol.

Further improvement involved the use of 2,2,2-trichloroethyl chloroformate to give a carbamate intermediate (**245,** R = CO$_2$CH$_2$CCl$_3$), which could be cleaved by reduction with zinc in acetic acid or methanol to produce normorphine (**247**) in 75% yield. This method of Montzka *et al.* (*252*) later found widespread application in morphine chemistry.

Even milder conditions are required for the decomposition of the vinyl urethane intermediate. Vinyl chloroformate, introduced by Olofson and co-workers (*192,194,253*), is inexpensive, more reactive than other chloroformates, and results in pure *N*-demethyl products in high yields. Decomposition of the vinyl carbamate starts with the addition of dry hydrogen chloride or bromide (or bromine), and the α-halogenoethyl derivative obtained is simply warmed in alcohol (generally in methanol) to provide

$$\overset{\backslash}{\underset{/}{N}} - \overset{\overset{O}{\|}}{C} - CH = CH_2 \quad \xrightarrow{HX} \quad \overset{\backslash}{\underset{/}{N}} - \overset{\overset{O}{\|}}{C} - CHX - CH_3$$

$$\searrow MeOH$$

$$\overset{\backslash \oplus}{\underset{/}{N}}H_2 X^{\ominus} + CO_2 + CH_3CH(OCH_3)_2$$

the salt of the secondary amine. In this demethylation process there is no need to protect the phenolic and alcoholic hydroxy groups of morphine; the *O,O,N*-trivinyloxycarbonyl derivative is formed, and during decomposition either all three acyl functíons are eliminated, or selective *N*-VOC removal can be achieved (*253*).

Olofson *et al.* (*254*) later synthesized, from phosgene and acetaldehyde, the α-chloroethyl chloroformate, which was found to be a similarly effective demethylating agent. When 14-acyloxy-morphinans were treated with this reagent, O-acyl to N-acyl migration completed the procedure after recovering the nor base from its hydrogen chloride salt (*194*) (see Section IV,C).

Both the von Braun and chloroformate methods can be applied only for alkaloids or derivatives not containing $\Delta^{8,14}$-double bond. For example, in the reaction of thebaine (**69**), either with cyanogen bromide (*255–257*) or with chloroformates (*258,259*), the allylamine structure leads to C-9–N bond fission, and, instead of N-demethylation, a urethane of type **249** is formed. However, if the diene structure is blocked by complex formation (e.g., thebaine–Fe(CO)$_3$], demethylation can be successfully performed both with the von Braun method (*260*) and with trichloroethyl chloroformate (*261*).

The same type of C-9–N bond cleavage can be effected by treating thebaine (**69**) either with p-toluenesulfonyl bromide under Schotten–Baumann conditions or with β-(trimethylsilyl)ethyl chloroformate in the presence of K$_2$CO$_3$ (*262*). In contrast to earlier assumptions (*258*), the urethane **254**, obtained in 94% yield, was shown to have the $\Delta^{7,14}$-dienone structure on the basis of its NMR spectrum. Treatment of **254** with trifluoroacetic acid at ambient temperature reversed the direction of the reaction, and a mixture of codeinone (**140**) and neopinone (**141**) was obtained

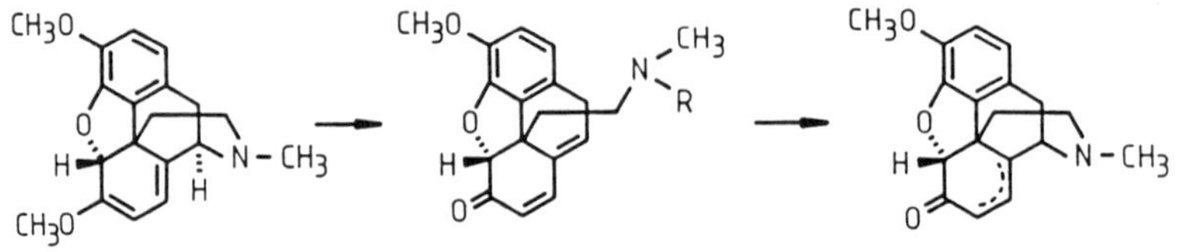

69

	R
253	Tosyl
254	$-CO_2CH_2CH_2TMS$

140 $\Delta^{7,8}$
141 $\Delta^{8,14}$

(*262*). The action of sodium methoxide solution on 7-bromoneopinone dimethylacetal–methoperchlorate (**255**) causes similar C-9–N cleavage and formation of two isomeric 10-methoxy dienes **256** and **257** (*263*).

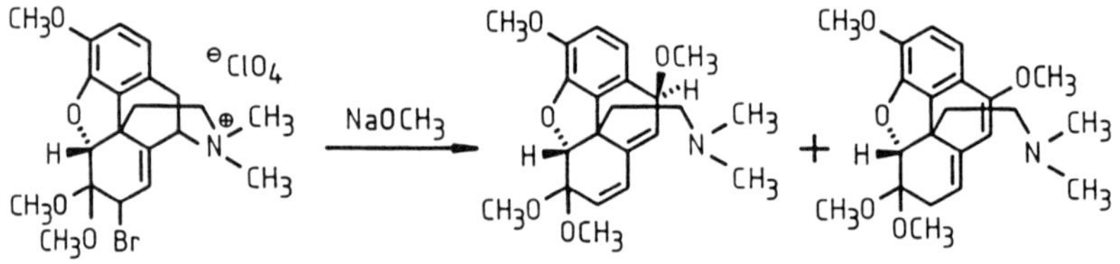

255 256 257

Diethyl azodicarboxylate (DAD) reacts with tertiary methylamines to afford a 1:1 adduct, from which the secondary amine, formaldehyde, and diethyl hydrazinedicarboxylate are formed on treatment with dilute acid.

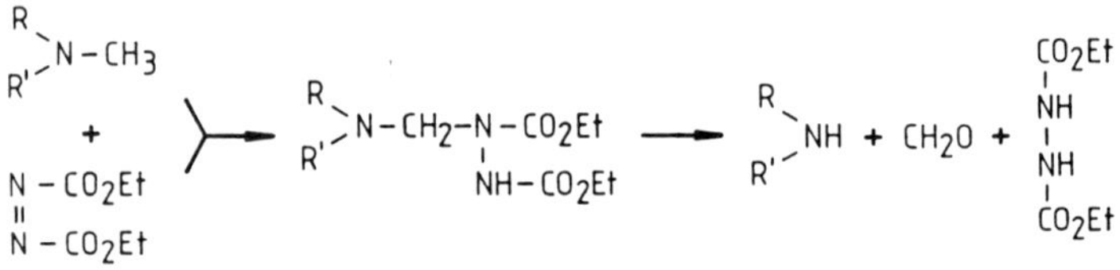

Accordingly, DAD can be regarded as a special demethylating agent (*264*), which, in contrast to the above-mentioned methods, is able to convert thebaine to *N*-demethylthebaine (**252**) directly (*265,266*). Because of the acid sensitivity of thebaine, the 1:1 adduct was hydrolyzed with pyridine hydrochloride in ethanol or with ammonium chloride in a water–ethanol system.

As thebaine (**69**) is an active diene component, excess DAD easily provides a Diels–Alder by-product. Application of two equivalents of DAD resulted in both *N*-demethylation and Diels–Alder reactions occur-

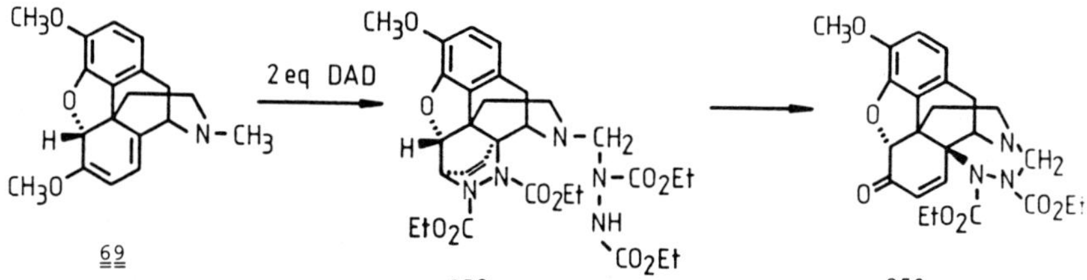

CH₃O
O
H
CH₃O
N—CH₃
69
2 eq DAD
CH₃O
O
H
N—CH₂
CO₂Et
N—CO₂Et
EtO₂C
N
CO₂Et
NH
CO₂Et
258
CH₃O
O
H
O
N—CH₂
N—N
EtO₂C
CO₂Et
259

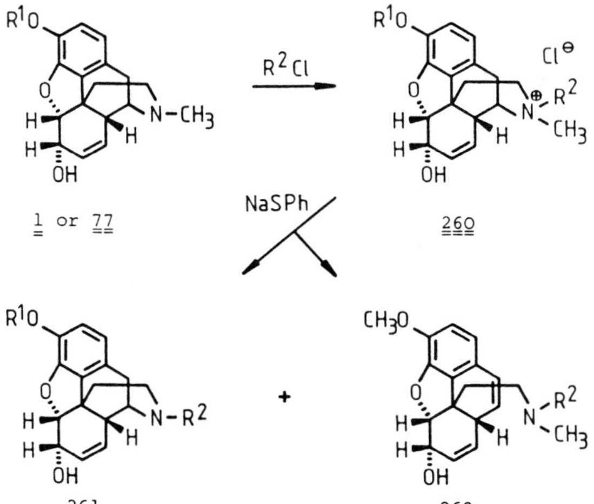

R¹O
O
H
H
OH
H
N—CH₃
1 or 77
R²Cl
R¹O
O
H
H
OH
Cl⊖
N⊕—R²
CH₃
260
NaSPh
R¹O
O
H
H
OH
H
N—R²
261
+
CH₃O
O
H
H
OH
H
CH₂—N—R²
CH₃
262
R¹=CH₃ or H   R²=n-C₃H₇, i-C₃H₇ or n-C₄H₉

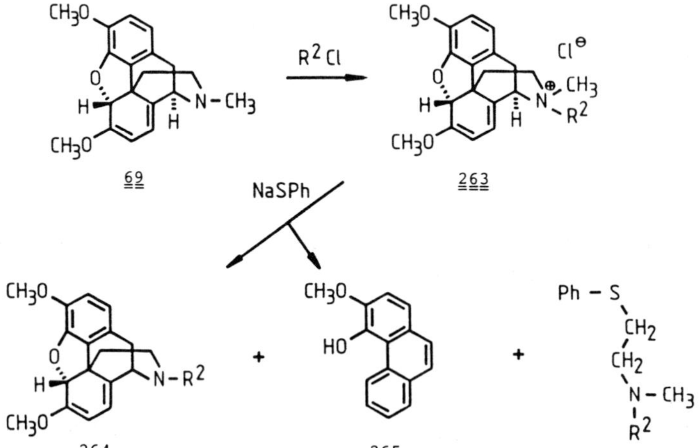

CH₃O
O
H
H
CH₃O
N—CH₃
69
R²Cl
CH₃O
O
H
H
CH₃O
Cl⊖
N⊕—CH₃
R²
263
NaSPh
CH₃O
O
H
CH₃O
N—R²
264
+
CH₃O
HO
265
+
Ph—S
CH₂
CH₂
N—CH₃
R²

ring simultaneously (*267*). The resulting adduct **258** afforded the 14-hydrazine-substituted codeinone derivative **259** on subsequent hydrolysis.

A number of research groups have investigated and utilized the N-demethylation of thebaine with DAD in connection with semisynthetic approaches to *N*-substituted *N*-demethylthebaine derivatives (*268–272*). Obviously, DAD can also be utilized for the *N*-demethylation of other morphinan derivatives containing an $\Delta^{8,14}$-double bond (*273–275*).

Besides these comprehensive and widely used methods, other N-demethylation procedures of morphines have also been published. Among the methods of lesser importance, a photochemical *N*-demethylation of codeine should be mentioned (*276*).

An unusual way to prepare *N*-alkylnormorphines concludes this section. Madyastha and co-workers (*277,278*) developed a convenient method for the replacement of the tertiary *N*-methyl by another alkyl group in two simple steps: quaternization of the alkaloid with the appropriate alkyl halide and preferential removal of the methyl group by a nucleophile. In this way, *N-n*-propyl, *N-i*-propyl, and *N-n*-butyl derivatives of morphine, codeine, and thebaine were prepared. For the demethylation of the quaternary methiodides (**260** or **263,** respectively) sodium thiophenolate was used, and in addition to the target molecules **261** and **264,** different amounts of the elimination products **262** and **265,** formed by C-9–N cleavage, were obtained as by-products.

By utilizing the results of the investigations on *N*-demethylation, Makleit and co-workers recently synthesized a series of *N*-alkylated (allyl, *n*-propyl, cyclopropylmethyl, cyclobutylmethyl, etc.) *N*-demethyl morphine derivatives. Thus, *N*-alkyl derivatives of isomorphine and isocodeine (*279*), dihydroisomorphine and dihydroisocodeine (*280*), 14β-hydroxymorphine and 14β-hydroxycodeine (*281*), as well as 14β-hydroxydihydromorphine (*282a*), and 14β-hydroxyisomorphine and -dihydroisomorphine (*282b*) have been stereohomogeneously synthesized.

B. Conversion of Tosyl and Mesyl Derivatives of Morphine Alkaloids

Starting from the 6-*O*-tosylates or 6-*O*-mesylates of codeine, morphine, or various derivatives, a great number of semisynthetic compounds can be obtained. Since the late 1960s, Makleit and co-workers have paid special attention to investigating and exploring the nucleophilic reactions of these derivatives. They found that although azidolysis of dihydrocodeine mesylate or tosylate (**266**) affords 6-deoxy-6-azido-dihydroisocodeine (**268**), whose reduction with $LiAlH_4$ gives the corresponding 6-amino-dihydroisocodeine (**269**), a similar reaction of the allylic codeine esters

	R^1	R^2
266	CH_3	Ts or Ms
267	Ac	Ts or Ms

	R^1	R^2
268	CH_3	N_3
269	CH_3	NH_2
270	H	N_3
271	H	NH_2
272	CH_3	SH
273	H	SH

	R^1	R^2
274	CH_3	Ts or Ms
275	Ac	Ts or Ms

	R^1	R^2
276	CH_3	N_3
277	CH_3	NH_2
278	H	N_3
279	H	NH_2
280	CH_3	SH
281	H	SH

(**274**) via a S_N2' mechanism results in 8-azido- (**276**) and 8-amino-8-deoxypseudocodeine (**277**), respectively (*283*).

The same results were attained by repeating the reaction series with 3-*O*-acetyl derivatives of morphine (**275**) and dihydromorphine (**267**). In the dihydro series, 6-azido and 6-amino-6-deoxy-dihydroisomorphine (**270** and **271**), and in the allylic morphine series, 8-azido and 8-amino derivatives **278** and **279,** could be obtained (*284*). The structures of the mercapto derivatives (**272, 273, 280, 281**), obtained by treatment of the above-mentioned tosylates or mesylates with potassium thioacetate as the nucleophile and subsequent hydrolysis, unambiguously support the earlier findings (S_N2 sequence in the dihydro and S_N2' in the allylic series) (*285*).

A valuable summary of the earlier results of nucleophilic substitution of codeine tosylate (**274**) with halogenides has been published (*286,287*). The fluoride ion (tetrabutylammonium fluoride, TBAF) attacks at C-6 (S_N2), and the 6-fluoro derivative **282** cannot be isomerized into 8-fluorocodide.

Substitution with chlorides primarily affords the 6-chloro derivative (**283**) as the kinetic product (S_N2), which can be easily isomerized to the 8-chloro compound **284** (S_Ni'). Treatment with bromide or iodide as the nucleophile results exclusively in the 8-halogeno derivatives **285** and **286**. The S_N2' mechanism is supposed to involve an S_N2 reaction and a subsequent fast S_Ni' step via a [3,3]-suprafacial sigmatropic rearrangement (*287*).

In the dihydrocodeine series, reaction of halogenides with mesylate **266** affords either 6-halogeno-dihydroisocodeine (**287a–c**) or deoxycodeine C

	R^1	R^2
274	CH_3	Ts or Ms
275	Ac	Ts or Ms

	R
289	CH_3
290	Ac

(**288**). With fluoride and chloride, **287** is obtained; iodide treatment leads exclusively to **288.** The result with bromides depends on the reaction conditions applied (*288*).

When thiocyanate anion is applied in the reaction of **274** or **275,** as expected on the basis of the above mechanistic summarization (*286,287*), the 8-deoxy-8-isothiocyanate derivatives **289** and **290** are obtained (S_N2 and S_Ni') (*289*). Starting from the 6-*O*-mesyl or 6-*O*-tosyl derivatives of morphine and codeine, an improved synthesis of isomorphine, isocodeine and their dihydro derivatives has been published (*290–292*).

Two further good procedures for C-6 epimerization have been published. Refluxing codeine in glacial acetic acid with dimethylformamide–neopentylacetal, followed by alkaline hydrolysis, affords isocodeine in excellent yield (*182*). Similarly good results have been attained in the C-6 epimerization of different morphine and codeine derivatives by applying the Mitsunobu reaction followed by hydrolysis of the initially obtained isobenzoates or thioacetates (*279,293,294*).

The nucleophilic substitution of 6-*O*-tosylates and 6-*O*-mesylates of 14-hydroxymorphines has also been studied. Although preparation of **291** and **292,** as well as their reactions with some nucleophiles (Cl^-, Br^-, I^-, H^-, AcO^-), had been reported earlier (*295,296*), some new aspects and derivatives have been published by Makleit *et al.* (*297*). By azidolysis of **291** either a 6- or 8-azido derivative, **293** or **294,** respectively, could be prepared depending on the reaction time. Of course, **293** can be easily isomerized to **294** by heating in dimethylformamide solution. Azido compounds obtained were converted to the corresponding amino derivatives **295** and **296,** respectively. As expected, from the dihydro derivative **292** only 6-azido (**300**) and 6-amino derivatives (**301**) can be produced;

291   R=Ts or Ms

	R
293	N_3
295	NH_2
297	Cl
299	F

	R
294	N_3
296	NH_2
298	Cl

292   R=Ts or Ms

	R
300	N_3
301	NH_2

302

8-amino-14-hydroxy-dihydropseudocodeine (**302**) is available from **294** or
296 by catalytic hydrogenation. The S_Ni'-type suprafacial sigmatropic
rearrangement of chlorocodide **297** to 8-chloropseudocodide (**298**) as well
as the preparation of 6-deoxy-6-fluoro-14-hydroxyisocodeine (**299**) have
also been published (*298*).

In a similar manner, 6-azido-14-hydroxymorphine and *N*-substituted
N-norazidomorphines have also been prepared (*299*), and, together with
other azidomorphines, their pharmacological properties were investigated
(*300,301*). A summary of experiences in the field of nucleophilic reactions
of mesyl and tosyl morphines has been presented by Makleit and co-
workers in a conference proceedings (*302*).

275

LAH

	R
303	H
304	CH_3

305

As a further demonstration of the utilization of tosylate reactions, a novel and convenient synthesis of deoxymorphine-E (**303**), deoxycodeine-E (**304**), and dihydrodeoxymorphine-D (**305**), starting from 3-*O*-acetyl-6-*O*-tosylmorphine (**275**), by LiAlH₄ reduction and subsequent diazomethane treatment or catalytic hydrogenation, respectively, has been reported (*303*). Nucleophilic reactions of pseudocodeine tosylate (*304*), as well as of neopine tosylate (*229*), have also been reported by Makleit and co-workers.

220	z
c	Cl
d	Br

306	X
a	Cl
b	Br
c	OTs

6-*O*-Tosyl derivatives of 14β-chloro- and 14β-bromocodeines (**220c,d**) occupy a unique place among the tosylates and mesylates investigated, as

220	X
a	Br
b	Cl

307	X
a	Br
b	Cl

308	X
a	βBr
b	βCl
c	αCl

306a

they contain a double allylic system. Chloride or bromide treatment of **220c** gives 6-halogeno-6-demethoxythebaine (**306a,b**); similar reaction of the more reactive **220d** provides 6-tosyloxy-6-demethoxythebaine (**306c**) as the main product. The mechanism of the procedure has been discussed in detail (*305,306*).

In a subsequent paper by Makleit and co-workers (*307*), nucleophilic reactions of another doubly allylic family, the $6\beta,14\beta$-dihalogenocodeines (**307**), prepared from 14β-halocodeines (**220**) with PCl_5, have been published. Only the leaving group linked to the tertiary C-14 is claimed to take part in the reaction, and either in the absence or in the presence of a nucleophilic partner the 6,7-dihalo-6-deoxyneopine derivative **308** is formed primarily, which can be easily transformed into 6-chloro-6-demethoxythebaine (**306a**).

C. REACTIONS OF THEBAINE

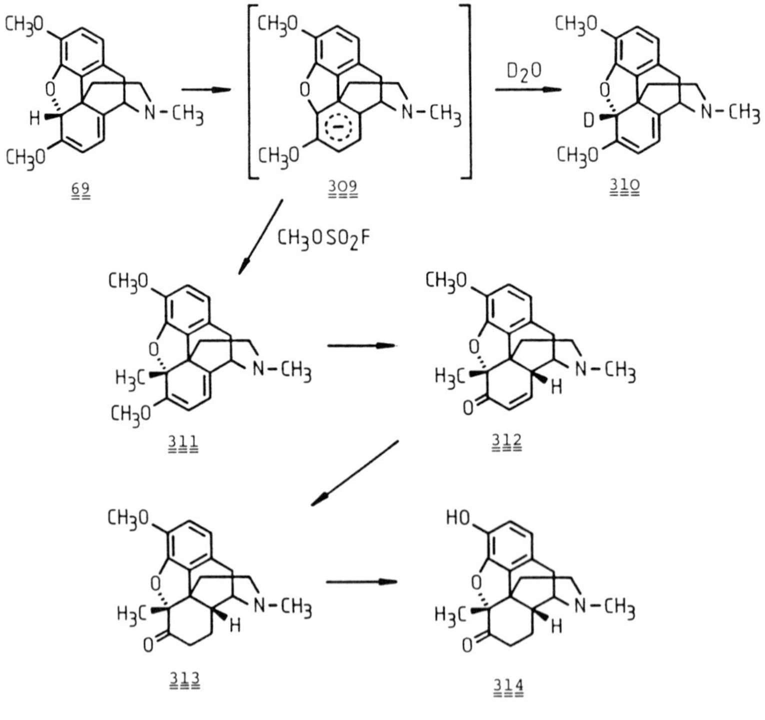

Although thebaine has been the subject of chemical studies since 1835, and at least 300 papers on the chemistry of the parent compound or its derivatives have been published, easy formation of an anion from thebaine was first observed only in 1979 by Boden (*308,309*). Treatment of a solution of thebaine (**69**) in THF with *n*-butyllithium at −78°C gave a deep red solution of thebaine anion (**309**). Quenching with D_2O yielded 5-monodeuterothebaine (**310**). Treatment of the anion with methyl fluorosulfonate led to 5-methylthebaine (**311**). From 5-methylthebaine, whose structure has been determined by ^{1}H-NMR spectroscopy, 5-methylcodeinone (**312**), 5-methyldihydrocodeinone (**313**), and metopon (**314**) have been easily produced.

In another paper by Gates *et al.* (*310*), conversion of 5-methylcodeinone (**312**) to 5-methylcodeine, 5-methylmorphine, and 5-methylheroin was also reported. Furthermore, treatment of thebaine anion with benzyl chloride or ethyl chloroformate to yield 5-benzyl- or 5-carbethoxythebaine, respectively, has been described.

Reaction of thebaine with lithium dimethylcuprate results in the formation of 7β-methyldihydro-β-thebaine (**315**) (*311*), which is a unique starting material, offering as it does an entry into a series of 7-methyl and 7-methyl-8-alkyl-morphinan-6-ones and isomorphinan-6-ones (*312*). Later, Kotick *et al.* extended their work to synthesize the 4,5α-epoxy counterparts of the previously reported 7-methyl and 7-methyl-8-alkyl derivatives (*313*).

The methoxydiene system of thebaine is normally attacked by electrophilic reagents at C-14 (*314*). Nitration of thebaine could be performed either with tetranitromethane (*315,316*) or with dinitrogen tetroxide (*317,318*) regioselectively. If the reaction with tetranitromethane is conducted in methanol, 14β-nitrocodeinone dimethylacetal (**316**) is the main product, which has been converted to 14β-nitrocodeinone (**317**), 14β-aminocodeinone (**318**), and 14β-aminocodeine (**319**). In benzene, in the presence of oxygen (air), however, the nitration took a different course. The major product was identified as the cyclic peroxide 8α,10α-epidioxy-8,14-dihydro-14β-nitrothebaine (**320**), the structure of which was elucidated by chemical degradation studies, as well as by IR and NMR spectroscopy, and verified by single crystal X-ray analysis (*315,316*).

When dinitrogen tetroxide is used for the nitration of thebaine, the major product, 14β-nitrocodeinone (**317**) (23% yield) is accompanied by 8-nitrothebaine (7%) (*317,318*). Osei-Gyimah and Archer made use of the

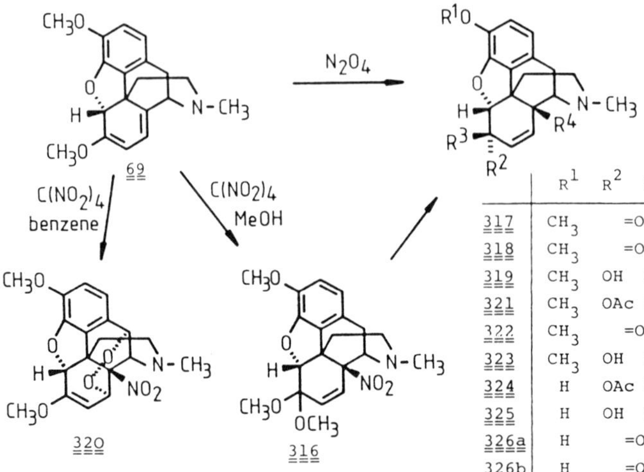

	R^1	R^2	R^3	R^4
317	CH_3	=O		NO_2
318	CH_3	=O		NH_2
319	CH_3	OH	H	NH_2
321	CH_3	OAc	H	NHAc
322	CH_3	=O		SCN
323	CH_3	OH	H	SH
324	H	OAc	H	NHAc
325	H	OH	H	SH
326a	H	=O		Br
326b	H	=O		Cl
156a	CH_3	=O		Br
156b	CH_3	=O		Cl

intermediate **317** in synthesizing 14β-acetamidocodeine-6-acetate (**321**) and the corresponding morphine derivative **324** by successive reductive steps and demethylation with BBr_3.

Reaction of thiocyanogen with thebaine similarly furnished 14β-thiocyanocodeinone (**322**), which was converted by $LiAlH_4$ reduction to 14β-mercaptocodeine (**323**), and by subsequent BBr_3 treatment to 14β-mercaptomorphine (**325**). By treating thebaine with *N*-chloro- or *N*-bromosuccinimide, (see also Refs. *167* and *222*), 14β-chloro- (**156b**) and 14β-bromocodeinone (**156a**), and, via subsequent BBr_3 treatment, the corresponding morphinone derivatives (**326a,b**) could be obtained (*318*).

As a conjugated diene, thebaine reacts with nitrosyl cyanide, generated *in situ* from silver cyanide and nitrosyl chloride, to afford a 6,14-cycloadduct **327.** Hydrogenation of this 1,2-oxazine derivative over Adams catalyst produced 14β-cyanoaminocodeinone (**328**) (*319*).

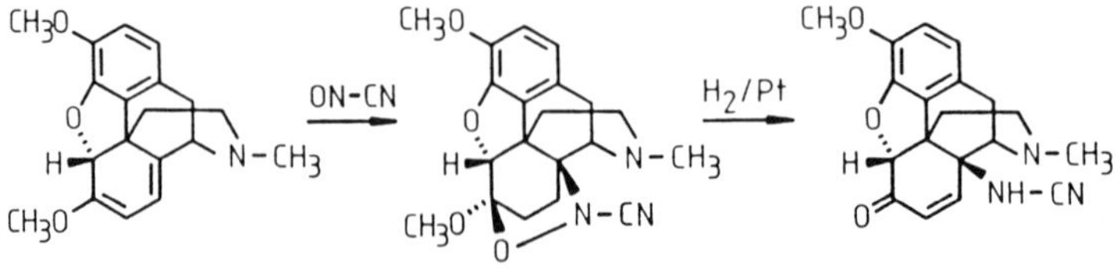

As has been demonstrated, electrophilic attack of thebaine takes place at position 14. Kirby and co-workers, nevertheless, reported on two exceptional cases. Nitrosation of thebaine with an excess of nitrosyl chloride in the presence of alcohols occurs predominantly at C-7, thus producing oxime **329** (*320*). Similarly, C-7 attack and formation of 7β-iodoneopinone dimethylketal (**330**) are observed when thebaine is reacted with excess of iodine in alcohol at ambient temperature (*321*). Treatment of **330** with silver acetate in acetic acid caused rearrangement with displacement of iodine to yield the indolinocodeine derivative **332.** This silver-catalyzed procedure, offering an entry to the indolinomorphine series, is likely to involve an aziridinium intermediate **331.** The same aziridinium perchlorate was obtained and isolated when 14β-bromocodeinone dimethylacetal (**333**) was treated with silver perchlorate (*322*). The α configuration for the 9-substituent was supported by NMR data.

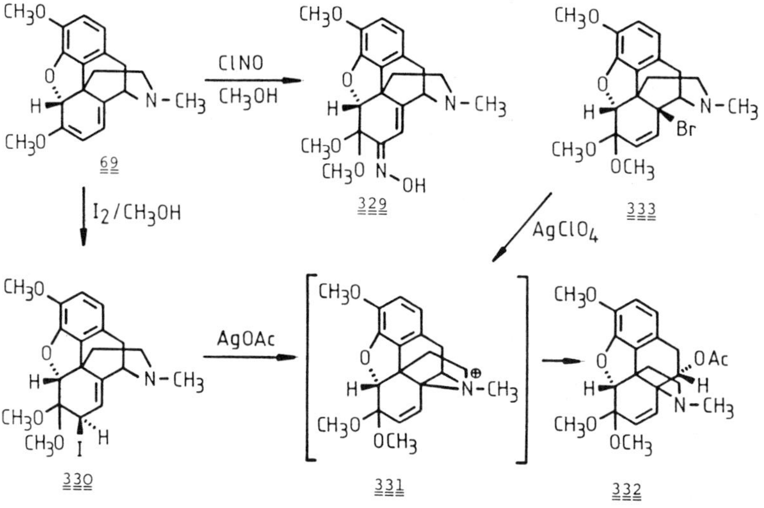

Reduction of thebaine with potassium–liquid ammonia gives a 1:1 mixture of β-dihydrothebaine (**334**) and diene **335** in 95% combined yield. Application of other metals (e.g., Ca, Li, K/FeCl$_3$, and Na) results in exclusive formation of ϕ-dihydrothebaine (**335**) (*323,324*). Oxidation of thebaine to 14β-hydroxycodeinone (**173**) has been discussed within the transformations section (see Section IV,C).

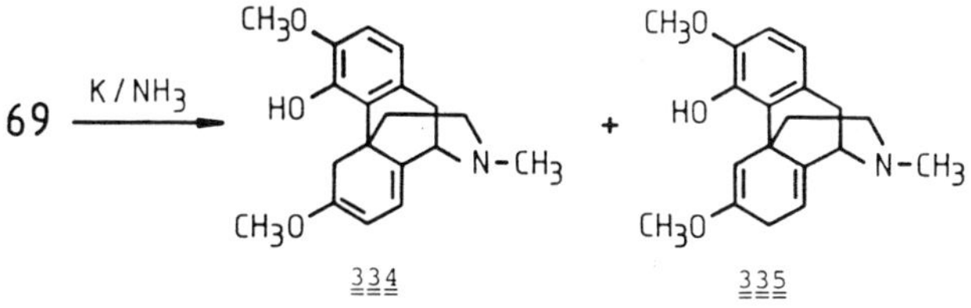

D. Reactions of Codeine, Morphine, Codeinone, and Dihydrocodeinone

The majority of the reactions of morphine and codeine have already been mentioned in earlier sections or were investigated earlier and reviewed in preceding volumes of this treatise. Here, only some new reactions are presented in brief.

A regio- and stereoselective C-10 alkylation of a protected codeine η-chromium tricarbonyl complex has been performed by two research groups. The original idea comes from Arzeno and co-workers (*325*), who protected codeine as the *O*-silyl ether **336** and treated it with hexacarbonyl chromium to afford a single η-complex **337**. This complex can be deprotonated to yield an anion, which, on being treated with methyl iodide, gives the protected complex of (10*S*)-methylcodeine (**338**). Exposure to light and TBAF then supplies the free base **339**. The stereochemistry of C-10 in **339** was defined by X-ray crystallography. Later, Mathews and Sainsbury investigated the scope and limitations of the procedure, and they synthesized other (10*S*) alkylated codeines (*326*). The relative, and hence the absolute, configuration of C-10 was established by NOE studies.

	R			R
77	H		**337**	H
336	Bu^t(CH$_3$)$_2$Si		**338**	CH$_3$

Other C-alkylations in the skeleton of codeinone derivatives have also been presented. Sargent and Jacobson treated codeinone (**140**) with dimethylsulfoxonium methylide, and, instead of the expected *exo*-7,8-cyclopropyl derivative, oxirane **340** was obtained, though in rather poor yield. This compound was transformed by LiAlH$_4$ reduction into 6-methylisocodeine (**342**) (*327*).

Similar results have been published by Bognár and co-workers (*328*), who converted dihydrocodeinone (**53b**) to both of the 6-methyl diastereomers, **343** and **347**. 6-Methyldihydrocodeine (**343**) has been prepared via the method of Small and Rapoport (*329*) by addition of methyl lithium on the carbonyl function; the iso compound **347** was obtained via the oxirane **345**, produced by three different routes from **53b**. Stereoar-

	R^1	R^2
341	CH_3	OH
342	OH	CH_3

rangements were elucidated partly by IR spectroscopy and partly by chemical correlation: isomer **343** proved to be identical with a sample produced by catalytic hydrogenation of 6-methylcodeine (**341**), using a route published by Findlay and Small (*330*), and the structure was elucidated by NMR spectroscopy (*327*).

In an essentially identical approach, both C-6 epimers of 6-methyl-14β-hydroxydihydrocodeine, **344** and **348,** were also synthesized from oxycodone (**174**); configurations and conformations were determined by IR analysis (*328*). The same research group reported on the selective ethynylation at C-6 of codeinone, 1-bromocodeinone, and dihydrocodeinone (*331*).

	R
53b	H
174	OH

	R
343	H
344	OH

3 diff. methods

	R
345	H
346	OH

	R
347	H
348	OH

From 6β-ethinylcodeine, 6β-acetyl-, 6β-vinyl-, or 6β-ethylcodeine could also be produced.

Reaction of codeinone (**140**) with diazomethane in the presence of palladium(II) acetate yielded a mixture of starting material and 7β,8β-methanodihydrocodeinone (**349**) (*332*). The β steric arrangement of the cyclopropane ring in **349** was established by cleavage with aqueous HCl to afford the 8β-chloromethyl compound **350,** which was followed by carbonyl reduction and reductive dehalogenation to afford 8β-methyldihydrocodeine (**351**), identical with the major product

	R¹	R²	X
350		=O	Cl
351	H	OH	H

obtained in the borohydride reduction of the known (*333,334*) 8β-methyldihydrocodeinone (**352;** $R^2 = CH_3$). Similar cyclopropyl derivatives (**349** type) have also been prepared from *N*-cycloalkylmethyl-norcodeinones in the same way (*332*).

Conjugate addition of lithium dialkylcuprates to codeinone (**140**) gave a series of 8β-alkyldihydrocodeinones (**352**). In some cases, the 8α isomer (**353**) could also be isolated in low yield. 8β-Acyldihydrocodeinones (**354**) were prepared by the addition of acyl carbanion equivalents [protected cyanohydrin method or use of lithium bis(α-ethoxyvinyl)cuprate] to **140** followed by hydrolysis. Similar treatment of 14β-hydroxycodeinone (**173**) with lithium dimethylcuprate gave a 4:1 isomeric mixture of the corresponding 8α- and 8β-methyl-14β-hydroxycodeinones (**355** and **356,** respectively). Protection of the hydroxy group via silylation influences the product ratio (*334*). Some of these compounds were modified (e.g., to *N*-substituted nordihydrocodeinone and morphinone derivatives) and the analgesic and narcotic effects investigated (*333,334*).

Leland and Kotick (*335*) treated dihydrocodeinone (**53b**) and its 8β-alkyl derivatives **352** with formaldehyde in the presence of calcium hydroxide in aqueous dioxane and obtained 7,7-bis(hydroxymethyl)-6β-ols (**357,** R = H or alkyl). Ditosylation of **357** followed by $Li(CH_2CH_3)_3$ BH reduction gave either 7α-methyl-6β,7β-oxetane compounds **358** or 7,7-dimethyl-6β-ols **359.** Two-step C-6 epimerization of **359** via oxidation–reduction supplies

	R
140	H
173	OH

	R¹	R²	R³
352	H	alkyl	H
353	H	H	alkyl
354	H	R–CO–	H
355	OH	H	CH₃
356	OH	CH₃	H
53b	H	H	H

357

358

359

360

361

the 7,7-dimethyl-6α-ols **361.** The 6-oxo intermediates **360** were converted to *N*-cycloalkylmethyl nor derivatives and then to the corresponding 3-hydroxy compounds (dihydromorphine series); the analgesic–narcotic antagonist activities of these new molecules have been studied.

Mannich reaction of dihydrocodeinone (**53b**) was first reported by Rapoport and Small (*336*). The only product they could isolate was characterized as the dimer **362.** The reaction was later thoroughly investigated by Polazzi (*337*), who used *N,N,N′,N′*-tetramethyl-methanediamine as the Mannich reagent and obtained the same dimer. The structure, however, was revised to be the Diels–Alder dimer **363** of 7-methylenedihydrocodeinone.

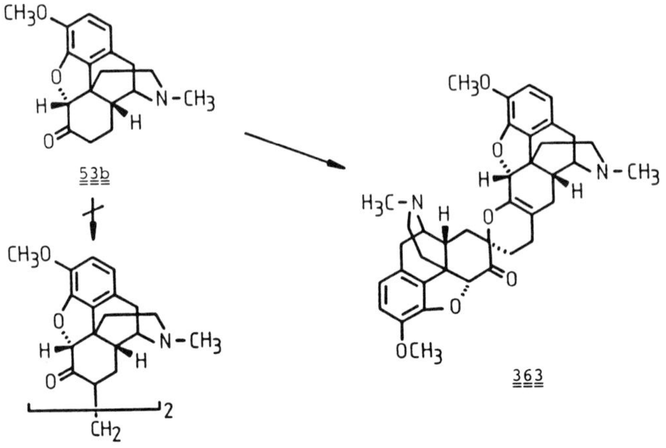

Chloromethylation of dihydrocodeinone (*53b*), followed by reduction with zinc, yielded 1-methyldihydrocodeinone (**53e**) (*338*). The position of the methyl group at C-1 was established by further reduction of **53e** to 1-methyldihydrothebainone (**39b**), which had been prepared previously by total synthesis (*54,55*).

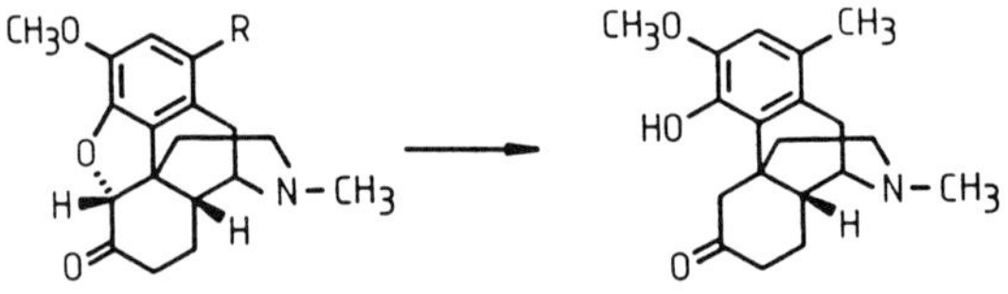

53	R
b	H
e	CH₃

Coupling reactions of morphine derivatives with diazonium salts have been studied by Auterhoff and Tittjung (*339*). They found that morphine derivatives in the 3-hydroxy series, as well as dihydrocodeines, give exclusively C-2 azido coupled products. Codeine reacts at C-2 and C-8 to afford two regioisomeric azido derivatives. Hydrocodone and oxycodone also gave two products: one is coupled at C-2 and the other at C-7 (α position to C = O).

It has already been mentioned (see Section IV,A) that, whereas hydride reduction of codeinones normally produces 6α-hydroxy derivatives (codeine series), formamidinesulfinic acid (FSA) reduces 6-oxo morphines with opposite stereoselectivity. In this way, isocodeine or isomorphine derivatives are obtained (*340,341*). As in certain cases FSA treatment resulted in opening of the 4,5-oxide ring (*157,158*), Chatterjie *et al.* investigated the scope and mechanism of this reaction (*158*). During their study, they found that FSA in alkaline medium reduces aromatic nitro groups to amines very smoothly, so that 2-nitrocodeine (**364**) or 2-nitromorphine (**365**) could be transformed to the corresponding 2-amino derivatives **366** and **367** in relatively high yield without oxide ring scission (*342*).

	R
364	CH_3
365	H

	R
366	CH_3
367	H

The Mitsunobu reaction with morphines has already been noted in Section V,B; some applications, however, are worth presenting here as well. This reaction, discovered in the mid-1960s (*343*), enables alcohols to alkylate H-acids in the presence of diethyl azodicarboxylate and triphenylphosphine according to the following equation:

$$ROH + HX + EtO_2C - N = N - CO_2Et + Ph_3P \rightarrow RX +$$
$$Ph_3P = O + (NHCO_2Et)_2$$

The reaction is totally stereoselective and results in a change of configuration (inversion, epimerization) of a chiral alcohol. Thus, Mitsunobu reaction of morphine, codeine, dihydromorphine, and dihydrocodeine with thioacetic acid (*293,294*) or benzoic acid (*279*) resulted in C-6 epimerization and formation of esters of the iso series, hydrolysis of which afforded isomorphine (**372**), isocodeine (**371**), dihydroisomorphine (**378**), and dihydroisocodeine (**377**), respectively. Similarly, the Mitsunobu reaction of codeine and morphine derivatives with phthalimide or succinimide leads to 6β-aminomorphines and codeines on treating the 6β-phthalimido compounds with hydrazine (*344*).

Claisen–Eschenmoser reaction of codeine with *N*,*N*-dimethylacetamide dimethylacetal leads to 8α-(dimethylcarbamoylmethyl)-8,14-dihydro-6-demethoxythebaine (**379a**) in 80% yield. Utilizing this intermediate, Fleischhacker and Richter also synthesized a series of 8α-substituted 8,14-dihydro-6-demethoxythebaines (**379b–d**) and some oripavine derivatives **379e** and **379f** (*345*). Direct Claisen–Eschenmoser

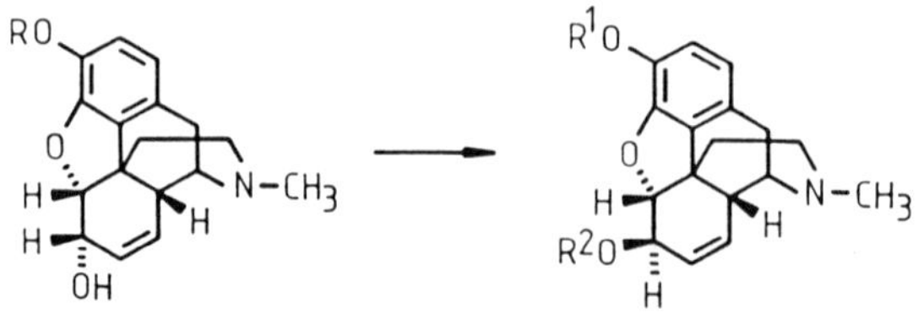

	R
77	CH$_3$
368	Ac

	R^1	R^2
369	CH$_3$	PhCO or CH$_3$CS
370	Ac	PhCO or CH$_3$CS
371	CH$_3$	H
372	H	H

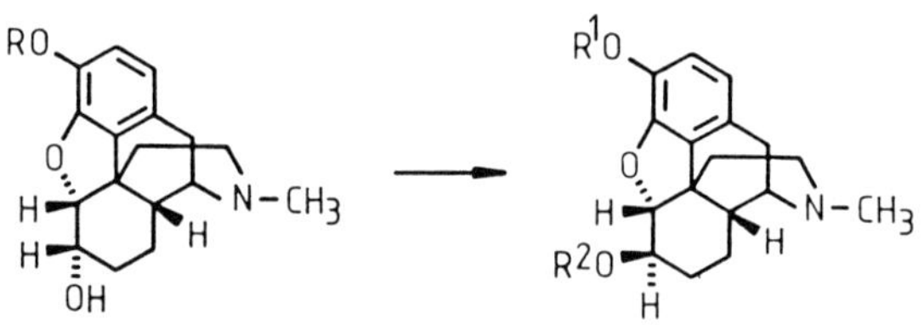

	R
373	CH$_3$
374	Ac

	R^1	R^2
375	CH$_3$	PhCO or CH$_3$CS
376	Ac	PhCO or CH$_3$CS
377	CH$_3$	H
378	H	H

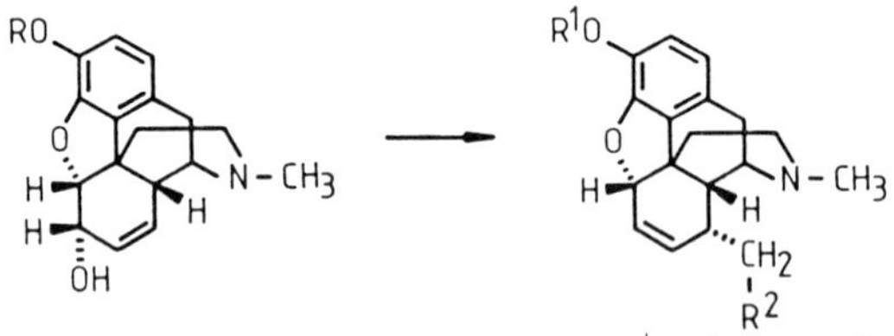

77

379	R^1	R^2
a	CH$_3$	CON(CH$_3$)$_2$
b	CH$_3$	CH$_2$N(CH$_3$)$_2$
c	CH$_3$	CH$_2$OH
d	CH$_3$	CH$_3$
e	H	CON(CH$_3$)$_2$
f	H	CH$_2$N(CH$_3$)$_2$

reaction of morphine did not lead to oripavine derivatives **379e** and **379f;** instead, O-methylation took place, and codeine and its rearranged product **379a** were isolated. 8α-Substituted 8,14-dihydro-6-demethoxyoripavines (**379e,f**) have been obtained by demethylation of the corresponding thebaine derivatives **379a** and **379b.**

Carroll and co-workers (*346*) reported an improved synthesis of morphine-3-glucuronide, the major metabolite of morphine and heroin in human beings and animals. The modified procedure, which involves the addition of the solid bromosugar to a concentrated solution of the lithium salt of morphine in methanol at room temperature, furnished the metabolite in 53% yield.

There are two publications since the last review in this treatise that report on the reductive elimination of the phenolic hydroxyl group from the C-3 position of morphines. Lewis and Readhead prepared the diethoxyphosphate esters of oripavine derivatives and performed the reductive cleavage with sodium in liquid ammonia (*347*). The well-known method utilizing hydrogenolytic cleavage of phenyltetrazolyl ethers was applied to morphine by Bognár *et al.* (*348*). During the reduction, the $\Delta^{7,8}$-double bond was also hydrogenated to afford 3-deoxydihydromorphine (**381**).

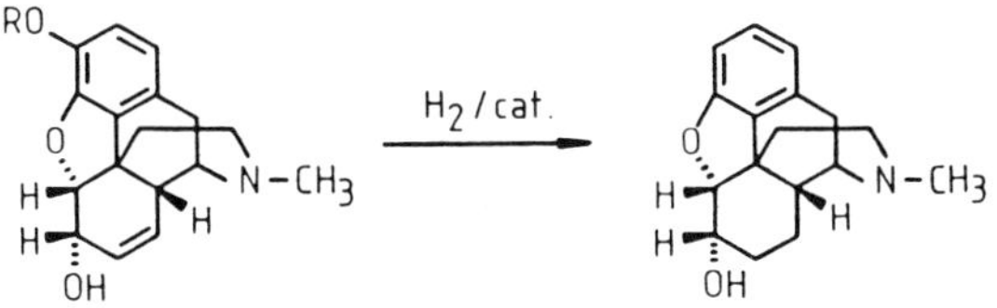

	R
1	H
380	$C_7H_5N_4$

381

Beckmann and Schmidt rearrangements of 6-oxodihydromorphine alkaloids have also been studied (*349,350*). It was found that in the two types of rearrangements, the isomeric lactams **382** and **383,** respectively, could be obtained as the main products.

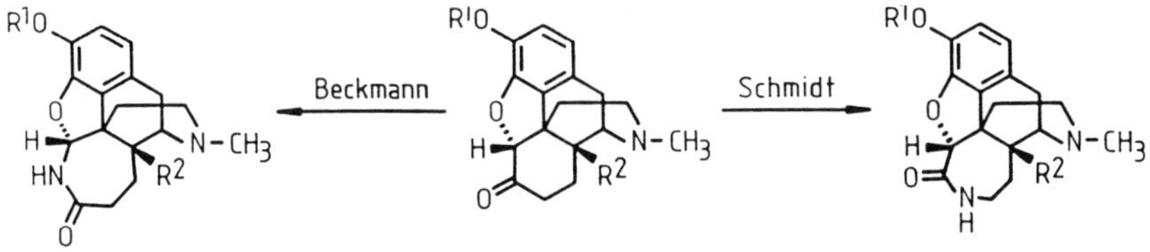

382

R^1 = H or CH_3
R^2 = H or OH

383

E. Reactions of 14β-Bromocodeine Derivatives

As was already mentioned in Sections IV,D and V,C, 14β-bromocodei-
none (**156a**) can be produced from thebaine with *N*-bromosuccinimide.
This active intermediate, containing a conjugated carbonyl group and an
allylic bromide function, can undergo a number of versatile transformations. Here, some recent applications are presented to demonstrate its
reactivity.

Reduction of 14β-halocodeinones (**156**) has been summarized in Section
IV,D. It is worth mentioning, nevertheless, that sodium borohydride reduction of 14β-halocodeinones (**156**) results in the stereoselective formation of 14β-halocodeines (**220**) (*222*), whereas catalytic hydrogenation of
156a leads to neopinone (**141**) (see p. 183).

Oxidation of 14β-bromocodeinone dimethylacetal (**333**) was performed
by shaking an alcoholic solution in air, in the presence of sodium hydrox-

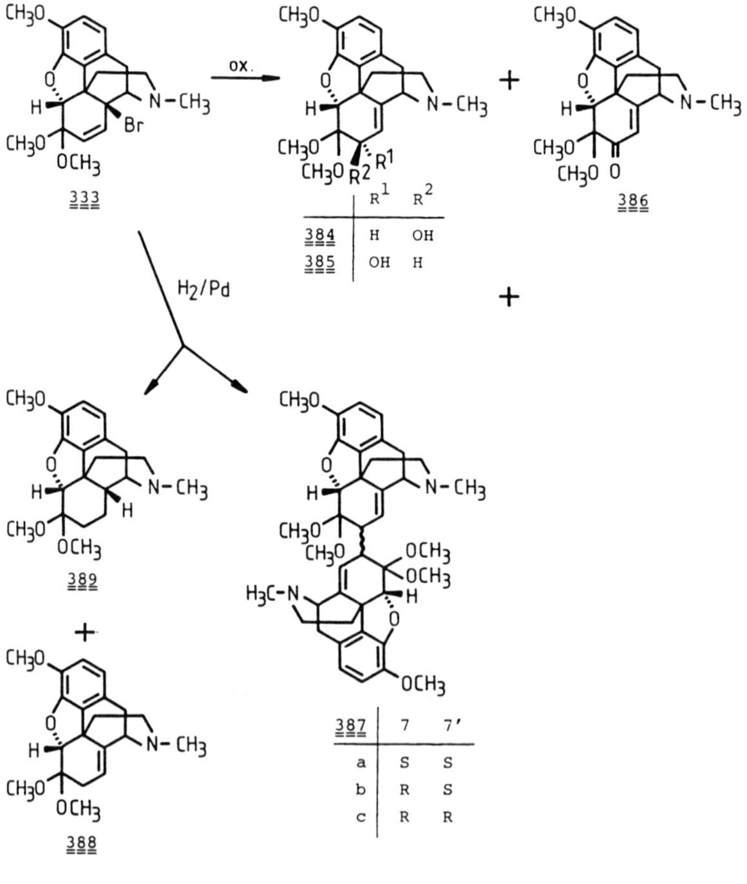

	R^1	R^2
384	H	OH
385	OH	H

	7	7'
a	S	S
b	R	S
c	R	R

333   **384**   **385**   **386**   **387**   **388**   **389**

ide and a specially prepared hydrogenation catalyst, to yield 7β-hydroxyneopinone dimethylacetal (**384**) as the main product (75%). From the mother liquor, the C-7 epimer carbinol **385**, the 7-oxoneopinone derivative (**386**), and three diastereomeric dimers, **387a–387c,** could be isolated by preparative TLC (*351*).

The same three dimers, **387a–387c,** together with neopinone dimethylacetal (**388**), a small amount of 7β-hydroxy derivative **384**, and dihydrocodeinone dimethylacetal (**389**) were obtained when dimethylacetal **333** was hydrogenated over Pd catalyst. The ratio of the six products could be influenced by the reaction conditions applied, and the structure and stereo arrangement of all products were elucidated (*352*).

By investigating the Claisen–Eschenmoser rearrangement of the 7β-hydroxy derivative **384,** Fleischhacker and Richter (*353*) synthesized a series of new 14β-substituted derivatives (e.g., **390** and **391**). The two latter products underwent some unexpected conversions to lactones **392** and **393,** as well as to the cyclic ethers **394** and **395,** respectively, under the conditions of acetal hydrolysis. The mechanism of these reactions has been discussed (*354*).

Methanolysis of 14β-bromocodeinone dimethylacetal (**333**) has been thoroughly investigated by Fleischhacker and co-workers. In the presence

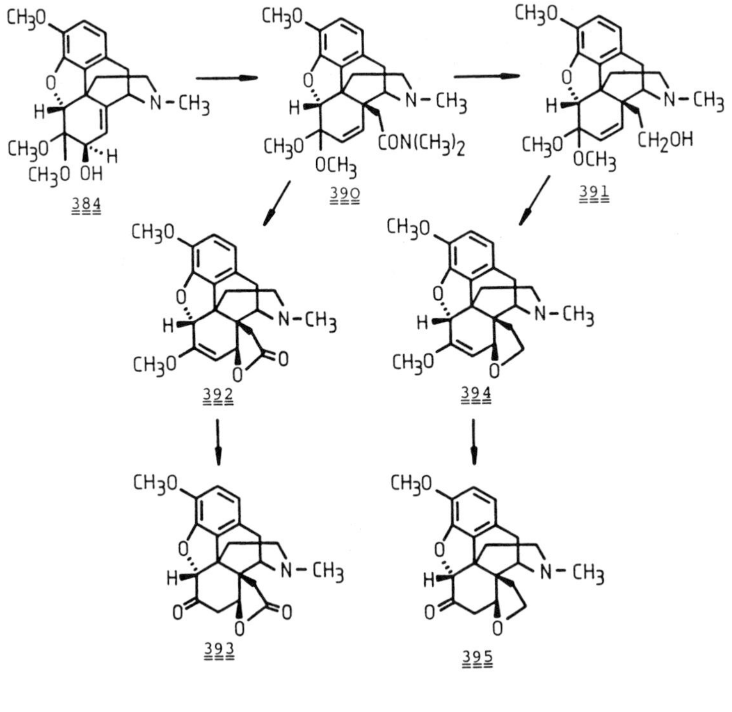

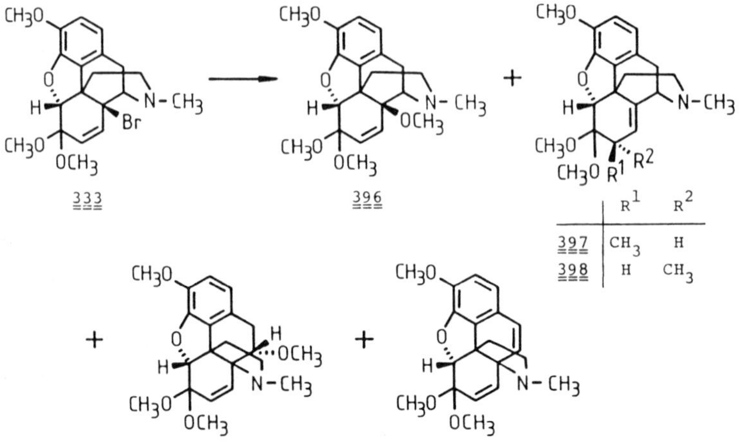

333 → **396** + **397** / **398**

	R¹	R²
397	CH_3	H
398	H	CH_3

399a + **400**

of sodium carbonate, this reaction provided five products; isolation and structure elucidation of 14β-, 7β-, and 7α-methoxy derivatives **396–398**, as well as of two indolinocodeinone derivatives **399a** and **400**, have been described (*355,356*).

Onda and co-workers investigated some reactions of 14β-bromocodeine (**220a**) and found it to be more reactive than the 6-oxo derivative. Hydrolysis, for example, resulted in a mixture containing 9α-hydroxyindoli-

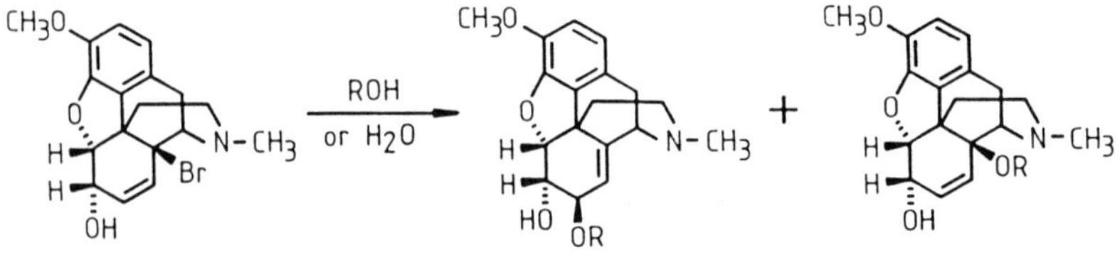

220a → (ROH or H_2O) **401** + **402**

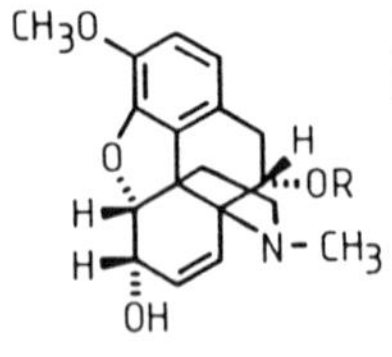

399

399, 401, 402	R
a	CH_3
b	H

nocodeine (**399b**), 7β-hydroxyneopinone (**401b**), and 14β-hydroxycodeine (**402b**) (*357,358*). Similar results were obtained under solvolytic conditions (e.g., with CH_3OH, **401a, 402a,** and **399a** are the main products).

VI. Opiate-Mediated Analgesia

In the previous sections, we have dealt with the synthetic methods suitable for preparing morphine and morphine congeners. With all of this chemistry available, a great number of derivatives have been made either by semisynthetic or total synthetic approaches, aiming at analogs of morphine with an improved pharmacological profile. Several reviews covering this subject have been published (*359–362*). Discussion of all the derivatives and their pharmacology is beyond the scope and limitation of this chapter. Thus we confine ourselves to the intriguing problem of how opiates and opioid peptides interact with their corresponding receptor.

This issue has aroused the interest of many scientists in the last decades, and the overview of Gottschlich (*360*) deserves special attention, for it serves here as the main source of the present brief discussion. It was a reasonable assumption that the opiates, like other pharmacologically active compounds, mediate their effects by interacting with specific receptors. Two short-chain opioid peptides, the enkephalins, were detected in mammalian brain and the amino acid sequence published in 1975 (*363*) (Fig. 1). These peptides were considered to be the endogenous ligands of the opiate receptor. Since then, several further opioid peptides have been characterized. In 1980 Morley (*364*) reviewed the pharmacological and binding properties of nearly 200 enkephalin analogs. The results for some hydroxylated and nonhydroxylated derivatives are shown in Fig. 2. The presence of the phenolic moiety of tyrosine is essential. The introduction of a phenolic hydroxyl group into the phenylalanine unit leads, however, to an almost complete loss of pharmacological activity.

This phenomenon is also observed with morphine and morphine analogs. When the phenolic hydroxyl group is eliminated, the affinity for the opiate receptor is substantially decreased (*365*), indicating a similarity between the N terminus of the enkaphalins and a certain part of the

Tyr-Gly-Gly-Phe-Leu(=Leu enkephalin)

Tyr-Gly-Gly-Phe-Met(=Met enkephalin)

Fig. 1. Enkephalins.

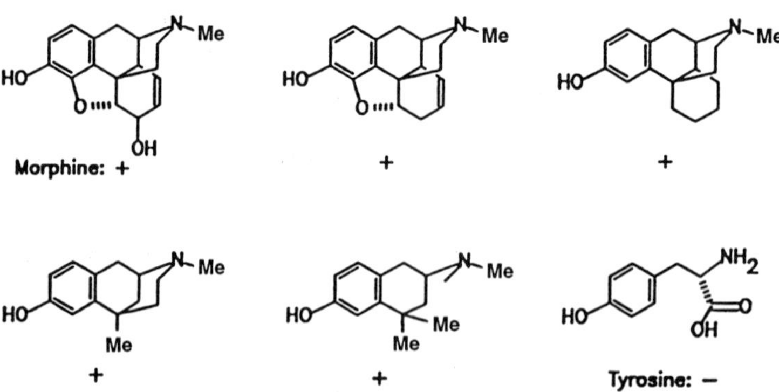

Fig. 2. Importance of the phenolic hydroxyl group for the biological activity of enkephalins.

morphine nucleus. Perhaps the structure of morphine mimics the N-terminal portion of the enkephalins. However, tyrosine itself does not behave like an opioid. Thus, we may conclude that further structural requirements for opioid activity reside in the morphine molecule. Which of them are really necessary?

To answer the question we have to examine a series of compounds (Fig. 3). From Fig. 3 we may conclude that a certain rigid arrangement of important structural elements is indispensable. The above observations and others (*366–368*) led to the interpretation that morphine mimics the N-terminal part of the opioid peptides. A nonpeptidic opioid compound with a phenyl ring loses its opioid activity after a hydroxyl group has been introduced, as shown in Fig. 4. Allylprodine, a congener of the analgesic Dolantin, becomes inactive if a hydroxyl group is present. Thus, allylprodine probably represents a structural equivalent of the phenylalanine moiety of the enkephalins.

Further extensive studies of different molecules proved that variation of the structural bridges connecting the essential elements of opiates and

Fig. 3. Opioid activity of partial structures of morphine (+, inactive; −, active).

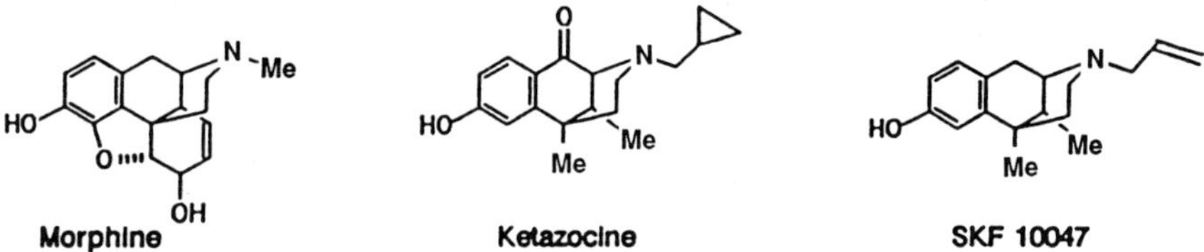

FIG. 4. Allylproline and Dolantine.

opioid peptides, within certain limits, makes little difference to the opiate receptor. The phenylpiperidines, like Dolantin, owe their opioid character to this extraordinary toleration by the receptor.

It can therefore be concluded that (1) the enkephalins need three essential structural elements, namely, a basic nitrogen, a "tyrosine equivalent," and a "phenylalanine equivalent," in order to possess opioid activity, and (2) the rigid opiates, on the other hand, are biologically active with only two of these moieties (i.e., morphine and analogs possessing the basic nitrogen and "tyrosine equivalent" sections and phenylpiperidines having the basic nitrogen and "phenylalanine equivalent" elements).

A. The "Opiate Receptor" as a Family of Receptors

In 1976, Martin *et al.* (*369*) demonstrated that opiates can be separated into three groups according to their pharmacological profiles (see Fig. 5): the μ-opiates/receptors (after morphine); the κ-opiates/receptors (after ketazocine, belonging to the class of benzomorphans, and, in contrast to morphine, having sedative activities); and the σ-opiates/receptors (after SKF 10047, another benzomorphan, with σ-opiates possessing hallucinogenic properties). In the 1970s, two further types of opiate receptors, the δ- and ε-receptors were detected, but nowadays the μ-, κ-, and to some extent the δ-opiates are regarded as the most interesting representatives of

FIG. 5. Morphine, ketazocine, and SKF 10047 as prototypes of opiates interacting with μ-, κ-, and σ-receptors.

Try-Gly-Gly-Phe-Leu-Arg-Arg-Ile-Arg-Pro-Lys-Leu-Lys-Trp-Asp-Asn-Gln

FIG. 6. Dynorphin.

the analgesics. The above-discussed considerations apply only to the μ-receptors, but what about the κ-opiates/receptors?

A few years after enkephalins had been discovered, a new mammalian opioid peptide (dynorphin, Fig. 6) having high affinity for the κ-receptor was identified by Goldstein *et al.* (*370*). Why the C-terminal extension of Leu-enkephalin, present in dynorphin, transforms the μ-active enkephalins to κ-active agents is not clearly understood yet. The situation is also complex with respect to the nonpeptidic ligands of the κ-opiate receptor. Some benzomorphans and morphinans with high κ-activity are shown in Fig. 7. The elimination of the phenolic hydroxyl from these derivatives markedly reduces the biological activity. However, the question of why changing the N-substituent leads to such a shift from μ- to κ-activity, cannot be answered for sure and remains open for further speculation.

A further intriguing question is, What makes an opiate agonist act sometimes as an antagonist? The κ-agonists of the benzomorphan/morphinan type also have affinity for the μ-opiate receptor. The interection of these compounds with the μ-receptor, however, results in its blockade instead of its activation.

An accepted concept in this connection is that opioid activity requires a close interaction, under physiological conditions, between the cationic nitrogen of the opiate and an "anionic site" of the receptor (Fig. 8). If morphine is methylated, the resulting compound can no longer approach the anionic site as close as morphine because of steric reasons. *N*-Methylmorphine shows only limited activity as an opiate agonist and can

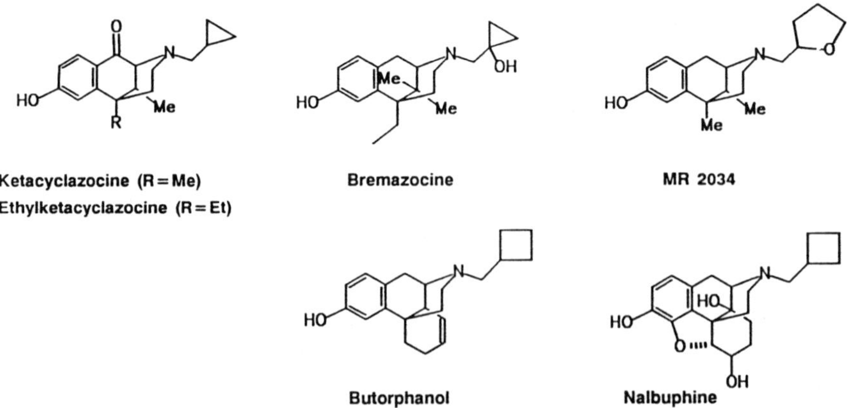

FIG. 7. Compounds active at K-opioid receptors.

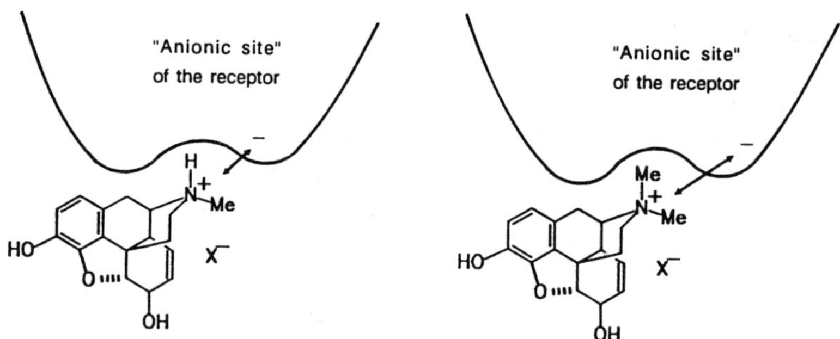

FIG. 8. Morphine and *N*-methylmorphine interacting with the anionic site of the opiate receptor.

prevent a stronger agonist from interacting with the receptor, that is, can behave like an antagonist (*371*).

In the mono-*N*-alkylated derivatives, the substituent can occupy two different steric positions, axial or equatorial, with respect to the piperidine ring (Fig. 9). The nalbuphine isomer **A** can interact with the anionic site of the receptor easily and acts as an agonist. The cyclobutylmethyl N-substituent is responsible for the κ-specificity, for unknown reasons. On the contrary, the stereoisomer **B** does not interact and hence possesses antagonistic activity.

B. BIVALENT LIGANDS OF OPIATE RECEPTORS

A brand-new approach to more potent and selective opioid agonists and antagonists is largely connected with the name of Portoghese (*372*). If opiate receptors, located in a membrane, keep a well-defined mean distance from one another, a bivalent ligand should react favorably, compared to two single opiate molecules, with the opiate receptors owing to entropic factors. Such bivalent ligands have been synthesized using

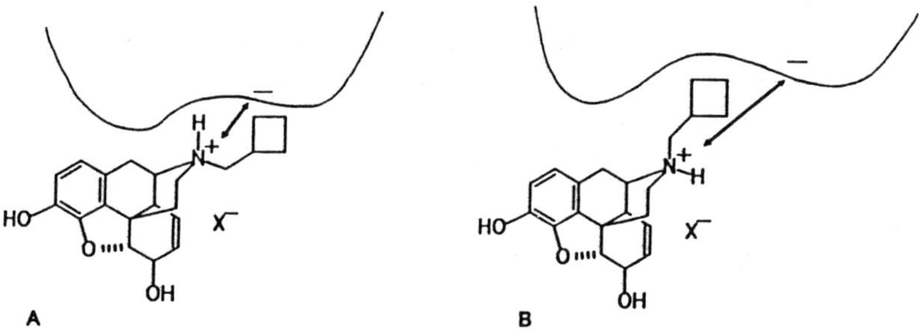

FIG. 9. Two configurations of nalbuphine.

enkephalins and morphinelike molecules as templates. The optimal distance between the pharmacophores might also differ according to the opiate subtypes.

The results obtained with certain types of bivalent opiate antagonists are given in Fig. 10. The bivalent compound **b** (called TENA) shows outstanding κ-antagonism. Bivalent ligands containing a very short and rigid spacer, like norbinaltorphimine (Fig. 11), even exceeded TENA as a κ-selective antagonist. However, it is still a remarkable problem that the most important compound synthesized according to the "bivalent concept" turned out not to act in the way that was expected on the basis of the original theoretical considerations.

Recently, the discovery of the first nonpeptide δ-antagonist, naltrindole (NTI) (Fig. 12), was reported (*373*). It is speculated that NTI is a bivalent ligand and that the two aromatic rings align in a manner similar to that of the aromatic rings of the enkephalins when bound to δ-receptors. Of the opioid receptors, the functions of the δ-receptor are the least well understood (*374,375*). The discovery of the highly potent and selective agent NTI provides researchers with an important new tool for studying δ-receptor function.

Further intensive study of opioid receptors appears to be very important for the development of new therapeutic agents. The phenomenon of pain has occupied human thought from the earliest times. To relieve pain is one of the most challenging areas of medicinal chemistry, and it will remain so for a long time.

VII. Biogenesis

The biogenesis of morphine alkaloids has been the subject of numerous studies. The theories of Robert Robinson on the origin of these alkaloids, from tyrosine via benzylisoquinoline alkaloids (*74–77*), have been broadly confirmed. However, recent labeling experiments resulted in some modification on the overall sequence shown in the last review in this treatise (*1*).

The first revision concerns the formation of reticuline (**54b**) (see Fig. 13). Zenk and co-workers seriously questioned the intermediacy of (*S*)-norlaudanosoline (**405**) in the biogenesis of (*S*)-reticuline [(*S*)-**54b**] when a common origin for the trioxygenated bases of the coclaurine type (**407**) and reticuline was shown in *Aunoma reticula* plants (*376*). Later, the same research group demonstrated that it is not (*S*)-norlaudanosoline [(*S*)-**405**], but rather (*S*)-norcoclaurine [(*S*)-**406**], originating from stereospecific condensation of dopamine (**403**) and 4-hydroxyphenylacetaldehyde (**404**), which represents the first true alkaloidal intermediate in the biogenesis of

Relative antagonistic activity:

	μ	κ
a	2.1	1.0
b	2.0	10.4
c	3.5	1.6

A =

FIG. 10. Antagonistic activity of bivalent opiates.

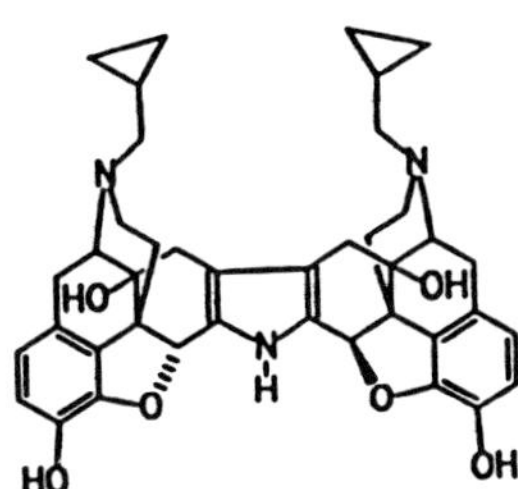

FIG. 11. Norbinaltorphimine.

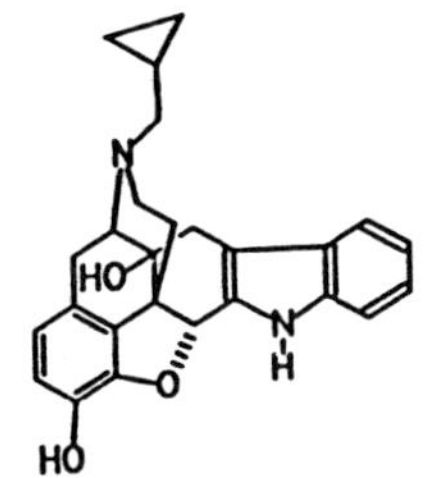

FIG. 12. Naltrindole (NTI).

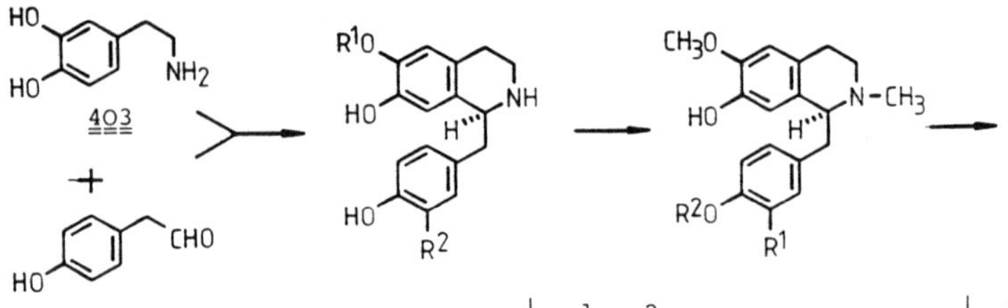

		R^1	R^2			R^1	R^2
405	(S)-norlau-danosoline	H	OH	**408**	(S)-N-methyl-coclaurine	H	H
406	(S)-norcoc-laurine	H	H	**409**	(S)-3′-hydr-oxy-N-methyl-coclaurine	OH	H
407	(S)-coclau-rine	CH$_3$	H	**54b**	(S)-reticuline	OH	CH$_3$

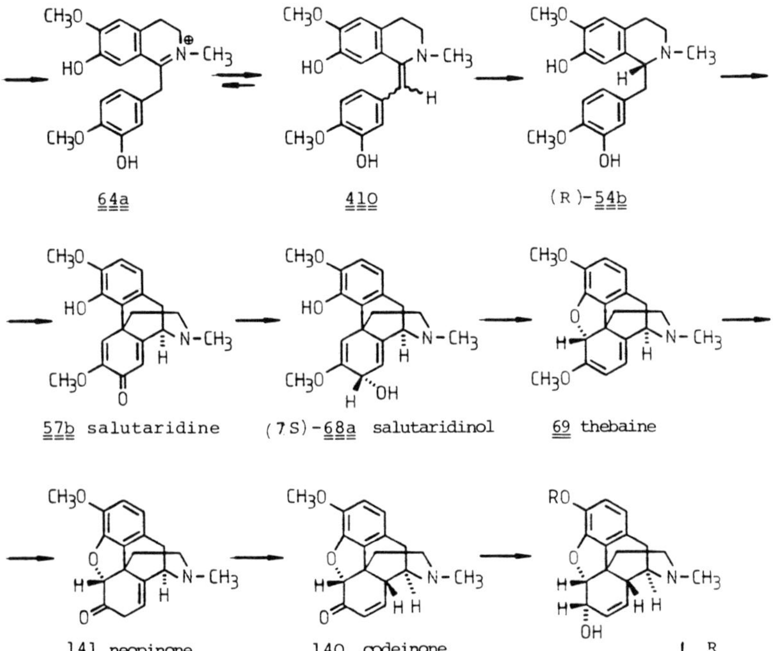

64a   **410**   (R)-**54b**

57b salutaridine   (7S)-**68a** salutaridinol   **69** thebaine

141 neopinone   **140** codeinone

		R
77	codeine	CH$_3$
1	morphine	H

FIG. 13. Biosynthesis of morphine in plants.

reticuline-derived alkaloids (*377,378*). The transformation of (*S*)-**406** to (*S*)-reticuline [(*S*)-**54b**] entails one hydroxylation, one *N*-methylation, and two *O*-methylation steps.

In a following publication (*379*) Stadler and Zenk clarified the sequence and the stereoselectivity of the enzymatic steps with ^{13}C-labeling studies, and as a result of this research, the "first half" of the proposed biosynthetic scheme has been revised as follows. The trihydroxylated base (*S*)-**406** is rapidly converted to (*S*)-coclaurine [(*S*)-**407**] by a relatively unselective 6-*O*-methyltransferase enzyme (*380*). Evidence suggests that enzymatic *N*-methylation of (*S*)-**407** to (*S*)-**408** by a stereochemically unspecific *N*-methyltransferase (*381*) precedes the 3′-hydroxylation of ring C to furnish a tetra-oxygenated benzylisoquinoline intermediate (*S*)-**409,** a step which represents the hitherto missing link between (*S*)-**406** and the reticuline moiety and is catalyzed by a relatively unspecific phenolase enzyme (*382*). Regioselective 4′-*O*-methylation converts (*S*)-**409** to (*S*)-**54b** and requires an absolutely stereoselective 4′-*O*-methyltransferase, the isolation and characterization of which had been successfully accomplished (*383*).

The absolute configuration of (*S*)-reticuline has to be altered in the course of the biogenesis. This is supposed to occur by a series of oxidation–reduction steps via the 1,2-dehydroreticulinium salt **64a** (*384*). He and Brossi suggested that the cytosolic NADPH$_2$-dependent enzyme, which converts the 1,2-dehydroreticulinium compound under alkaline conditions (pH 7.8), may be operating on the enamine **410** rather than on a quaternary immonium salt (*385*).

In another publication, Zenk and co-workers investigated the structure of the two salutaridinols (**68a**) formed by NaBH$_4$ reduction of salutaridine (**57b**) (*161*) (see also Section IV,B). They verified that enzymatically (*in vitro*) only salutaridinol-I [(7*S*)-**68a**] is formed and can be converted to thebaine, codeine, and morphine, as was found by Barton *et al.* (*78*). Its configuration, however, on the basis of X-ray crystallography, is (7*S*), opposite of that which had been previously determined (*79*).* The modified biogenetic pathway disclosed by the above recent studies is summarized in Fig. 13.

Finally, it is worth mentioning that Brossi and co-workers have identified the formation of endogenous codeine and morphine in rat brain. To demonstrate the occurrence of a biosynthetic pathway in mammals, (+)-salutaridine, (−)-thebaine, and (−)-codeine were administered to rats intravenously. In all cases, a marked increase in the codeine and morphine levels in rat tissues was observed. This provides evidence for a biosynthe-

---

* Subsequent X-ray analysis of the original isolation of nudaurine (*386*) (2,3-methylenedioxy substitution pattern at the aromatic ring), stored for almost 30 years at −20°C without significant decomposition, has revealed that the product possesses the 7*S* absolute configuration (*387*) in contrast to earlier assumptions (*388*).

tic pathway similar to that delineated in Fig. 13 for the "plant production" (*389,390*).

Noted Added in Proof

Overman *et al.* published a paper in which they described an asymmetric synthesis of both enantiomers of opium alkaloids and morphinans via a chiral synthon obtained by enantioselective reduction of 2-allyl-cyclohex-2-en-1-one. Asymmetric induction was carried out by using catechol boran as reducing agent in the presence of (R)- and (S)-oxazaborolidine catalyst (*391*).

References

1. K. W. Bentley, *in* "The Alkaloids" (R. H. F. Manske, ed.), Vol. 13, Chap. 1. Academic Press, New York and London, 1971.
2. H. L. Holmes, *in* "The Alkaloids" (R. H. F. Manske and H. L. Holmes, eds.), Vol. 2, Chaps. 1 and 2. Academic Press, New York and London, 1952.
3. G. Stork, *in* "The Alkaloids" (R. H. F. Manske, ed.), Vol. 6, Chap. 7. Academic Press, New York and London, 1960.
4. K. W. Bentley, "The Chemistry of Morphine Alkaloids." Oxford Univ. Press (Clarendon), Oxford, 1954.
5. K. W. Bentley, *in* "The Isoquinoline Alkaloids," Chap. 7. Macmillan (Pergamon), Oxford, 1965.
6. T. Kametani, "The Chemistry of Isoquinoline Alkaloids," Vol. 1. Elsevier, New York, 1969.
7. T. Kametani and K. Fukumoto, *J. Heterocycl. Chem.* **8,** 341 (1971).
8. K. L. Stuart, *Chem. Rev.* **21,** 47 (1971).
9. A. Brossi, *Trends Pharmacol. Sci.* **50,** 239 (1982).
10. K. C. Rice, *in* "The Chemistry and Biology of Isoquinoline Alkaloids" (J. D. Phillipson, M. F. Roberts, and M. Zenk, eds.), p. 191. Springer-Verlag, Berlin, 1985.
11. G. Blaskó and G. A. Cordell, *Heterocycles* **27,** 1269 (1988).
12. "Specialist Periodical Report. The Alkaloids," The Royal Society of Chemistry, Burlington House, London, **1,** 122 (1971); **2,** 128 (1972); **3,** 157 (1973); **4,** 154 (1974); **5,** 141 (1975); **6,** 131 (1976); **7,** 115 (1977); **8,** 111 (1978); **9,** 115 (1979); **10,** 110 (1980); **11,** 100 (1981); **12,** 119 (1982); **13,** 144 (1983).
13. K. W. Bentley, *Nat. Prod. Rep.* **1,** 360 (1984); **2,** 88 (1985); **3,** 161 (1986); **4,** 690 (1987); **5,** 282 (1988); **6,** 405 (1989); **7,** 259 (1990); **8,** 354 (1991); **9,** 379 (1992).
14. K. P. Tiwari, R. N. Choudhary, and G. D. Pandey, *Phytochemistry* **20,** 863 (1981).
15. A. L. Djakoure, *Ann. Univ. Abidjan, Ser. C* **17,** 105 (1981).
16. R. A. Barnes and O. H. Soeiro, *Phytochemistry* **20,** 543 (1981).
17. M. Shamma, P. Chinnasamy, S. F. Hussain, and F. Khan, *Phytochemistry* **15,** 1802 (1976).
18. R. L. Khosa, V. K. Lal, A. K. Wahi, and A. B. Ray, *Chem. Ind. (London)*, 662 (1980).

19. V. Vecchietti, C. Casagrande, G. Ferrari, B. Danieli, and G. Palmisano, *J. Chem. Soc., Perkin Trans. 1*, 578 (1981).
20. R. Hocquemiller, A. Oeztekin, F. Roblot, and A. Cavé, *J. Nat. Prod.* **47**, 342 (1984).
21. A. Oeztekin, R. Hocquemiller, and A. Cavé, *J. Nat. Prod.* **47**, 557 (1984).
22. B. Gözler, P. Ozic, A. J. Freyer, and M. Shamma, *J. Nat. Prod.* **53**, 986 (1990).
23. V. Vecchietti, C. Casagrande, and G. Ferrari, *Tetrahedron Lett.*, 1631 (1976).
24. V. Vecchietti, C. Casagrande, and G. Ferrari, *Farmaco Ed. Sci.* **32**, 767 (1977).
25. D. J. Slatkin, N. Doorenbos, J. E. Knapp, and P. L. Schiff, *Lloydia* **37**, 488 (1974).
26. T. N. Ilinskaya, M. E. Perelson, I. I. Fadeeva, and O. N. Tolkachev, *Khim. Prir. Soedin.*, 650 (1973).
27. M. E. Perelson, I. I. Fadeeva, and T. M. Ilinskaya, *Khim. Prir. Soedin.*, 188 (1975); *Chem. Abstr.* **83**, 79434 (1975).
28. B. Charles, H. Guinaudeau, J. Bruneton, and P. Cabalion, *Can. J. Chem.* **67**, 1257 (1989).
29. A. Patra, *Phytochemistry* **26**, 2391 (1987).
30. A. I. Spiff, V. Zabel, W. H. Watson, M. A. Zemaitis, A. M. Ateya, D. J. Slatkin, J. E. Knapp, and P. L. Schiff, *J. Nat. Prod.* **44**, 160 (1981).
31. F. Tillequin and M. Koch, *Heterocycles* **23**, 1357 (1985).
32. E. Tojo, D. Dominguez, and L. Castedo, *J. Nat. Prod.* **52**, 415 (1989).
33. E. Tojo, D. Dominguez, and L. Castedo, *Phytochemistry* **30**, 1005 (1991).
34. G. Sariyar, A. J. Freyer, H. Guinaudeau, and M. Shamma, *J. Nat. Prod.* **53**, 1383 (1990).
35. S. Pfeifer and D. Thomas, *Pharmazie* **27**, 48 (1972).
36. G. Sariyar and A. Oeztekin, *Plant. Med. Phytother.* **15**, 160 (1981).
37. H. Hokawa, S. Tsuruoka, K. Takeya, N. Mori, T. Sonobe, S. Kosemura, and T. Hamanaka, *Chem. Pharm. Bull.* **35**, 1660 (1987).
38. H. Hokawa, K. Takeya, N. Mori, T. Sonobe, S. Kosemura, N. Okamura, T. Ogawa, M. Ogoshi, K. Yamakawa, and T. Hamanaka, *Jpn. Kokai Tokkyo Koho* JP 62,289,565; *Chem. Abstr.* **109**, 73724 (1988).
39. J. S. Glasby, "Encyclopedia of the Alkaloids," Vol. 2, p. 948. Plenum, New York, 1975.
40. C. Chambers and K. L. Stuart, *Chem. Commun.*, 328 (1968).
41. T. Kametani, K. Fukumoto, F. Satoh, and H. Yagi, *J. Chem. Soc. C*, 520 (1969).
42. I. R. C. Bick, H. M. Leow, N. W. Preston, and J. J. Wright, *Aust. J. Chem.* **26**, 455 (1973).
43. A. N. Tackie, D. Dwuma-Badu, J. E. Knapp, D. J. Slatkin, and P. L. Schiff, *Phytochemistry* **19**, 2884 (1974).
44. D. Dwuma-Badu, J. S. K. Ayim, S. F. Withers, N. O. Ogyemang, A. M. Ateya, M. M. El-Azizi, J. E. Knapp, D. J. Slatkin, and P. L. Schiff, *J. Nat. Prod.* **43**, 123 (1980).
45. H. Meshulam and D. Lavie, *Phytochemistry* **19**, 2633 (1980).
46. M. Juichi, Y. Fujitani, and H. Furukawa, *Yakugaku Zasshi* **104**, 946 (1984).
47. S.-T. Lu, Y.-C. Wu and S.-P. Leon, *Phytochemistry* **24**, 1829 (1985).
48. A. Shafiee, A. Ghanbarpour, and S. Akhlaghi, *J. Nat. Prod.* **48**, 855 (1985).
49. R. Grewe, H. Pohlmann,. and M. Schnoor, *Chem. Ber.* **84**, 527 (1951).
50. R. Grewe and W. Friedrichsen, *Chem. Ber.* **100**, 1550 (1967).
51. R. Grewe and W. Friedrichsen, *Chem. Ber.* **100**, 1 (1967).
52. G. C. Morrison, R. O. Waite, and J. Shavel, Jr., *Tetrahedron Lett.*, 4055 (1967).
53. M. Gates and G. Tschudi, *J. Am. Chem. Soc.* **78**, 1380 (1956).
54. H. C. Beyerman, E. Buurman, and L. Maat, *J. Chem. Soc., Chem. Commun.*, 918 (1972).

55. H. C. Beyerman, E. Buurman, L. Maat, and C. Olieman, *Recl. Trav. Chim. Pays-Bas* **95**, 184 (1976).
56. H. C. Beyerman, T. S. Lie, L. Maat, H. S. Bosman, E. Buurman, E. J. M. Bijsterveld, and H. J. M. Sinnige, *Recl. Trav. Chim. Pays-Bas* **95**, 24 (1976).
57. H. C. Beyerman, E. Buurman, T. S. Lie, and L. Maat, *Recl. Trav. Chim. Pays-Bas* **95**, 43 (1976).
58. H. C. Beyerman, L. van Bommel, L. Maat, and C. Olieman, *Recl. Trav. Chim. Pays-Bas* **95**, 312 (1976).
59. H. C. Beyerman, J. van Berkel, T. S. Lie, L. Maat, J. S. M. Wessels, H. H. Bosman, E. Buurman, E. J. M. Bijsterveld, and H. J. M. Sinnige, *Recl. Trav. Chim. Pays-Bas* **97**, 127 (1978).
60. T. S. Lie, L. Maat, and H. C. Beyerman, *Recl. Trav. Chim. Pays-Bas* **98**, 419 (1979).
61. A. Brossi, F.-L. Hsu, and K. C. Rice, *J. Org. Chem.* **47**, 5214 (1982).
62. F.-L. Hsu, K. C. Rice, and A. Brossi, *Helv. Chim. Acta,* **63**, 2042 (1980).
63. F.-L. Hsu, A. E. Jacobson, K. C. Rice, and A. Brossi, *Heterocycles* **13**, 259 (1979).
64. F.-L. Hsu, K. C. Rice, and A. Brossi, *Helv. Chim. Acta* **65**, 1576 (1982).
65. A. Brossi, *Trends Pharmacol. Sci.* **3**, 239 (1982).
66. A. Brossi, *Proc. 4th Asian Symp. Med. Plants Spices* (UNESCO Publ.), 261 (1981).
67. M. Gates and M. S. Shepard, *J. Am. Chem. Soc.* **84**, 4125 (1962).
68. Merck and Co., Inc., Neth. Pat. 7,107,921 (Dec. 13, 1971); Br. Pat., 1,330,581 (Sept. 19, 1973).
69. J. I. DeGraw, J. C. Christensen, V. H. Brown, and M. J. Cory, *J. Heterocycl. Chem.* **11**, 363 (1974).
70. K. C. Rice, *J. Org. Chem.* **45**, 3135 (1980).
71. U.S. Pat. Appl., 165,690 (Dec. 19, 1980).
72. D. D. Weller and H. J. Rapoport, *J. Med. Chem.* **19**, 1171 (1976).
73. K. C. Rice, *J. Med. Chem.* **20**, 164 (1977).
74. J. M. Gulland and R. Robinson, *Mem. Proc. Manchester Lit. Philos. Soc.* **69**, 79 (1925).
75. R. Robinson and S. Sugasawa, *J. Chem. Soc.,* 3163 (1931).
76. D. H. R. Barton and T. Cohen, *Festschr. Prof. Dr. Arthur Stoll, Siebzigsten Geburtstag,* 117 (1957).
77. D. H. R. Barton, G. W. Kirby, W. Steglich, and G. M. Thomas, *Proc. Chem. Soc. (London),* 203 (1963).
78. D. H. R. Barton, G. W. Kirby, W. Steglich, G. M. Thomas, A. R. Battersby, T. A. Dobson, and H. Ramuz, *J. Chem. Soc.,* 2423 (1965).
79. D. H. R. Barton, D. S. Bhakuni, R. James, and G. W. Kirby, *J. Chem. Soc. C,* 128 (1967).
80. H. I. Parker, G. Blaschke, and H. Rapoport, *J. Am. Chem. Soc.* **94**, 1276 (1972).
81. K. C. Rice and A. Brossi, *J. Org. Chem.* **45**, 592 (1980).
82. Cs. Szántay, G. Dörnyei, G. Blaskó, M. Bárczai-Beke, and P. Péchy, *Arch. Pharm. (Weinheim, Ger.)* **314**, 983 (1981).
83. Cs. Szántay, G. Blaskó, M. Bárczai-Beke, G. Dörnyei, and P. Péchy, *Planta Med.* **48**, 207 (1983).
84. Cs. Szántay, G. Blaskó, M. Bárczai-Beke, G. Dörnyei, and P. Péchy, *Magy. Kem. Lapja* **40**, 37 (1985).
85. G. Dörnyei, M. Bárczai-Beke, G. Blaskó, P. Péchy, and Cs. Szántay, *Tetrahedron Lett.* **28**, 2913 (1982).
86. T. Kametani and K. Fukumoto, *J. Pharm. Soc. Jpn.* (*Yakugaku Zasshi*) **83**, 1031 (1963).
87. N. H. Martin, S. L. Champion, and P. B. Belt, *Tetrahedron Lett.* **21**, 2613 (1980).
88. Cs. Szántay, G. Blaskó, M. Bárczai-Beke, G. Dörnyei, and L. Radics, *Heterocycles* **14**, 1127 (1980).

89. T. Kametani and K. Fukumoto, *J. Heterocycl. Chem.* **8,** 341 (1971), and references cited therein.
90. W. W.-C. Chan and P. Maitland, *J. Chem. Soc. C,* 753 (1966).
91. A. H. Jackson and J. A. Martin, *J. Chem. Soc. C,* 2061 (1966).
92. T. Kametani, K. Fukumoto, A. Kozuka, H. Yagi, and M. Koizumi, *J. Chem. Soc. C,* 2034 (1969).
93. T. Kametani, A. Kozuka, and K. Fukumoto, *J. Chem. Soc. C,* 1021 (1971).
94. T. Kametani, K. Fukumoto, K. Kigasawa, and K. Wakisaga, *Chem. Pharm. Bull.* **19,** 714 (1971).
95. B. Franck, G. Dunkelman, and H. J. Lubs, *Angew. Chem., Int. Ed. Engl.* **6,** 1075 (1967); *Angew. Chem.* **79,** 1066 (1967).
96. B. Franck, J. Lubs, and G. Dunkelman, *Angew. Chem.* **79,** 989 (1967).
97. M. A. Schwartz, *Synth. Commun.* **3,** 33 (1973).
98. T. Kametani, M. Ihara, and T. Honda, *J. Chem. Soc. C,* 1060 (1970).
99. M. P. Cava and K. T. Buck, *Tetrahedron* **25,** 2795 (1969).
100. T. Kametani, T. Sugahara, and K. Fukumoto, *Tetrahedron* **25,** 3667 (1969).
101. S. M. Kupchan, O. P. Dhingra, and Ch-K. Kim, *J. Org. Chem.* **41,** 4049 (1976).
102. M. A. Schwartz and I. S. Mami, *J. Am. Chem. Soc.* **97,** 1239 (1975).
103. M. A. Schwartz *Proc. 11th IUPAC Int. Symp. Chem. Nat. Prod.* **4,** (Part 2), 274 (1978); *Chem. Abstr.* **92,** 59050 (1980).
104. M. A. Schwartz, U. S. Pat. 4,003,903 (1977); *Chem. Abstr.* **86,** 155848g (1977).
105. M. A. Schwartz and R. A. Wallace, *Tetrahedron Lett.,* 3257 (1979).
106. Cs. Szántay, G. Blaskó, M. Bárczai-Beke, P. Péchy, and G. Dörnyei, *Tetrahedron Lett.* **21,** 3509 (1980).
107. Cs. Szántay, M. Bárczai-Beke, P. Péchy, G. Blaskó, and G. Dörnyei, *J. Org. Chem.* **47,** 594 (1982).
108. G. Blaskó, G. Dörnyei, M. Bárczai-Beke, P. Péchy, and Cs. Szántay, *J. Org. Chem.* **49,** 1439 (1984).
109. J. D. White, G. Caravatti, T. B. Kline, E. Edström, K. C. Rice, and A. Brossi, *Tetrahedron* **39,** 2393 (1983).
110. The versatile application of hypervalent iodine reagents has recently been reviewed; see H. Waldmann, *Nachr. Chem. Tech. Lab.* **39,** 828 (1991), and references cited therein.
111. M. A. Schwartz and M. F. Zoda, *J. Org. Chem.* **46,** 4623 (1981).
112. H. Blumberg and H. B. Dayton, *in* ''Agonist and Antagonist Actions of Narcotic Analgesic Drugs'' (H. W. Kosterlitz, J. O. J. Collier, and J. E. Villareal, eds.), p. 110. Macmillan, London, 1972.
113. E. L. May, *in* ''Psychopharmacological Agents'' (M. Gordon, ed.), Vol. 4, p. 35. Academic Press, New York, 1967.
114. D. G. Vanderlaan and M. A. Schwartz, *J. Org. Chem.* **50,** 743 (1985).
115. T. Kametani, S. Takano, and T. Kobari, *J. Chem. Soc. C,* 1030 (1971).
116. T. Kametani, Y. Ohta, M. Takemura, M. Ihara, and K. Fukumoto, *Heterocycles* **6,** 415 (1977).
117. M. H. Zenk, R. Gerardy, and L. Stadler, *J. Chem. Soc., Chem. Commun.,* 1725 (1989).
118. T. Amann and M. H. Zenk, *Tetrahedron Lett.* **32,** 3675 (1991).
119. L. L. Miller and R. F. Stewart, *J. Org. Chem.* **43,** 1589 (1978), and references cited therein.
120. J. M. Bobbit, I. Noguchi, R. S. Ware, K. Ng. Chiong, and S. J. Huang, *J. Org. Chem.* **40,** 2924 (1975), and references cited therein.
121. E. Kotani and S. Tobinaga, *Tetrahedron Lett.,* 4759 (1973).
122. L. L. Miller, F. R. Stermitz, J. Y. Becker, and V. Ramachandran, *J. Am. Chem. Soc.* **97,** 2922 (1975).

123. T. Kametani, M. Koizumi, and K. Fukumoto, *Chem. Pharm. Bull.* **17,** 2245 (1969).
124. T. Kametani, T. Sugahara, H. Yagi, and K. Fukumoto, *J. Chem. Soc. C,* 1063 (1969).
125. T. Kametani, K. Fukumoto, and T. Sugahara, *J. Chem. Soc. C,* 801 (1969).
126. T. Kametani, K. Fukumoto, F. Satoh, and H. Yagi, *J. Chem. Soc. C,* 3084 (1968).
127. T. Kametani, M. Ihara, K. Fukumoto, and H. Yagi, *J. Chem. Soc. C,* 2030 (1969).
128. D. R. Elmaleh, F. E. Granchelli, and J. L. Neumeyer, *J. Heterocycl. Chem.* **16,** 87 (1979).
129. T. Kametani, T. Sugahara, and K. Fukumoto, *Chem. Pharm. Bull.* **22,** 966 (1974).
130. T. Kametani, K. Fukumoto, S. Shibuya, H. Neomoto, T. Nakano, T. Shugahara, T. Takahashi, Y. Aizawa, and M. Toriyama, *J. Chem. Soc. Perkin Trans. 1,* 1435 (1972).
131. L. Stella, B. Raynier, and J. Sufzur, *Tetrahedron Lett.,* 2721 (1977).
132. R. D. Gless and H. Rapoport, *J. Org. Chem.* **44,** 1324 (1979).
133. W. H. Moos, R. D. Gless, and H. Rapoport, *J. Org. Chem.* **48,** 227 (1983).
134. D. A. Evans, C. H. Mitch, R. C. Thomas, D. M. Zimmermann, and R. L. Robey, *J. Am. Chem. Soc.* **102,** 5955 (1980).
135. D. A. Evans and C. H. Mitch, *Tetrahedron Lett.* **23,** 285 (1982).
136. J. E. McMurry, V. Farina, W. J. Scott, A. H. Davidson, D. R. Summers, and A. Shenvi, *J. Org. Chem.* **49,** 3803 (1984).
137. D. D. Weller and H. Rapoport, *J. Am. Chem. Soc.* **98,** 6650 (1976).
138. D. D. Weller, R. D. Gless, and H. Rapoport, *J. Org. Chem.* **42,** 1485 (1977).
139. W. H. Moos, R. D. Gless, and H. Rapoport, *J. Org. Chem.* **46,** 5064 (1981).
140. W. H. Moos, R. D. Gless, H. Rapoport, *J. Org. Chem.* **47,** 1831 (1982).
141. A. G. Schultz, R. D. Lucci, J. J. Napier, H. Kinoshita, R. Ravichandran, P. Shannon, and Y. K. Yee, *J. Org. Chem.* **50,** 217 (1985).
142. E. Ciganek, *J. Am. Chem. Soc.* **103,** 6261 (1981).
143. D. D. Weller and D. L. Weller, *Tetrahedron Lett.* **23,** 5239 (1982).
144. D. D. Weller, E. P. Stirchak, and D. L. Weller, *J. Org. Chem.* **48,** 4597 (1983).
145. J. E. Toth, P. R. Hamman, and P. L. Fuchs, *J. Org. Chem.* **53,** 4694 (1988).
146. R. B. Barber and H. Rapoport, *J. Med. Chem.* **19,** 1175 (1976).
147. L. F. Small and G. L. Browning, *J. Org. Chem.* **3,** 618 (1939).
148. M. Gates and G. M. K. Hughes, *Chem. Ind. (London),* 1506 (1956).
149. M. Gates and M. S. Sheppard, *J. Am. Chem. Soc.* **84,** 4125 (1962).
150. C. Schöpf and T. Pfeifer, *Justus Liebigs Ann. Chem.* **483,** 157 (1930).
151. C. Schöpf, T. Pfeifer, and H. Hirsch, *Justus Liebigs Ann. Chem.* **492,** 213 (1932).
152. K. Goto and I. Yamamoto, *Proc. Jpn. Acad.* **32,** 45 (1956).
153. K. Goto and I. Yamamoto, *Proc. Jpn. Acad.* **36,** 145 (1960).
154. K. Goto, I. Yamamoto, and T. Yamazaki, *Proc. Jpn. Acad.* **36,** 282 (1960).
155. C. Olieman, L. Maat, and H. C. Beyerman, *Recl. Trav. Chim. Pays-Bas* **97,** 31 (1978).
156. I. Iijima and K. C. Rice, *Heterocycles* **6,** 1157 (1977).
157. G. A. Brine, K. G. Boldt, M. L. Coleman, D. J. Bradley, and F. I. Carrol, *J. Org. Chem.* **43,** 1555 (1978).
158. N. Chatterjie, J. G. Umans, C. E. Inturrisi, W. T. C. Chen, D. D. Clarke, S. P. Bhatnagar, and U. Weiss, *J. Org. Chem.* **43,** 1003 (1978).
159. J. A. Lawson and J. I. DeGraw, *J. Med. Chem.* **20,** 165 (1977).
160. D. M. Smirnov, E. L. Sigal, K. Ya. Marechek, and V. P. Zakharov, *Khim. Farm. Zh.* **6,** 31 (1972); *Chem. Abstr.* **77,** 101,952g (1972).
161. H. Lotter, J. Gollwitzer, and M. H. Zenk, *Tetrahedron Lett.* **33,** 2443 (1992).
162. P. Sohar and E. F. Schönewaldt (Merck and Co., Inc.), Ger. Offen. 2,518,519; *Chem. Abstr.* **86,** 55618p (1977).

163. Merck and Co., Inc., Neth. Appl. 75 05, 109; *Chem. Abstr.* **86,** 155,849h (1977).
164. F. L. Bjeldanes and H. Rapoport, *J. Org. Chem.* **37,** 1453 (1972).
165. W. Fleischhacker, F. Vieböck, and F. Zeidler, *Monatsch. Chem.* **101,** 1215 (1970).
166. D. E. Rearick and M. Gates, *Tetrahedron Lett.,* 507 (1970).
167. H. Conroy, *J. Am. Chem. Soc.* **77,** 5960 (1955).
168. G. Horváth and S. Makleit, *Acta Chim. Acad. Sci. Hung.* **106,** 37 (1981).
169. V. V. Kiselev and R. A. Konovalova, *J. Gen. Chem. USSR (Engl. Transl.)* **18,** 142 and 855 (1948).
170. D. Neubauer, *Arch. Pharm. (Weinheim, Ger.)* **298,** 47 (1965).
171. N. Sharglii and L. Lalezari, *Nature (London)* **213,** 1244 (1967).
172. M. D. Gates, Jr., U.S. Pat. 2,773,832 (1957).
173. M. D. Gates, *J. Am. Chem. Soc.* **75,** 4340 (1953).
174. F. Krausz, U.S. Pat. 3,112,323 (1963).
175. J. P. Gavard, F. Krausz, and T. Rüll, *Bull. Soc. Chim. Fr.,* 486 (1965).
176. Fabr. de Prod. Quim. y Farm. Abello S. A., Belg. 839,732 (1976).
177. Fabr. de Prod. Quim. y Farm Abello S. A., Fr. Demande 2,343,288 (1977).
178. F. Calvo (Fabr. de Prod. Quim. y Farm. Abello S. A.), U.S. Pat. 4,052,402 (1977).
179. W. G. Dauben, C. P. Baskin, and H. C. H. A. van Riel, *J. Org. Chem.* **44,** 1567 (1979).
180. W. G. Dauben, C. P. Baskin, and H. C. H. A. van Riel, U.S. Pat. Appl. 115,411 (1980).
181. C. C. Hodges and H. Rapoport, *Phytochemistry* **19,** 1681 (1980).
182. R. B. Barber and H. Rapoport, *J. Med. Chem.* **18,** 1074 (1975).
183. R. K. Razdan, P. Herlihy, H. C. Dalzell, and D. E. Portlock, *J. Org. Chem.* **44,** 3730 (1979).
184. M. J. Lewenstein, Br. Pat. 995,493 (1964).
185. H. Blumberg, U.S. Pat. 3,332,950 (1967); *Chem. Abstr.* **67,** 100,301a (1967).
186. Sankyo Co., Belg. Pat. 615,009 (1962); *Chem. Abstr.* **57,** 15171c (1962).
187. I. Seki, *Takamine Kenkyusho Nempo* **12,** 56 (1960); *Chem. Abstr.* **55,** 8449b (1961).
188. M. Freund and E. Speyer, *J. Prakt. Chem.* **94,** 135 (1916).
189. I. Iijima, J. Minamikawa, A. E. Jacobson, A. Brossi, K. C. Rice, and W. Klee, *J. Med. Chem.* **21,** 398 (1978).
190. F. M. Hauser, T.-K. Chen, and F. I. Carroll, *J. Med. Chem.* **17,** 117 (1974).
191. I. Iijima, K. C. Rice, and A. Brossi, *Helv. Chim. Acta* **60,** 2135 (1977).
192. R. A. Olofson, R. C. Schnur, L. Bunes, and J. P. Pepe, *Tetrahedron Lett.* **18,** 1567 (1977).
193. Some yields are missing for the key steps in the published data (*185*).
194. R. A. Olofson and J. P. Pepe, *Tetrahedron Lett.* **18,** 1575 (1977).
195. I. Seki, *Chem. Pharm. Bull.* **18,** 671 (1970).
196. M. A. Schwartz and R. A. Wallace, *J. Med. Chem.* **24,** 1525 (1981).
197. K. C. Rice and E. L. May, *J. Heterocycl. Chem.* **14,** 665 (1977).
198. *N*-Ethoxycarbonylnorcodeinone (**188**) has been previously prepared by a very similar procedure; instead of MnO_2, Jones reagent (chromic acid–acetone) was used as oxidant [J. A. Lawson, J. I. DeGraw, and M. Anbar, *J. Heterocycl. Chem.* **13,** 593 (1976)].
199. H. Schmidhammer, A. E. Jacobson, L. Atwell, and A. Brossi, *Helv. Chim. Acta* **64,** 253 (1981).
200. J. Reden, M. F. Reich, K. C. Rice, A. E. Jacobson, A. Brossi, R. A. Streaty, and W. A. Klee, *J. Med. Chem.* **22,** 256 (1979).
201. I. Iijima, K. C. Rice, and J. V. Silvertone, *Heterocycles* **6,** 1157 (1977).
202. F.-L. Hsu, A. E. Jacobson, K. C. Rice, and A. Brossi, *Heterocycles* **13,** 259 (1979).
203. M. D. Rozwadowska, F.-L. Hsu, A. E. Jacobson, K. C. Rice, and A. Brossi, *Can. J. Chem.* **58,** 1855 (1980).

204. F.-L. Hsu, K. C. Rice, and A. Brossi, *Helv. Chim. Acta* **63,** 2042 (1980).
205. A. E. Jacobson, F.-L. Hsu, M. D. Rozwadowska, H. Schmidhammer, L. Atwell, A. Brossi, and F. Medzihradsky, *Helv. Chim. Acta* **64,** 1298 (1981).
206. H. Schmidhammer, A. E. Jacobson, L. Atwell, and A. Brossi, *Heterocycles* **16,** 1859 (1981).
207. R. Bognár, Gy. Gaál, P. Kerekes, G. Horváth, and M. T. Kovács, *Org. Prep. Proced. Int.* **6,** 305 (1974).
208. Y. K. Sawa and H. Tada, *Tetrahedron* **24,** 6185 (1968).
209. I. Monkovic, H. Wong, A. W. Pircio, Y. G. Perron, I. J. Pachter, and B. Belleau, *Can. J. Chem.* **53,** 3094 (1975).
210. I. Monkovic, T. T. Conway, H. Wong, Y. G. Perron, I. J. Pachter, and B. Belleau, *J. Am. Chem. Soc.* **95,** 7910 (1973).
211. S. Makleit, J. Knoll, R. Bognár, S. Berényi, G. Somogyi, and G. Kiss, *Acta Chim. Acad. Sci. Hung.* **93,** 169 (1977).
212. S. Makleit, S. Berényi, R. Bognár, and S. Elek, *Acta Chim. Acad. Sci. Hung.* **93,** 161 (1977).
213. S. Berényi, S. Makleit, S. Hosztafi, Zs. Fürst, T. Friedmann, and J. Knoll, *Med. Chem. Res.,* 185 (1991).
214. K. Goto, "Sinomenine: An Optical Antipode of Morphine Alkaloids," p. 96. Kitasato Institute, Tokyo, 1964.
215. K. Goto and I. Yamamoto, *Proc. Jpn. Acad.* **30,** 769 (1954).
216. K. Goto and I. Yamamoto, *Proc. Jpn. Acad.* **33,** 477 (1957).
217. K. Goto and I. Yamamoto, *Proc. Jpn. Acad.* **34,** 60 (1958).
218. I. Iijima, J. Minamikawa, A. E. Jacobson, A. Brossi, and K. C. Rice, *J. Org. Chem.* **43,** 1462 (1978).
219. S. Okuda, S. Yamaguchi, and K. Tsuda, *Chem. Pharm. Bull.* **13,** 1092 (1965).
220. K. Abe, M. Onda, and S. Okuda, *Chem. Pharm. Bull.* **17,** 1847 (1969).
221. S. W. Wunderly and E. Brochmann-Hanssen, *J. Org. Chem.* **42,** 4277 (1977).
222. S. Makleit, S. Berényi and R. Bognár, *Acta Chim. Acad. Sci. Hung. (Engl.)* **94,** 165 (1977); *Chem. Abstr.* **83,** 170,362d (1978).
223. W. Fleischhacker and H. Markut, *Monatsh. Chem.* **102,** 587 (1971).
224. S. Berényi and S. Makleit, *Acta Chim. Acad. Sci. Hung. (Engl.)* **104,** 97 (1980); *Chem. Abstr.* **94,** 121,755 (1981).
225. H. Inoue, M. Takeda, and H. Kugita, *Chem. Pharm. Bull.* **18,** 1569 (1970).
226. H. Kugita and M. Takeda, *Chem. Pharm. Bull. Tokyo* **13,** 1422 (1965).
227. M. Takeda, H. Inoue, and H. Kugita, *Tetrahedron* **25,** 1851 (1969).
228. H. Kugita, M. Takeda, and H. Inoue, *Tetrahedron* **25,** 1839 (1969).
229. S. Berényi, S. Makleit, R. Bognár, and A. Tegdes, *Acta Chim. Acad. Sci. Hung. (Engl.)* **103,** 365 (1980); *Chem. Abstr.* **94,** 47,564 (1981).
230. S. Berényi, S. Hosztafi, S. Makleit, and I. Szeifert, *Acta Chim. Acad. Sci. Hung. (Engl.)* **110,** 363 (1982); *Chem. Abstr.* **98,** 198,490 (1983).
231. P. R. Crabbendam, L. Maat, and H. C. Beyerman, *Recl. Trav. Chim. Pays-Bas* **100,** 293 (1981).
232. C. H. Hutchins, G. K. Cooper, S. Pürro, and H. Rapoport, *J. Med. Chem.* **24,** 773 (1981).
233. K. W. Bentley, "The Chemistry of Morphine Alkaloids," p. 309. Oxford Univ. Press (Clarendon), Oxford, 1954.
234. S. Berényi, S. Hosztafi, S. Makleit, and I. Molnár *Acta Chim. Acad. Sci. Hung.* **113,** 51 (1983); *Chem. Abstr.* **99,** 140,197 (1983).
235. G. Stork, *in* "The Alkaloids" (R. H. F. Manske and H. L. Holmes, eds.), Vol. 2, p. 195. Academic Press, New York, 1952.

236. R. R. Lorenz, E. D. Parady, and W. H. Thielking (Sterling Drug Inc.) Ger. Offen. 2,758,954 (1977); *Chem. Abstr.* **89,** 834 (1978).
237. F. E. Granchelli, C. N. Filer, A. H. Soloway, and J. L. Neumeyer, *J. Org. Chem.* **45,** 2275 (1980).
238. C. N. Filer, D. Ahern, F. E. Granchelli, J. L. Neumeyer, and S.-J. Law, *J. Org. Chem.* **45,** 3465 (1980).
239. M. V. Koch, J. G. Cannon, and A. M. Burkman, *J. Med. Chem.* **11,** 977 (1968).
240. S. Berényi, S. Makleit, and F. Rantal, *Acta Chim. Acad. Sci. Hung.* **120,** 201 (1985).
241. Cs. Simon, S. Hosztafi, S. Makleit, and S. Berényi, *Synth. Commun.* **21,** 2309 (1991).
242. S. Berényi, S. Hosztafi, and S. Makleit, *J. Chem. Soc., Perkin Trans. 1,* 2693 (1992).
243. J. von Braun, *Chem. Ber.* **47,** 2312 (1914).
244. H. A. Hageman, *Org. React.* **7,** 198 (1953).
245. K. W. Bentley and D. G. Hardy, *J. Am. Chem. Soc.* **89,** 3281 (1967).
246. A. C. Currie, G. T. Newbold, and F. S. Spring, *J. Chem. Soc.,* 4693 (1961).
247. E. Mohácsi, *J. Heterocycl. Chem.* **24,** 471 (1987).
248. J. D. Hobson and J. G. McCluskey, *J. Chem. Soc.,* 2015 (1967).
249. M. M. Abdel-Monem and P. S. Portoghese, *J. Med. Chem.* **15,** 208 (1972).
250. K. C. Rice, *J. Org. Chem.* **40,** 1850 (1975).
251. G. A. Brine, K. G. Boldt, C. K. Hart, and F. I. Carroll, *Org. Prep. Proced. Int.* **8,** 103 (1976).
252. T. A. Montzka, J. D. Matiskella, and R. A. Partyka, *Tetrahedron Lett.,* 1325 (1974).
253. R. A. Olofson and R. C. Schnur, *Tetrahedron Lett.,* 1571 (1977).
254. R. A. Olofson, J. T. Martz, J. O. Senet, M. Piteau, and T. Malfroot, *J. Org. Chem.* **49,** 2081 (1984).
255. E. Speyer and H. Rosenfeld, *Ber.* **58,** 1125 (1925).
256. L. F. Small and J. E. Mallonee, *J. Org. Chem.* **5,** 350 (1940).
257. C. Bertgen, W. Fleischhacker, and F. Vieböck, *Chem. Ber.* **100,** 2992 (1967).
258. C. Bertgen, W. Fleischhacker, and F. Vieböck, *Chem. Ber.* **100,** 3002 (1967).
259. D. U. Lee and W. Wiegrebe, *Arch. Pharm. (Weinheim, Ger.)* **319,** 694 (1986).
260. A. J. Birch and H. Fitton, *Aust. J. Chem.* **22,** 971 (1969).
261. A. J. Birch, L. F. Kelly, and A. J. Liepa, *Tetrahedron Lett.* **26,** 501 (1985).
262. J. E. Tóth and P. L. Fuchs, *J. Org. Chem.* **51,** 2594 (1986).
263. G. Heinisch and F. Vieböck, *Monatsh. Chem.* **101,** 1253 (1970).
264. S. Hosztafi, *Sci. Pharm.* **55,** 61 (1987).
265. A. Pohland and H. R. Sullivan, Jr., U.S. Pat. 3,342,824 (Sept. 19, 1967); *Chem. Abstr.* **65,** 15441d (1966).
266. Eli Lilly and Co., Neth. Pat. Appl., 6,515,815 (1966); *Chem. Abstr.* **65,** 15,441 (1966).
267. H. Merz and K.-H. Pook, *Tetrahedron* **26,** 1727 (1970).
268. L. S. Schwab, *J. Med. Chem.* **23,** 689 (1980).
269. S. Hosztafi, S. Makleit, and R. Bognár, *Acta Chim. Acad. Sci. Hung.* **103,** 371 (1980).
270. L. Maat, J. A. Peters, and M. A. Prazeres, *Recl. Trav. Chim. Pays-Bas* **103,** 296 (1984).
271. S. Hosztafi, T. Timár, and S. Makleit, *Magy. Kém. Folyóirat* **91,** 126 (1985).
272. L. Maat, J. A. Peters, and M. A. Prazeres, *Recl. Trav. Chim. Pays-Bas* **104,** 205 (1985).
273. W. Fleischhacker and B. Richter, *Sci. Pharm.* **57,** 351 (1989).
274. J. T. Linders, P. Briel, E. Fog, T. S. Lie, and L. Maat, *Recl. Trav. Chim. Pays-Bas* **108,** 268 (1989).
275. M. A. Prazeres, J. A. Peters, J. T. M. Linders, and L. Maat, *Recl. Trav. Chim. Pays-Bas* **105,** 554 (1986).
276. J. H. E. Lindner, H. J. Kuhn, and K. Gollnick, *Tetrahedron Lett.,* 1705 (1972).
277. T. S. Manoharan, K. M. Madyastha, B. Bali Singh, S. P. Bhatnagar, and U. Weiss, *Synthesis,* 809 (1983).

278. T. S. Manoharan, K. M. Madyastha, B. B. Singh, S. P. Bhatnagar, and U. Weiss, *Indian J. Chem.* **23/B,** 5 (1984).
279. Cs. Simon, S. Hosztafi, and S. Makleit, *Synth. Commun.* **21,** 407 (1991).
280. S. Hosztafi, Cs. Simon, and S. Makleit, *Synth. Commun.* **22,** 1673 (1992).
281. S. Hosztafi, Cs. Simon, and S. Makleit, *Synth. Commun.* **22,** 2527 (1992).
282a. S. Hosztafi, S. Berényi, G. Tóth, and S. Makleit, *Monatsh. Chem.* **123,** 435 (1992).
282b. S. Hosztafi, Cs. Simon, and S. Makleit, *Heterocycles,* **36,** 1509 (1993).
283. R. Bognár and S. Makleit, *Acta Chim. Acad. Sci. Hung.* **59,** 373 (1969).
284. R. Bognár, S. Makleit, and T. Mile, *Acta Chim. Acad. Sci. Hung.* **59,** 379 (1969).
285. R. Bognár, S. Makleit, T. Mile, and R. Radics, *Acta Chim. Acad. Sci. Hung.* **64,** 273 (1970).
286. R. Bognár, S. Makleit, and L. Radics, *Acta Chim. Acad. Sci. Hung.* **67,** 63 (1971), and references cited therein.
287. S. Makleit, T. Mile, and R. Bognár, *Acta Chim. Acad. Sci. Hung.* **89,** 275 (1976).
288. G. Somogyi, S. Makleit, and R. Bognár, *Acta Chim. Acad. Sci. Hung.* **97,** 339 (1978).
289. R. Bognár, S. Makleit, T. Mile, and L. Radics, *Monatsh. Chem.* **103,** 143 (1972).
290. S. Makleit and R. Bognár, *Acta Chim. Acad. Sci. Hung.* **64,** 281 (1970).
291. W. Fleischhacker and B. Richter, *Chem. Ber.* **113,** 3866 (1980).
292. G. W. Kirby and S. R. Massey, *J. Chem. Soc. C,* 3047 (1971).
293. I. Fujii, H. Togame, M. Yamamoto, K. Kanematsu, I. Takayanagi, and F. Konno, *Chem. Pharm. Bull.* **36,** 2282 (1988).
294. K. Kanematsu, R. Naito, Y. Simohigashi, M. Ohno, T. Ogasawara, M. Kurono, and K. Yagi, *Chem. Pharm. Bull.* **38,** 1438 (1990).
295. A. C. Currie, J. Gillon, G. T. Newbold, and F. S. Spring, *J. Chem. Soc.,* 773 (1960).
296. I. Seki, *Ann. Senkyo Res. Lab.* **17,** 1 (1965).
297. S. Makleit, L. Radics, R. Bognár, T. Mile, and É Oláh, *Acta Chim. Acad. Sci. Hung.* **74,** 99 (1972).
298. S. Makleit, L. Radics, R. Bognár, and T. Mile, *Acta Chim. Acad. Sci. Hung.* **74,** 111 (1972).
299. S. Makleit, J. Knoll, R. Bognár, S. Berényi, and G. Kiss, *Acta Chim. Acad. Sci. Hung.* **93,** 165 (1977).
300. S. Makleit, J. Knoll, R. Bognár, S. Berényi, G. Somogyi, and G. Kiss, *Acta Chim. Acad. Sci. Hung.* **93,** 169 (1977).
301. S. Makleit, J. Knoll, R. Bognár, S. Berényi, G. Somogyi, and G. Kiss, *Acta Chim. Acad. Sci. Hung.* **93,** 175 (1977).
302. R. Bognár, S. Makleit, J. Knoll, S. Berényi, and G. Horváth, *Commun. Dep. Chem. Bulg. Acad. Sci. (Izv. Khim.)* **8,** 203 (1975).
303. S. Makleit, S. Berényi, and R. Bognár, *Acta Chim. Acad. Sci. Hung.* **88,** 409 (1976).
304. S. Makleit, G. Somogyi, and R. Bognár, *Acta Chim. Acad. Sci. Hung.* **89,** 173 (1976).
305. S. Berényi, S. Makleit, and L. Szilágyi, *Acta Chim. Acad. Sci. Hung.* **117,** 307 (1984).
306. S. Berényi, S. Makleit, and F. Rantal, *Acta Chim. Acad. Sci. Hung.* **120,** 171 (1985).
307. S. Berényi, S. Makleit, and Á Sepsi, *Acta Chim. Acad. Sci. Hung.* **126,** 275 (1989).
308. R. M. Boden, Dissertation, Univ. of Rochester, New York (1979).
309. R. M. Boden, M. Gates, S. P. Ho, and P. Sundararaman, *J. Org. Chem.* **47,** 1347 (1982).
310. M. Gates, R. M. Boden, and P. Sundararaman, *J. Org. Chem.* **54,** 972 (1989).
311. D. L. Leland, J. O. Polazzi, and M. P. Kotick, *J. Org. Chem.* **45,** 4026 (1981).
312. D. L. Leland and M. P. Kotick, *J. Med. Chem.* **23,** 1427 (1980).
313. M. P. Kotick, D. L. Leland, J. O. Polazzi, J. F. Howes, and A. R. Bousquet, *J. Med. Chem.* **24,** 1445 (1981).

314. K. W. Bentley, *"The Chemistry of Morphine Alkaloids,"* p. 188. Oxford Univ. Press (Clarendon), Oxford, 1954, and references cited therein.
315. R. M. Allen, C. J. Gilmore, G. W. Kirby, and D. J. McDougall, *J. Chem. Soc., Chem. Commun.,* 22 (1980).
316. R. M. Allen, G. W. Kirby, and D. J. McDougall, *J. Chem. Soc. Perkin Trans. 1,* 1143 (1981).
317. S. Archer and P. Osei-Gyimah, *J. Heterocycl. Chem.* **16,** 389 (1979).
318. P. Osei-Gyimah and S. Archer, *J. Med. Chem.* **23,** 162 (1980).
319. P. Horsewood and G. W. Kirby, *Chem. Commun.,* 1139 (1971).
320. K. W. Bentley, G. W. Kirby, A. P. Price, and S. Singh, *Chem. Commun.,* 57 (1969).
321. R. M. Allen and G. W. Kirby, *Chem. Commun.,* 1346 (1970).
322. R. M. Allen and G. W. Kirby, *Chem. Commun.,* 1121 (1971).
323. R. K. Razdan, D. E. Portlock, H. C. Dalzell, and C. Malmberg, *J. Org. Chem.* **43,** 3604 (1978).
324. D. E. Portlock (Sharps Assoc.), Ger. Offen. 2,647,642 (1977); *Chem. Abstr.* **87,** 1025086 (1977).
325. H. B. Arzeno, D. H. R. Barton, S. G. Davies, X. Lusinchi, B. Meunier, and C. Pascard, *Nouv. J. Chim.* **4,** 369 (1980).
326. N. Mathews and M. Sainsbury, *J. Chem. Res. Suppl.* 82 (1988).
327. L. J. Sargent and A. E. Jacobson, *J. Med. Chem.* **15,** 843 (1972).
328. G. Horváth, P. Kerekes, Gy. Gaál, and R. Bognár, *Acta Chim. Acad. Sci. Hung.* **82,** 217 (1974).
329. L. F. Small and H. Rapoport, *J. Org. Chem.* **12,** 284 (1947).
330. S. P. Findlay and L. F. Small, *J. Am. Chem. Soc.* **72,** 3249 (1950).
331. P. Kerekes, R. Bognár, Gy. Gaál and G. Horváth, *Acta Chim. Acad. Sci. Hung.* **82,** 211 (1974).
332. M. P. Kotick, *J. Med. Chem.* **24,** 722 (1981).
333. M. P. Kotick, D. L. Leland, J. O. Polazzi, and R. N. Schut, *J. Med. Chem.* **23,** 166 (1980).
334. D. L. Leland, J. O. Polazzi, and M. P. Kotick, *J. Org. Chem.* **46,** 4012 (1981).
335. D. L. Leland and M. P. Kotick, *J. Med. Chem.* **24,** 717 (1981).
336. H. Rapoport and L. Small, *J. Org. Chem.* **12,** 834 (1947).
337. J. O. Polazzi, *J. Org. Chem.* **46,** 4262 (1981).
338. C. Olieman, L. Maat, and H. C. Beyerman, *Recl. Trav. Chim. Pays-Bas* **95,** 189 (1976).
339. H. Auterhoff and W. Tittjung, *Arch. Pharm. (Weinheim, Ger.)* **313,** 985 (1980).
340. N. Chatterjie, C. E. Inturrisi, H. B. Dayton, and H. Blumberg, *J. Med. Chem.* **18,** 490 (1975).
341. N. Chatterjie, J. G. Umans, and C. E. Inturrisi, *J. Org. Chem.* **41,** 3624 (1976).
342. N. Chatterjie, A. Minar, and D. D. Clarke, *Synth. Commun.* **9,** 647 (1979).
343. O. Mitsunobu, *Synthesis,* 1 (1981).
344. Cs. Simon, S. Hosztafi, and S. Makleit, *Synth. Commun.* **22,** 913 (1992).
345. W. Fleischhacker and B. Richter, *Chem. Ber.* **113,** 3866 (1980).
346. B. Berrang, C. E. Twine, G. L. Hennessee, and F. I. Carroll, *Synth. Commun.* **5,** 231 (1975).
347. J. W. Lewis and M. J. Readhead, *J. Med. Chem.* **13,** 525 (1970).
348. R. Bognár, Gy. Gaál, P. Kerekes, G. Horváth, and M. Kovács, *Org. Prep. Proced. Int.* **6,** 305 (1974).
349. I. Seki, *Chem. Pharm. Bull.* **18,** 1269 (1970).
350. R. Bognár, S. Makleit, L. Radics, and I. Seki, *Org. Prep. Proced. Int.* **5,** 49 (1973).
351. W. Fleischhacker, *Monatsh. Chem.* **102,** 558 (1971).

352. W. Fleischhacker and H. Markut, *Monatsh. Chem.* **102,** 569 (1971).
353. W. Fleischhacker and B. Richter, *Chem. Ber.* **112,** 2539 (1979).
354. W. Fleischhacker and B. Richter, *Chem. Ber.* **112,** 3054 (1979).
355. G. Heinisch, V. Klintz, and F. Vieböck, *Monatsh. Chem.* **102,** 530 (1971).
356. W. Fleischhacker, H. Markut, and F. Vieböck, *Monatsch. Chem.* **103,** 1066 (1972).
357. K. Abe, Y. Nakamura, M. Onda, and S. Okuda, *Tetrahedron* **27,** 4495 (1971).
358. K. Abe, M. Onda, and S. Okuda, *J. Chem. Soc., Perkin Trans. 1,* 316 (1973).
359. A. F. Casy and R. T. Parfitt, "Opioid Analgesics." Plenum, New York, 1986.
360. R. Gottschlich, *Kontakte,* 3 (1990).
361. D. M. Zimmerman and J. D. Leander, *J. Med. Chem.* **33,** 895 (1990).
362. E. J. Simon, *Med. Res. Rev.* **11,** 357 (1991).
363. H. W. Kosterlitz, T. W. Smith, J. Hughes, L. A. Fothergill, B. A. Morgan, and H. R. Morris, *Nature (London)* **258,** 577 (1975).
364. J. S. Morley, *Annu. Rev. Pharmacol. Toxicol.* **20,** 81 (1981).
365. A. Brossi, J. Reden, M. F. Reich, K. C. Rice, and A. E. Jacobson, *J. Med. Chem.* **22,** 256 (1979).
366. P. A. Crooks, T. Deeks, and R. D. Waigh, *J. Med. Chem.* **26,** 762 (1983).
367. P. S. Portoghese and K. Ramakrishnan, *J. Med. Chem.* **25,** 1423 (1982).
368. P. S. Portoghese, A. W. Lipowski, and S. W. Tam, *J. Med. Chem.* **29,** 1222 (1986).
369. W. R. Martin, C. G. Eades, J. A. Thompson, R. E. Huppler and P. E. Gilbert, *J. Pharmacol. Exp. Ther.* **197,** 517 (1976).
370. A. Goldstein, S. Tachibana, L. I. Lowney, M. Hunkapiller, and L. Hood, *Proc. Natl. Acad. Sci. U.S.A.* **76,** 6666 (1979).
371. A. L. Misra, N. L. Vadlamani, and R. B. Pontani, *J. Pharm. Pharmacol.* **30,** 187 (1978).
372. P. S. Portoghese, *Trends Pharmacol. Sci.* **10,** 230 (1989).
373. P. S. Portoghese, M. Sultana, H. Nagase, and A. E. Takemori, *J. Med. Chem.* **31,** 281 (1988).
374. P. S. Portoghese, H. Sultana, W. L. Nelson, P. Klein, and A. E. Takemori, *J. Med. Chem.* **35,** 4086 (1992).
375. H. I. Mosberg and H. B. Kroona, *J. Med. Chem.* **35,** 4498 (1992).
376. R. Stadler, T. M. Kutchan, S. Löffler, N. Nagakura, B. K. Cassels, and M. H. Zenk, *Tetrahedron Lett.* **28,** 1251 (1987).
377. S. Löffler, R. Stadler, N. Nagakura, and M. H. Zenk, *J. Chem. Soc., Chem. Commun.,* 1160 (1987).
378. R. Stadler, T. M. Kutchan, and M. H. Zenk, *Phytochemistry* **27,** 2557 (1988).
379. R. Stadler and M. H. Zenk, *Liebigs Ann. Chem.,* 555 (1990).
380. M. Rüffer, N. Nagakura, and M. H. Zenk, *Planta Med.* **49,** 131 (1983).
381. C.-K. Wat, P. Steffens, and M. H. Zenk, *Z. Naturforsch. C: Biosci.* **41,** 126 (1986).
382. S. Löffler, Dissertation, Univ. of Munich (1988).
383. T. Frenzel, Dissertation, Univ. of Munich (1987).
384. A. R. Battersby, D. M. Foulkes, and R. Binks, *J. Chem. Soc.,* 3323 (1965).
385. X. He and A. Brossi, *J. Nat. Prod.,* **56,** 973 (1993).
386. H. Flentje, W. Döpke and P. W. Jeffs, *Naturwissenschaften* **52,** 259 (1965).
387. J. Gollwitzer, H. Lotter, W. Döpke, and M. H. Zenk, *Nat. Prod. Lett.* **2,** 197 (1993).
388. D. H. R. Barton, R. James, G. W. Kirby, W. Döpke, and H. Flentje, *Chem. Ber.* **100,** 2457 (1967).
389. J. Donnerer, K. Oka, A. Brossi, K. C. Rice, and S. Spector, *Proc. Natl. Acad. Sci. U.S.A.* **83,** 4566 (1986).
390. A. Brossi, *Med. Res. Rev.* **12,** 1 (1992).
391. Ch. Y. Hong, N. Kado, and L. E. Overman, *J. Am. Chem. Soc.* **115,** 11028 (1993).

Chapter 3

LYCOPODIUM ALKALOIDS

William A. Ayer and Latchezar S. Trifonov

Department of Chemistry
University of Alberta
Edmonton, Alberta, Canada T6G 2G2

I. Introduction

The last review of the *Lycopodium* alkaloids published in this treatise covered the literature until 1985 (*1*). This review is intended to cover the developments in the field since that time and until the spring of 1993. A brief review covering the period 1986–1990 has appeared (*2*). The period is highlighted by the discovery that some *Lycopodium* alkaloids are potent inhibitors of acetylcholinesterase (*3*). Huperzine A (**1**) is reported to increase efficiency for learning and memory in animals, and it shows promise in the treatment of Alzheimer's disease and myasthenia gravis (*4,5*). This alkaloid has provided the focus for a good portion of the synthetic work in the period under review. The biosynthesis of the alkaloids is still not completely understood, and limited biosynthetic studies have been reported in the last few years. Plants of the genus *Lycopodium* have not been cultivated, and labeling experiments must be carried out in the field. Because the club mosses often are not easily accessible, very few feeding studies have been conducted. This is an area where plant tissue culture may prove extremely useful in future biosynthetic studies.

In this review, the alkaloids are divided into four groups: (i) the lycopodine group, those possessing the "lycopodane" skeleton (**2**); (ii) the lycodine group, the dinitrogenous alkaloids which contain a pyridine

233

THE ALKALOIDS, VOL. 45

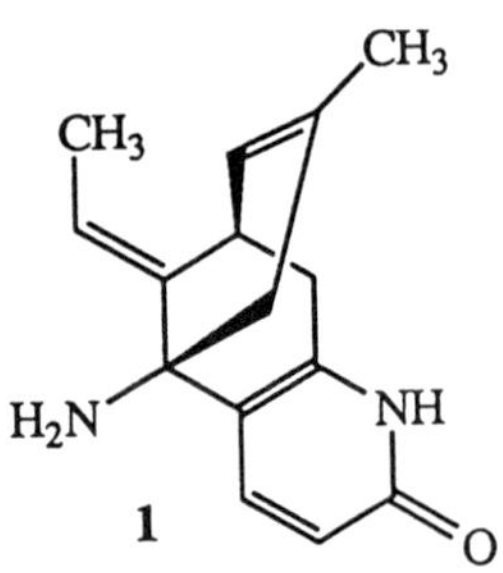

(occasionally reduced) or pyridone ring and which are exemplified by lycodine (**3**) and the previously mentioned huperzine A (**1**); (iii) the fawcettimine group, alkaloids which contain a five-membered B ring and which differ from the lycopodane group in that C-4 is bonded to C-12 rather than C-13 as exemplified by fawcettimine (**4**) (often the nitrogen to C-13 bond is severed in this group); and (iv) a miscellaneous group.

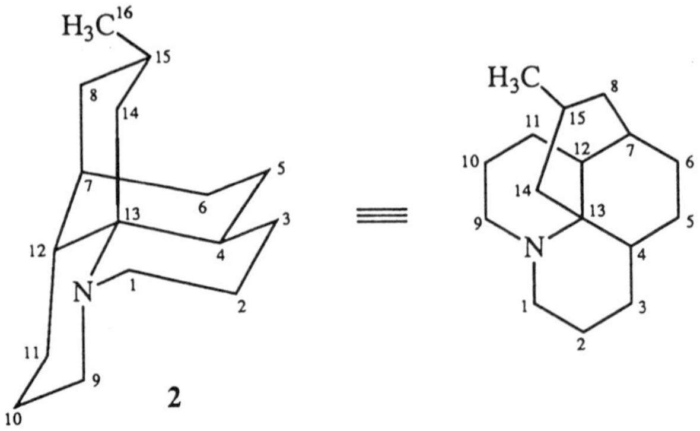

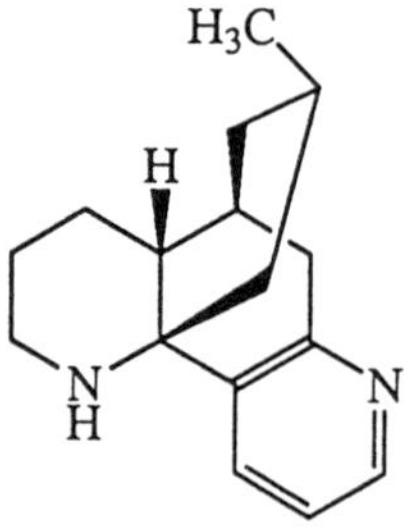

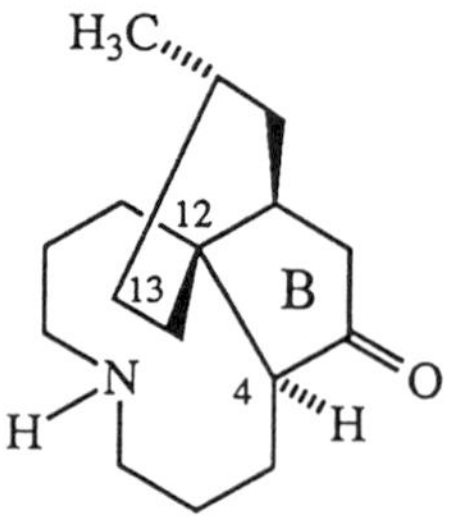

There are over 500 species in the genus *Lycopodium* (family Lycopodiaceae) (*6*), but the alkaloid content has been studied in fewer than 40 species. Most of the species are low, evergreen, coarsely mosslike plants, which are commonly known as club mosses. They are nonflowering plants which reproduce by means of spores rather than seeds. In many species, the spore-bearing bodies, known as strobili, appear as club-shaped growths at the tips of the mosslike branches, hence the name club mosses. The taxonomy of the genus and the family is still in a state of flux. Some botanists have subdivided the genus into four genera (*Lycopodium, Diphasiastrum, Lycopodiella,* and *Huperzia*), and some have placed *Huperzia* in a separate family (*7*). We prefer to retain the single genus name since this allows us to use a name which is familiar to most and typifies plants that can be easily recognized as being closely related.

II. Lycopodine Group

In the late 1940s and early 1950s, R. H. F. Manske (the founder of this treatise) and L. Marion examined the alkaloid content of several species of *Lycopodium*. During these early studies, the isolation and characterization of 35 alkaloids were reported. These were assigned numbers from L1 to L35. Most of these alkaloids have since been investigated and their structures assigned (*1*). Recently, many of the original samples of Manske and Marion have been reexamined by GC–MS and GC–IR, and the identity of all of the "L" alkaloids, except for L10, is now established (*8*). It has been shown that several of the original "L" alkaloids (often characterized as the hydroperchlorates) were mixtures, and only two new alkaloids, 5,15-oxidolycopodane (**5**), a component of alkaloids L28 and L31, and acetyl-annofoline (**6**), a component of L17, were discovered.

GC–MS was also used to reexamine the alkaloids of *L. clavatum* var. *borbonicum* and *L. deuterodensum* (*9*). Flabelliformine (**8**) and alkaloid L20 (**9**) were identified as constituents of *L. clavatum* var. *borbonicum*. *Lycopodium deuterodensum*, previously examined by Manske in 1953, contained lycopodine (**10**, ~94% of the total alkaloids), along with flabelline (**7**), flabelliformine (**8**), lycodoline (**11**), anhydrolycodoline (**12**), and clavolonine (**13**). In agreement with these results, alkaloid L35, reported from this species by Manske, has been shown (*8*) to be a mixture containing lycopodine (**10**), flabelliformine (**8**) and lycodoline (**11**). Lycodoline (**11**) also has been isolated from *H. serrata* (*10*).

The GC–MS method was used to examine two new species, *L. australianum* [shown to contain lycodine (**3**), cernuine (see Section V), and a new alkaloid whose structure was not established] and *L. fastigiatum* (*9*). The latter species contains several known members of the lycopodane (**2**) group, along with the new alkaloid fastigiatine (see below) and its *N*-demethyl derivative.

The structure of paniculine (**14**) has been confirmed by an X-ray study of the hydrobromide (*11*), and detailed ^{1}H- and ^{13}C-NMR assignments for **14** and related compounds have been reported (*12*). Paniculine (**14**) is an interesting member of the lycopodane group in that it is the only alkaloid functionalized at C-10. In the related pentacyclic alopecurane (**15**) skeleton, C-10 is the terminus of a carbon–carbon bond.

A new alkaloid with the 7,8-*seco*-lycopodane skeleton, lyconnotinol (**16**), has been isolated from *L. obscurum* (*13*). The structure of **16** was confirmed by its preparation from lyconnotine (**17**), the only other known alkaloid possessing this particular skeleton.

An elegant new synthesis of lycopodine (**10**) has been reported by Kraus and Hon (*14,15*). 5-Methylcyclohexenone was transformed into

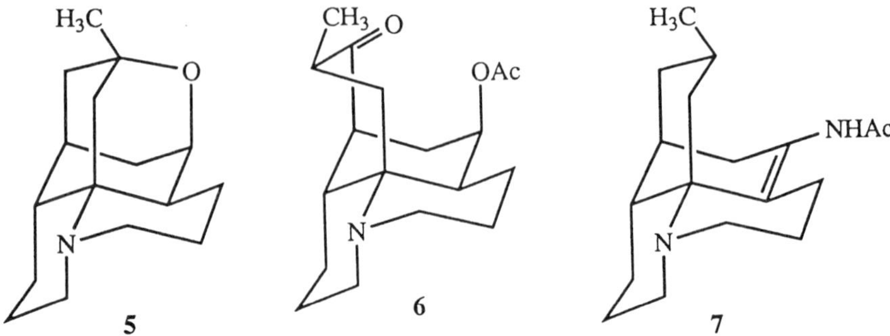

5

6

7

8 R = OH, $R^1 = R^2 = R^3 = H$
9 $R = R^2 = R^3 = H, R^1 = OH$
10 $R = R^1 = R^2 = R^3 = H$
11 $R = R^1 = R^3 = H, R^2 = OH$
12 $R = R^1 = R^3 = H, \Delta^{11,12}$
13 $R = R^1 = R^2 = H, R^3 = OH$

14

15

16 $R = CH_2OH$
17 $R = COOCH_3$

2-allyl-5-methylcyclohexenone (**18**) (Scheme 1) using a thiol addition, alkylation, and elimination sequence. Michael addition of acetoacetate to **18** and internal aldol condensation followed by decarbomethoxylation provided **19,** which was subjected to Brown hydration. The resulting primary hydroxyl was converted to the benzenesulfonate, and the bridgehead hydroxyl was transformed to the bromide **20.** Treatment of **20** with silver triflate, followed immediately by an excess of 1-amino-3-benzyloxypropane, gave the tertiary amine **21.** Alternatively, generation of the bridgehead enone and addition of 3-amino-1-propanol provides **22,** which also may be prepared by catalytic hydrogenolysis of **21.** The crystalline keto alcohol **22,** an intermediate in Heathcock's synthesis, was converted to racemic lycopodine in two steps using Heathcock's method (*16*). The overall yield from **18** to lycopodine was 25%.

Annotinine (**23**) and annotine (**24**) may be considered as 8,15- or 7,8-seco-lycopodane derivatives with an additional C — C bond between C-12

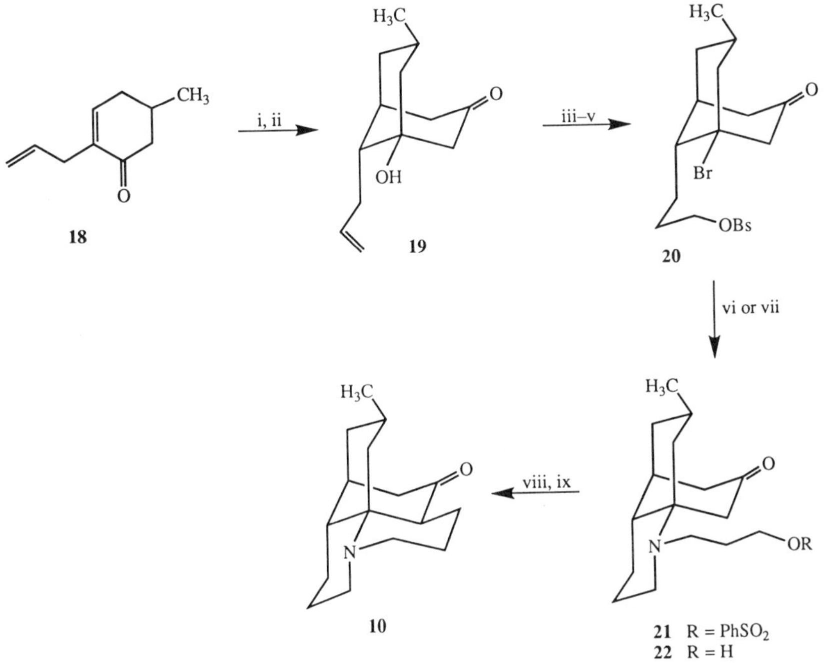

SCHEME 1. Reagents: i, MeCOCH$_2$CO$_2$Et, NaOMe, MeOH; ii, KOH, heat; iii, BH$_3$–THF; iv, PhSO$_2$Cl; v, PBr$_3$; vi, AgOTf, H$_2$N(CH$_2$)$_3$OBn; vii, H$_2$N(CH$_2$)$_3$OH, DBU; viii, 'BuOK, Ph$_2$CO, benzene, heat; ix, H$_2$, Pt.

23   24

and C-15 or C-7 and C-15, respectively. A detailed ^{1}H- and ^{13}C-NMR study of these two structurally unique alkaloids has been reported (*17*). These studies confirm the constitution of annotine (**24**), deduced earlier from degradation studies, spectroscopic information, and biogenetic considerations, and allowed the assignment of stereochemistry at C-4 and C-12. The conformation of the molecules deduced from the NMR studies shows good correlation with those determined by molecular mechanics calculations.

III. Lycodine Group

The lycodine group consists of the N_2 alkaloids containing a pyridine (as in lycodine, **3**), an α-pyridone (as in huperzine A, **1**), or a hydrogenated pyridine ring (as in the flabellidane group, **25**). Lycodine (**3**) has been detected in *L. deuterodensum*, which contains lycopodine (**10**) as the major alkaloid (*9*).

A biogenetically interesting new C_{19} alkaloid possessing the lycodine (**3**) nucleus with three additional carbon atoms attached at C-1 has been described (*13*). The new alkaloid has structure **26** and is named hydroxypropyllycodine because of its relationship to lycodine (**3**). The structure was elucidated mainly by spectroscopic methods, particularly NMR and mass spectrometry. The mass spectrum shows an intense peak at *m/z* 199, attributed to ion **A** formed by loss of the "bridging" atoms, a common fragmentation observed in related *Lycopodium* alkaloids, followed by McLafferty rearrangement with loss of acetaldehyde from the hydroxypropyl side chain. Hydroxypropyllycodine (**26**) is the first *Lycopodium*

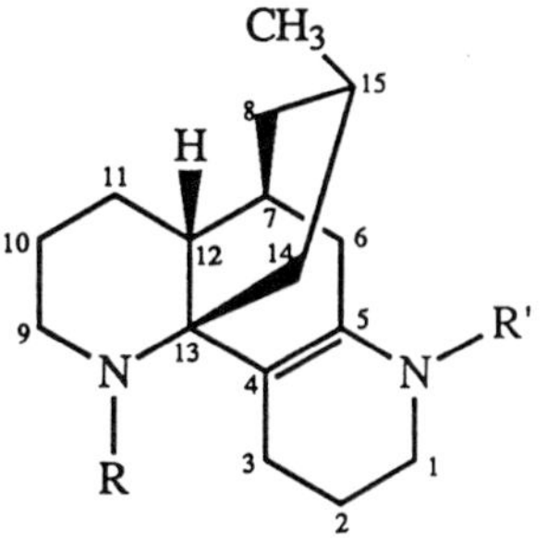

25

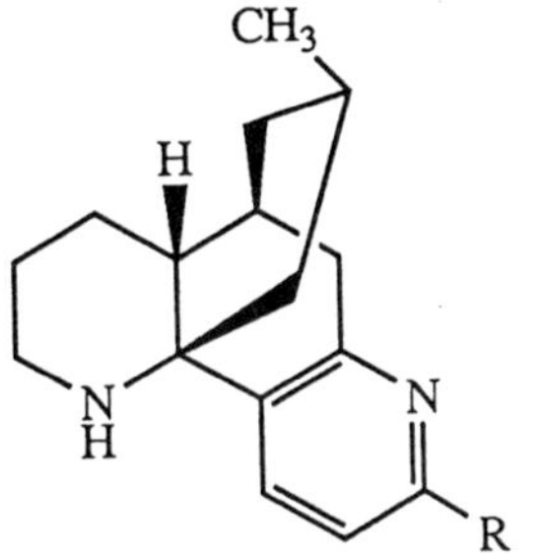

3 R = H
26 R = CH₂CH(OH)CH₃

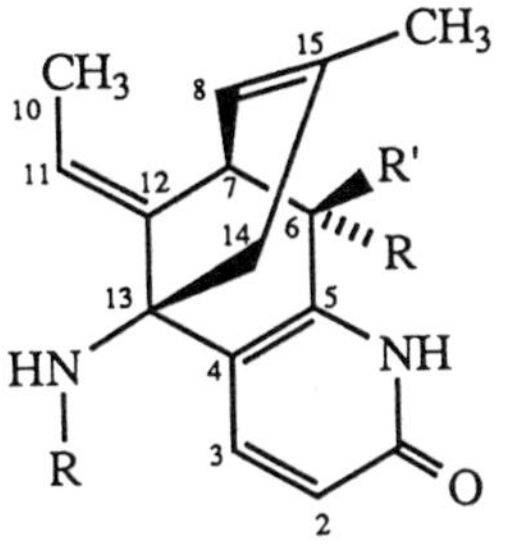

1 R = R' = R" = H
27 R = Ac, R' = R" = H
28 R = R" = H, R' = OH
29 R = H, R' + R" = O

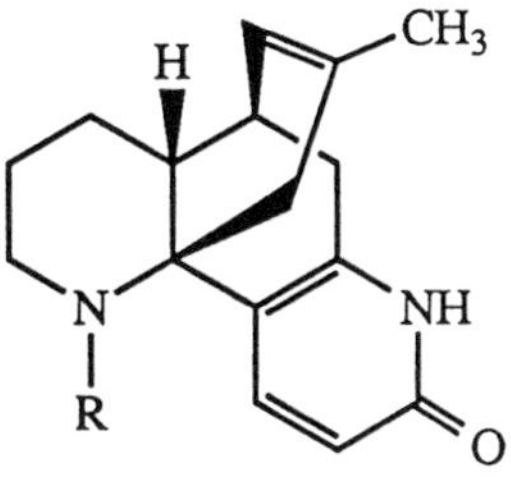

30 R = H
31 R = CH₃

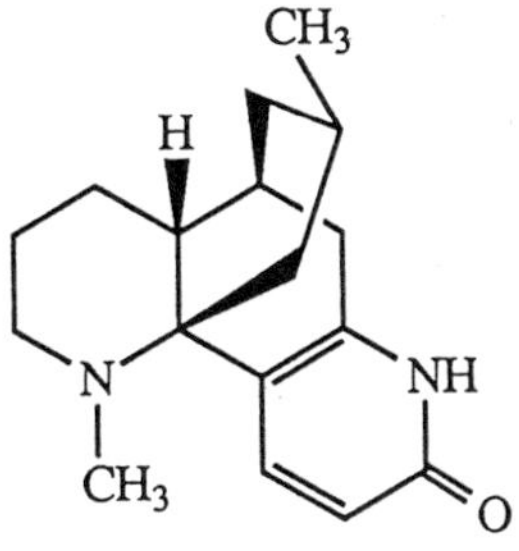

32

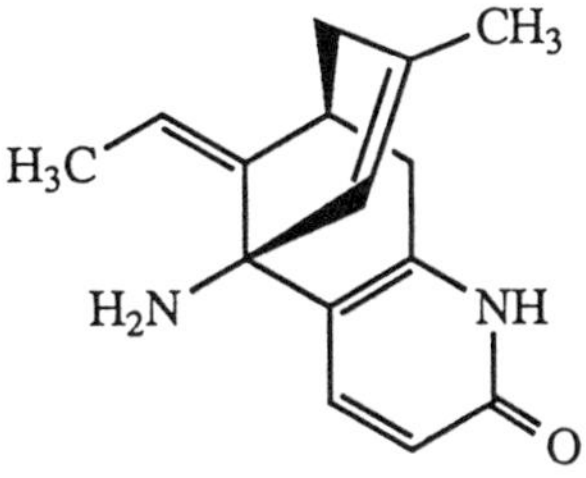

33

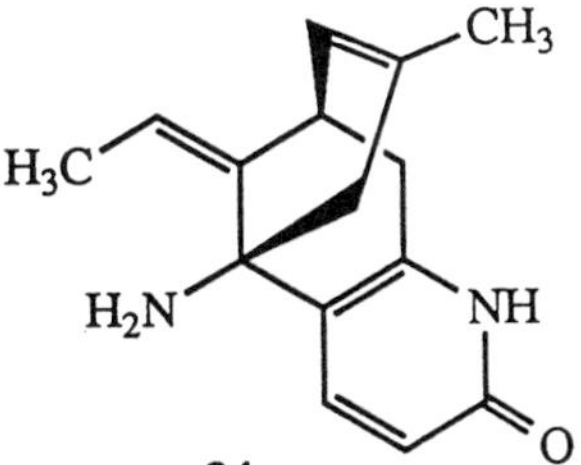

34

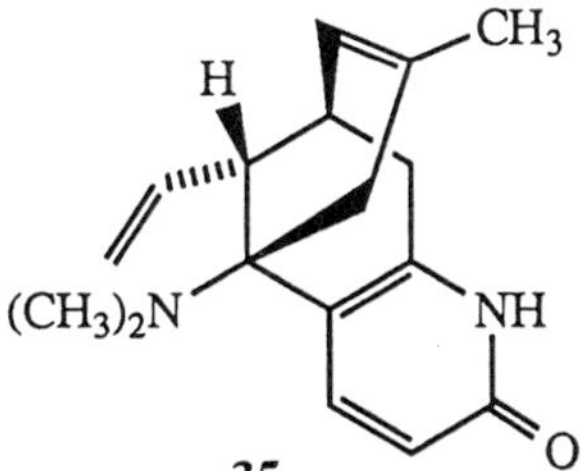

35

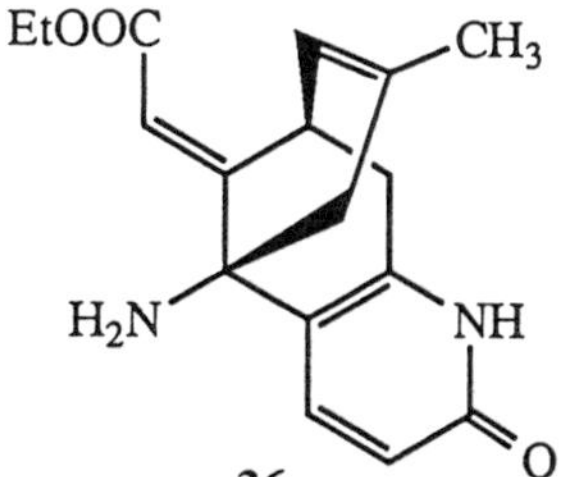

36

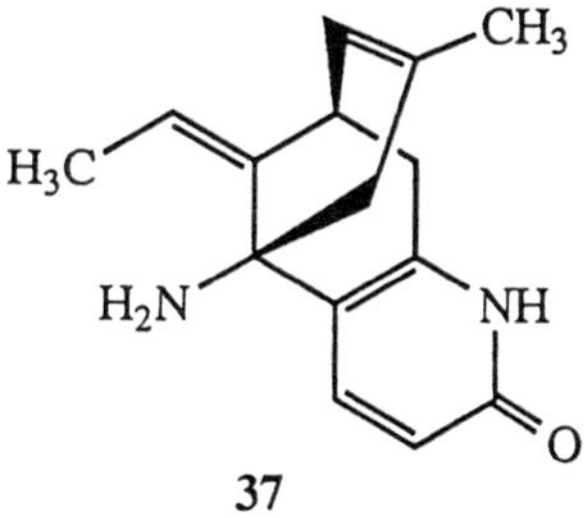

37

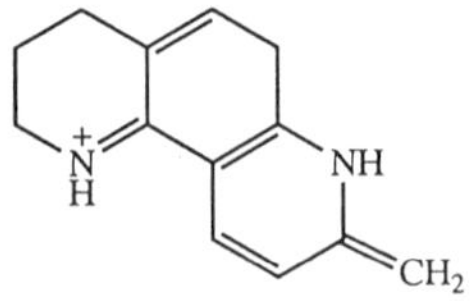

A

alkaloid encountered which possesses a 19-carbon skeleton, and, as illustrated in Scheme 2, it may be regarded as arising from the combination of
two pelletierine units and one acetoacetate unit. This is consistent with the
current (*1,16*), though still tentative, biogenetic hypotheses for this group
of alkaloids.

Huperzine A (**1**) and huperzine B (**30**) are important new alkaloids isolated from the Chinese folk medicinal plant *Huperzia serrata* (=*Lycopodium serratum*) (*18*). They exhibit strong anticholinesterase activity (*19*),
and huperzine A (**1**) has been shown to improve animal performance in
Y-maze experiments and to be useful in treating myasthenia gravis in
humans (*4*). Huperzine A is currently in clinical testing in China for the
treatment of Alzheimer's disease (*20*). The structure of huperzine A was
determined on the basis of extensive NMR, IR, and UV data, dehydrogenation to give 6-methyl-α-pyridone, preparation of mono-, di-, and tri-*N*-
methyl derivatives, and hydrogenation to dihydro and tetrahydro derivatives (*18*). The hydrogen on C-8 appears as a doublet (5 Hz) and is coupled
to a proton at δ 3.56 (H-7). Irradiation of the C-10 methyl hydrogens causes

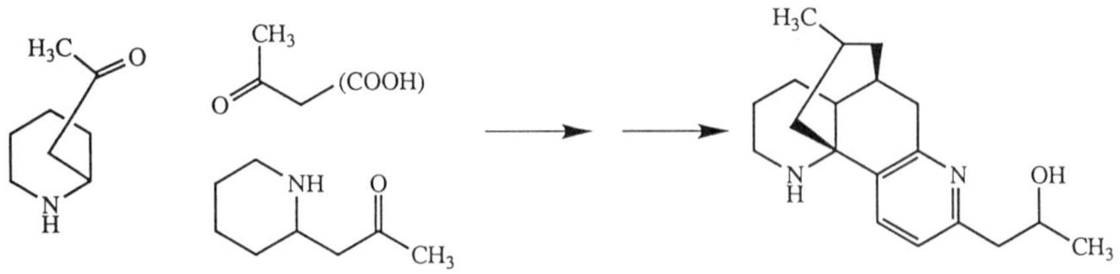

26

SCHEME 2.

nuclear Overhauser effect (NOE) enhancement of the δ 3.56 signal, allowing assignment of the stereochemistry of the exocyclic ethylidene group. The alkenic hydrogen at C-11 and the C-3 pyridone hydrogen are both shifted upfield in *N*-acetylhuperzine A (**27**), confirming the proximity of these hydrogens to the amino nitrogen. Huperzine B (**30**) lacks the ethylidene group signals in the NMR spectrum, but clearly shows the features of the C-8, C-15 endocyclic double bond and the α-pyridone ring system. Dehydrogenation of huperzine B (Pd/C, 300°C) furnishes 7-methylquinoline and 6-methyl-α-pyridone, accounting for all of the carbon and nitrogen atoms in **30.** Eschweiler–Clarke methylation gives *N*-methylhuperzine B (**31**), the spectral characteristics of which closely resemble those of β-obscurine (**32**). Catalytic hydrogenation of **31** gives 15-epi-β-obscurine, characterized by the high-field chemical shift (δ 0.60) of the C-15 methyl (anisotropically shielded by the pyridone ring). The fact that hydrogenation occurs exclusively from the C-12 side suggests that the stereochemistry at C-12 is the same as in β-obscurine. This conclusion is supported by NOE enhancement experiments.

The structure of huperzine A (**1**) is similar in many respects to that reported for selagine (**33**) (*1*), an alkaloid of *L. selago*. To obtain sufficient quantities of selagine for pharmacological testing, the alkaloids of *L. selago* indigenous to northwestern Canada were examined (*21*). The major pyridone alkaloid isolated proved to be identical with an authentic sample of huperzine A (**1**). The optical rotation and melting point reported for selagine (*22*) differed significantly from those of huperzine A (selagine, $[\alpha]_D$ −99°, mp 224–226°C; huperzine A, $[\alpha]_D$ −147°, mp 214–215°C), and it was thus important to obtain a sample of the compound isolated from *L. selago* and assigned structure **33.** The only remaining sample in the collection of the late Professor Wiesner was in the form of a KBr pellet used for the determination, in 1960, of the IR spectrum. Extraction of the pellet provided an authentic sample of selagine which proved to be identical with huperzine A. It is thus apparent that the structure **33** of selagine had been misassigned and that selagine is identical with huperzine A (**1**). Since structure **1** was first assigned the name huperzine A, that name has been retained and the name selagine has been abandoned. It has also been shown that isoselagine, originally assigned structure **34** (*23*), is also identical with huperzine A (**1**) (*21,24*). In the search for an authentic sample of selagine, a sample of a minor alkaloid isolated by the Wiesner group was uncovered. This has been shown to be 6β-hydroxyhuperzine (**28**) (*21*). Oxidation of **28** with Jones' reagent or with manganese dioxide gives the ketone **29** (*21*).

N-Methylhuperzine B (**31**) has been reported as a minor alkaloid of *Huperzia serrata,* and a new alkaloid, huperzinine (**35**), has been obtained from this source (*25*). Another new alkaloid, named phlemariuine M, has been isolated from *L. fordii* and assigned structure **36** (*26*). The ^{1}H- and ^{13}C-NMR spectra reported for phlemariuine M are consistent with the proposed structure, but other physical properties (melting point, IR, UV) have not been reported. Another paper from China that also discusses the alkaloids of *P. fordii* (*27*) reports the isolation of a compound $C_{16}H_{29}N_2O$ which was named fordimine and assigned structure **30** (huperzine B). It is not clear whether fordimine, which was claimed to be a new alkaloid, is thought to be a stereoisomer of huperzine B.

The pronounced physiological activity of huperzine A (**1**) has provided the impetus for renewed interest in the synthesis of these tricyclic *Lycopodium* alkaloids, and 1989 saw the completion of two very similar (*28,29*) syntheses of huperzine A. In one approach (*28*), the α-pyridone **39** (Scheme 3) was prepared by treating the enamine derived from the mono-ethylene ketal (**38**) of cyclohexane-1,4-dione with acrylamide. The resulting mixture of unsaturated lactams was N-benzylated and then dehydrogenated by α-selenation and oxidative elimination to give **39**. The pyridone ring in **39** proved rather sensitive in subsequent steps, and was thus protected by conversion to the methoxypyridine **40**. Hydrolysis of the ketal and carbomethoxylation gave the enolized β-ketoester **41**. Base-catalyzed addition of **41** to methacrolein and subsequent aldolization gave the bridged ketoalcohol **42**. This was transformed, via the mesylate, to olefin **43**. The ethylidene group of requisite stereochemistry was introduced by Wittig reaction followed by isomerization. Hydrolysis provided acid **44**, which was subjected to Curtius rearrangement to give urethane **45**. Treatment of **45** with trimethylsilyl iodide effected both *N*- and *O*-deprotection and afforded racemic huperzine A (**1**) (*28*). It was later found that the debenzylated derivative of pyridone **39** could be prepared from the monoketal of cyclohexane-1,4-dione, in one step, by treatment with methyl propiolate in methanolic ammonia under pressure (*30*). The synthesis of unnatural ($\pm$)-(*Z*)-huperzine (**37**) using essentially the route outlined in Scheme 3, but omitting the double bond isomerization step, has also been reported (*31*). The *Z* isomer **37** is not as active an acetylcholinesterase inhibitor as huperzine A itself, at least when the racemic forms are compared (*30*). Several other benzenoid and heterocyclic analogs of huperzine A, all less active than the alkaloid itself, have been prepared (*32–34*).

The second total synthesis of huperzine A (*29*) also proceeds through the bridged hydroxyketone **42** (Scheme 3), prepared by treating **41** with methacrolein in the presence of methanolic sodium methoxide. Mesylation and

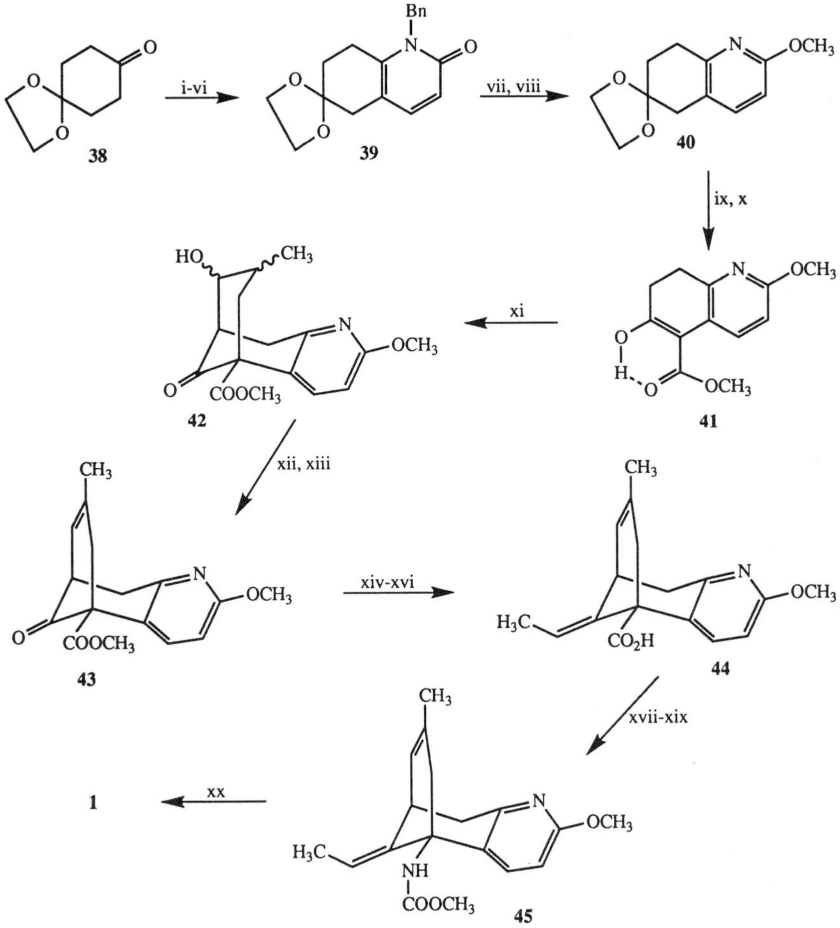

SCHEME 3. Reagents: i, pyrrolidine, PhH, *p*-TsOH; ii, acrylamide, dioxane, heat; iii, H_2O, dioxane, heat; iv, KH, BnCl, THF; v, LDA, PhSeCl, THF, $-78°C$; vi, $NaIO_4$, Et_3N, MeOH, heat; vii, H_2, Pd/C; viii, Ag_2CO_3, MeI, $CHCl_3$; ix, aq. HCl, Me_2CO; x, KH, $(MeO)_2CO$, heat; xi, methacrolein, tetramethylguanidine, CH_2Cl_2; xii, MsCl, Et_3N, DMAP, CH_2Cl_2; xiii, NaOAc, HOAc, heat; xiv, $Ph_3P = CHCH_3$, THF; xv, PhSH, AIBN, heat; xvi, aq. NaOH, THF, MeOH, heat; xvii, $SOCl_2$, toluene; xviii, NaN_3, heat; xix, MeOH, heat; xx, TMSI, $CHCl_3$, heat.

elimination gave the olefin **43**. Wittig olefination gave a mixture of the (*Z*) and (*E*)-ethylidene derivatives in which the desired *E* isomer was the minor component. The mixture of *Z/E* isomers was treated with methanolic potassium hydroxide, which brought about hydrolysis of the less sterically hindered *E* isomer, to provide acid **44**. Curtius rearrangement and transformation of the methoxypyridine to the pyridone provided huperzine A.

Recent *in vitro* and *in vivo* bioassays have shown that (−)-huperzine A (**1**) is almost twice as active as the racemic mixture and has 33-fold greater activity than that of (+)-huperzine A (*35*). Therefore, an elegant, molecular modeling-based synthesis of the more active (−)-isomer was developed utilizing the ester **46** (Scheme 4) bearing the easily removable 8-phenylmenthol chiral auxiliary. The latter induces a 4:1 ratio at room temperature (and 9:1 at −20°C) of ketones **48** with the major isomer possessing the desired (7*R*,13*S*) absolute stereochemistry, the ratio being virtually identical to that predicted by quantum mechanic calculations.

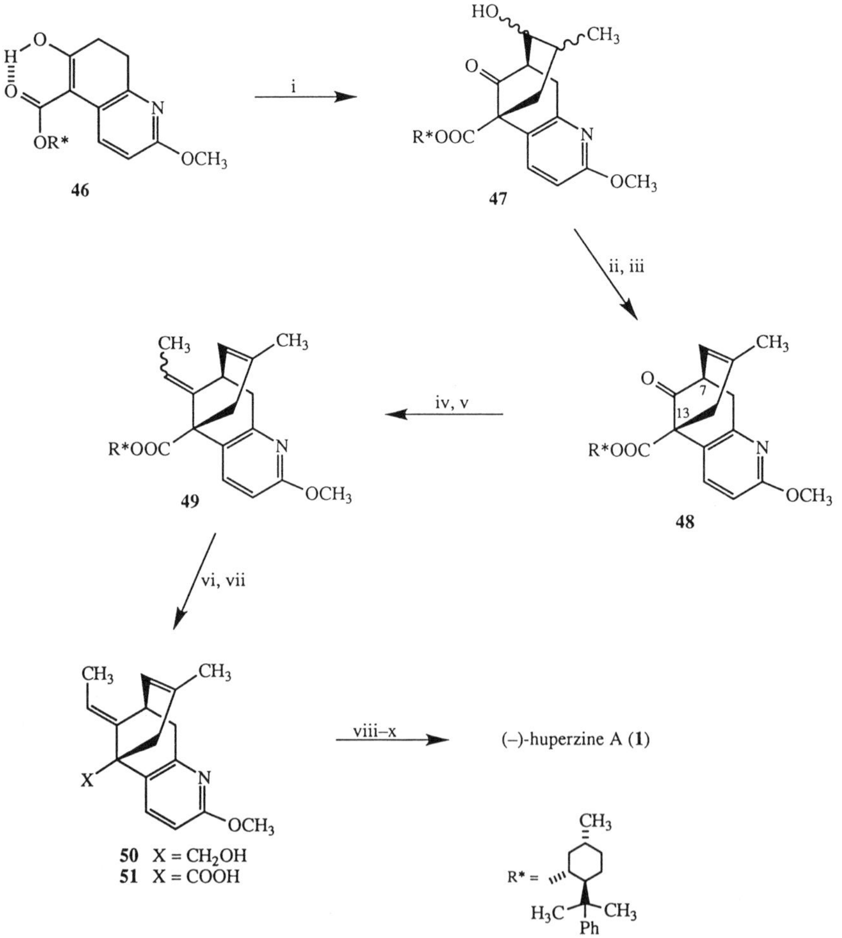

SCHEME 4. Reagents: i–iii, v, same as in Scheme 3; iv, separation of diastereomers; vi, LAH/THF; vii, Jones oxidation; viii, $SOCl_2$, $PhCH_3$; ix, NaN_3, $PhCH_3$; x, MeOH, heat.

The isomer has been further converted to a mixture of dienes **49** (*Z/E* ratio 10:1 or 1:1.4 after isomerization) as illustrated in Scheme 4. Lithium aluminum hydride reduction of the *E* isomer to remove the chiral auxiliary, oxidation to acid **51,** and Curtius rearrangement lead to optically pure (−)-huperzine A (**1**).

A detailed mass spectrometric study of the structural isomers hupersinine (**35**), *N,N*-dimethylhuperzine A, and *N*-methylhuperzine B (**31**), including tandem mass spectrometry (MS/MS), has shown that the alkaloids may be differentiated by this technique (*36*). Thus loss of C_3H_5 (C-10 to C-12) is the predominant fragmentation of huperzine, whereas loss of C_4H_7 (C-8, C-14, C-15, and C-16) occurs less frequently. Loss of C_4H_9, however, is the major fragmentation for alkaloids which lack a double bond in the bridge, such as β-obscurine (**32**). This study revealed that β-obscurine was present as an impurity in the early samples of *N*-methylhuperzine B and is responsible for the major fragments in the mass spectrum of the mixture of these two alkaloids.

Two structurally interesting new members of the lycodine group, fastigiatine (**52**) and des-*N*-methylfastigiatine (**53**) were initially detected in the GC–MS examination of the crude alkaloids of *L. fastigiatum,* a species of *Lycopodium* indigenous to New Zealand (*37*). The structure of fastigiatine

52 R = CH₃
53 R = H

54

55

was deduced from MS and NMR data and is supported by an X-ray analysis (*38*). The pentacyclic structure is related to the tetracyclic flabellidane (**54**) group with an extra carbon–carbon bond between C-4 and C-10. A C-4, C-10 bond is also found in the alopecurane (**55**, same as **12**), group of alkaloids (*1*). Des-*N*-methylfastigiatine (**53**) was correlated with **52** by *N*-methylation (*37*). It is suggested that fastigiatine may be derived biogenetically from a C-10 substituted *N*-methylflabellidine as shown in structure **54**.

IV. Fawcettimine Group

The alkaloids of the fawcettimine group possess either the fawcettimane (**56**) or the fawcettidane (**57**) skeleton. The two subgroups are logically treated together, since fawcettimine, the prototype alkaloid, exists as an equilibrium mixture of the keto amine form **58** and the carbinolamine form **59**. A characteristic of this group not shared with the other groups is the presence of a five-membered ring B. Recent structural and synthetic studies by Heathcock and co-workers have provided valuable new information, discussed later, on this interesting keto amine–carbinolamine tautomerization.

Several new members of the fawcettimine group have been isolated from *L. obscurum* (*13*). The structures of lobscurinol (**60**), epilobscurinol (**61**), and acetyllobscurinol (**62**) were determined mainly on the basis of extensive correlation spectroscopy (COSY) ^{1}H-NMR studies which allowed the complete analysis of the spin system corresponding to carbons 9 to 11, and of another spin system correlating H-14 with the alkenic methyl and on through the C-8 hydrogens all the way to C-2. The α,β-unsaturated carbonyl system present in the alkaloids gives rise to absorption at 234 nm in the UV and at 1658 cm^{-1} in the IR. The stereochemistry at C-5 was assigned on the basis of coupling constants and NOE enhancements. In particular, the hydrogen at C-5 in **61** and its *O*-acetyl derivative shows vicinal coupling with both C-6 H atoms (each 7 Hz), allylic coupling (2 Hz) with H-3, and **W** coupling with H-7. H-5 in **60** and **62** shows observable coupling to only one of the H-6 atoms (5 Hz), smaller allylic coupling (1 Hz), and no coupling to H-7.

Oxidation of lobscurinol provided the diketone **63**, which was named lobscurinine. Lobscurinine was not isolated among the alkaloids of *L. obscurum*, but its interesting ammonia adduct, obscurinine (**64**), was

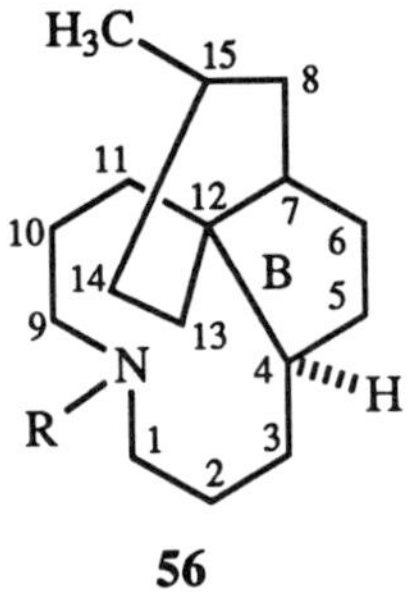

56

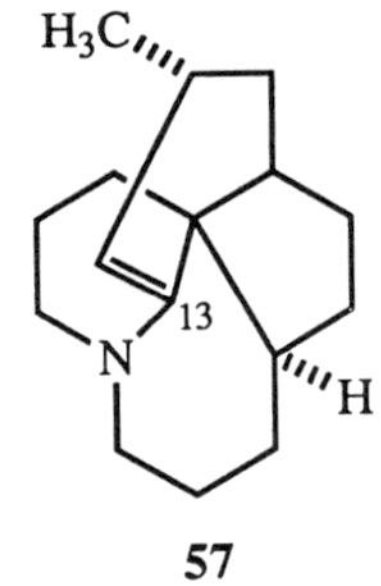

57

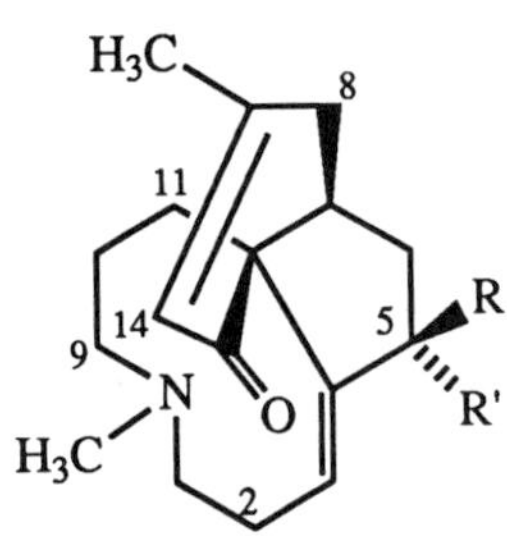

58

59

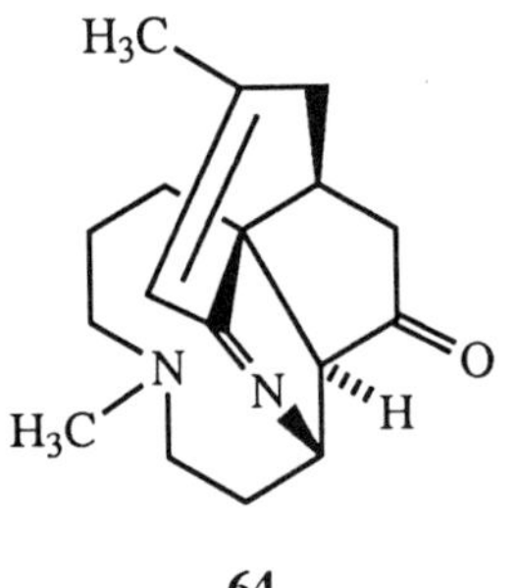

60 R = H, R' = OH
61 R = OH, R' = H
62 R = H, R' = OAc
63 R + R' = O

64

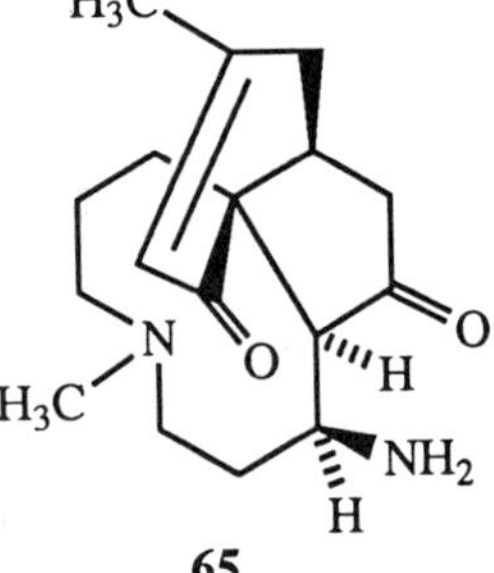

65

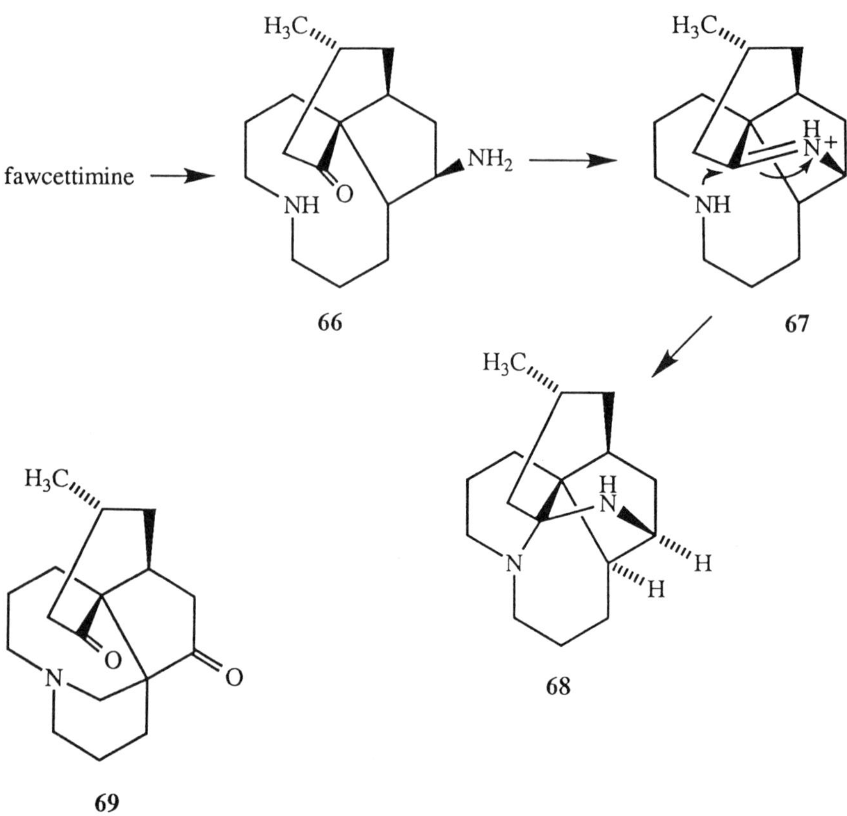

fawcettimine
66
67
68
69
70 R = OH, R' = H
71 R + R' = O

present. The structure of obscurinine, initially believed to be a new alkaloid, was determined by X-ray crystallography (*39*). Because in both
isolation procedures (*13,39*) aqueous acidic extracts of the alkaloids were
basified with ammonium hydroxide, it seems probable that obscurinine is
an artifact of the isolation process. Indeed, treatment of a chloroform
solution of lobscurinine (**63**), prepared by oxidation of lobscurinol (**60**),
with aqueous ammonia led to the formation of obscurinine (**64**) in virtually
quantitative yield (*13*), presumably by internal Schiff base formation in the
ammonia adduct **65** of lobscurinine. This partial synthesis of obscurinine
provides confirmation of the structures of lobscurinol (**60**) and epilobscurinol (**61**).

Macleanine (**68**) is a unique dinitrogeneous alkaloid which fits into the
fawcettimine group. The alkaloid was isolated from the mother liquors of a
large-scale (50 kg) extraction of *Lycopodium serratum* (=*Huperzia serrata*) carried out in China in order to isolate large quantities of huperzine A
(**1**), the pharmacologically important acetylcholinesterase inhibitor. The
structure of macleanine was determined mainly by high resolution NMR
studies and was confirmed by an X-ray crystallographic study of the
hydroperchlorate (*40*). Biogenetically, macleanine may be derived from
fawcettimine by reductive amination of the C-5 carbonyl group to give the
amine **66,** followed by formation of the internal immonium ion **67** and
transannular addition of the secondary amino function. Lycoflexine (**69**),
the interesting formaldehyde adduct of fawcettimine previously isolated
from *L. clavatum* var. *inflexum,* has been identified by GC–MS as a
constituent of *L. deuterodensum* (*9*).

Two elegant and strategically different total syntheses of magellanine
(**70**) and magellaninone (**71**) have been reported (*41,42*). The syntheses
illustrate the power of modern synthetic methods in the construction of
stereochemically complex natural products. They also show once again
the impetus provided by structurally unique natural products for the development of instructive and insightful synthetic sequences. Throughout the
past 25 years, the *Lycopodium* alkaloids have proved to be a rich ground
for the sharpening of these synthetic tools, providing several different
challenging carbon skeletons (*16*). The first of these two new syntheses, by
Overman and co-workers (*41*), provides the optically active alkaloids and
begins with the readily available (1*R*,5*S*)-bicycloheptenone **72** (Scheme 5)
which was ring-expanded to the methylthio ketone **73**. This was transformed stepwise to the lithio anion **76** which was coupled with the (*S*)-
cyclopentanone **77** to provide, after modification of functional groups, the
key intermediate **79**. This intermediate, on treatment with stannic
chloride, underwent an elegant Prins cyclization–pinacol rearrangement

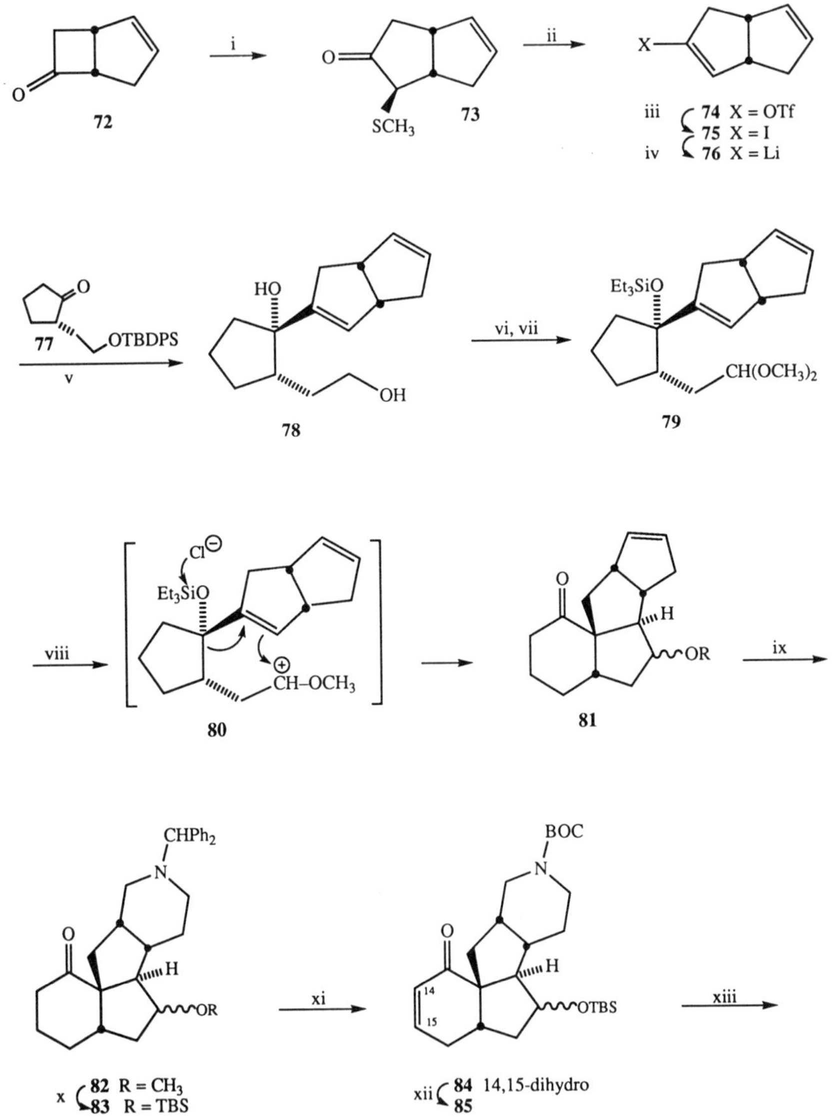

Scheme 5. Reagents: i, LiCH(SCH$_3$)$_2$, THF, 0°C; Cu(OTf)$_2$, PhH, (*i*-Pr)$_2$NEt, PhH, 50°C; ii, Li, NH$_3$–THF, −40°C; TMSCl, THF; CH$_3$Li, THF, −78 → 0°C; PhN(Tf)$_2$, −78 → 23°C; iii, [(CH$_3$)$_3$Sn]$_2$, Pd(Ph$_3$P)$_4$, LiCl, THF, 60°; *N*-iodosuccinimide, THF, 0°C; iv, 'BuLi, Et$_2$O, −110°C; v, Bu$_4$NF, THF; vi, Et$_3$SiCl, imidazole, DMAP, DMF, 50°C; Swern oxidation; vii, (CH$_3$O)$_3$CH, PPTS, CH$_2$Cl$_2$, room temperature; viii, SnCl$_4$, CH$_2$Cl$_2$, −78 → −23°C; ix, OsO$_4$ (cat.), NaIO$_4$, dioxane–H$_2$O, room temperature; Ph$_2$CHNH$_3$Cl, NaBH$_3$CN, *i*-PrOH, room temperature; x, Cl$_3$SiCH$_3$, NaI, CH$_3$CN, 80°C; TBSCl, imidazole,

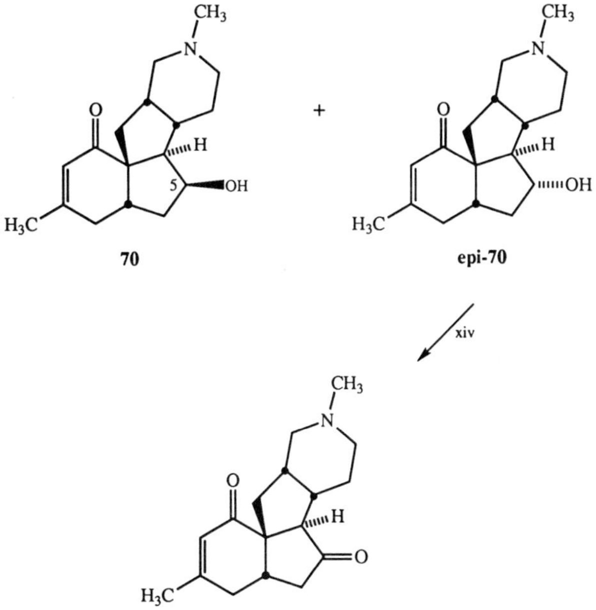

DMF, room temperature; xi, H_2, $Pd(OH)_2$, EtOAc, room temperature; $(BOC)_2O$, Et_3N, DMAP, CH_3CN, room temperature; xii, LDA, $(CH_3)_3SiCl$, THF, $-78°C$; $Pd(OAc)_2$, CH_3CN, 80°C; xiii, $Li(CH_3)_2Cu$, TMEDA, $(CH_3)_3SiCl$, $-78 \rightarrow 0°C$; $Pd(OAc)_2$; CH_3CN, 80°C; CF_3COOH, room temperature; HCHO, $NaBH_3CN$, CH_3CN, room temperature; HF, CH_3CN; xiv, Jones oxidation, room temperature.

sequence to provide the tetracyclic ketone **81** possessing all of the essential stereogenic centers of (−)-magellanine. Double bond cleavage of the mixture of epimers, insertion of the nitrogen, *N*-methylation, and methylation of C-15 afforded a separable mixture of (−)-magellanine (**70**) and 5-epimagellanine. Jones oxidation of the epi compound provided (+)-magellaninone (**71**).

The Paquette synthesis of racemic magellanine (**70**) and magellaninone (**71**) (*42*), appearing very shortly after that of Overman, utilizes as a key step the Michael addition of ethyl 5-ethoxy-3-oxo-4-pentenoate to the α,β-unsaturated ketone derived from **86.** Sequential internal Michael addition furnished the cycloannulated product **87** (Scheme 6). The ester **87** was smoothly transformed to the unsaturated ketone **88,** and the cyclopentanone was reduced selectively. Protection of the hydroxyl and

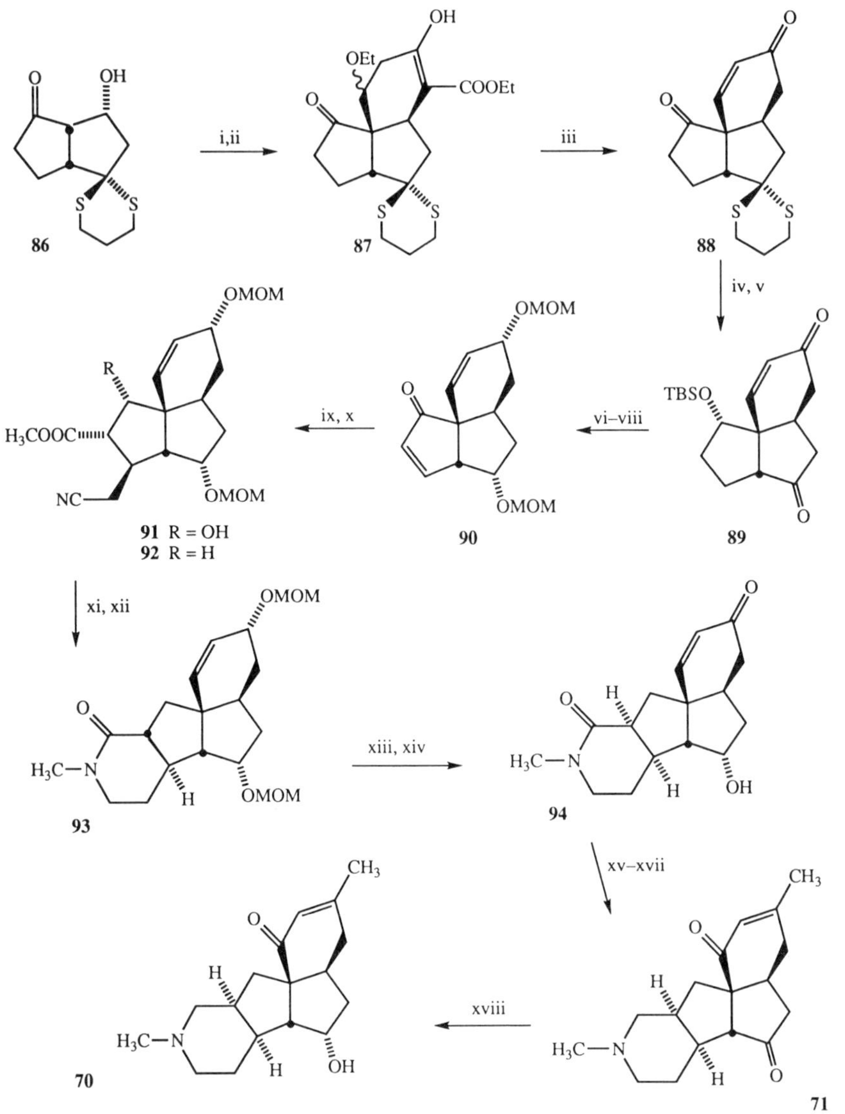

SCHEME 6. Reagents: i, MsCl, Et₃N; ii, (*E*)-EtOCH = CHCOCH₂COOEt, K₂CO₃, Al₂O₃, THF/EtOH, room temperature; iii, TsOH, C₆H₆, heat, NaCl, DMF; iv, NaBH₄, EtOH, CH₂Cl₂, 0°C; TBSOTf; imid, CH₂Cl₂, room temperature; v, Tl(NO₃)₃, MeOH, THF; vi, DIBAL-H, CH₂Cl₂, −78°C; vii, MOMCl, (*i*-Pr)₂NEt, CH₂Cl₂; Bu₄NF, HMPA, 3-Å sieves, room temperature; PCC on Al₂O₃, CH₂Cl₂; viii, LiN(SiMe₃)₂, THF; PhSeCl; H₂O₂, Py; ix, LiCH(CN)SiMe₃, HMPA, THF; KF, aq. CH₃CN; LDA, NCCOOCH₃; x, NaBH₄, MeOH, −20°C; xi, NaBH₄, CoCl₂, MeOH; KOH, MeOH; H₃O⁺; xii, NaH, MeI, THF; xiii, LDA, THF, −78 → 10°C; H₂O at −78°C; KCl, H₂O, THF; xiv, MnO₂, CHCl₃; xv, CH₃Li, THF, −78°C; xvi, LiAlH₄, THF, heat; xvii, Jones oxidation; xviii, NaBH₄, EtOH; Ph₃P, DEAD, HCOOH, THF; 10% KOH, MeOH.

deprotection of the thioketal provided the diketone **89.** This was transformed to the unsaturated ketone **90** by standard transformations. The piperidine ring was constructed in a novel way. Conjugate addition of lithio(trimethylsilyl)acetonitrile to **90** followed by C-acylation provided a keto ester, which on reduction gave **91.** The hydroxyl group was removed via the selenocarbonate. Selective reduction of the cyano group, lactam formation, and *N*-methylation provided **93.** Base-catalyzed *trans–cis* ring isomerization taking advantage of the lactam functionality, deprotection of the hydroxyl groups, and selective MnO_2 oxidation furnished the desired precursor **94,** which possesses all of the stereogenic centers of magellaninone. Introduction of the *C*-methyl group and appropriate adjustment of the oxidation level provided racemic magellaninone (**71**), which was in turn transformed into magellanine (**70**).

Another interesting approach to the synthesis of the complex tetracyclic skeleton of magellaninone (**71**) and related alkaloids has been described (*43*). The starting material (Scheme 7) is the pentacyclic dione **95** derived from the Diels–Alder adduct of cyclopentadiene and *p*-benzoquinone. This was transformed in a series of steps to the triquinanedienone **96.** Conjugate addition of 4-bromomagnesium-1-butene furnished as the major product the *cis–anti–cis* addition compound, which was transformed by phenylselenylation–selenoxide elimination to the unsaturated ketone **97.** Wacker-type oxidation of the side chain gave enedione **98.** This, on base-catalyzed ring closure, provided the tetracyclic diketone **99** containing the required stereochemistry at all five ring junction stereogenic centers. The cyclopentene ring was then transformed to an *N*-methylpiperidine ring by ozonolysis, reduction, mesylation, and treatment with methylamine, to provide the tetracyclic diketoamine **100,** which lacks the methyl group at C-15 and the oxygen substituent at C-13 of megallaninone.

In a slightly modified scheme, Metha and Rao (*44*) developed a synthesis of the intermediate **107** (Scheme 8), which is regarded as a possible synthetic precursor of magellanine (**70**), magellaninone (**71**), and paniculatine (**101**). In this synthesis, the regiospecific conversion of one of the carbonyl groups of **106** to an exomethylene functionality has been achieved.

The same authors developed a method of constructing the skeleton of fawcettimine (**4**) and serratinine (**102**), starting from the readily available intermediate **108** (Scheme 9) and following an approach based on carbocycle–heterocycle equivalence (*45*). Thus, 1,4-addition of the Grignard reagent to **108** in the presence of CuBr and $BF_3 \cdot OEt_2$ led to a mixture of *cis* and *trans* isomers which was isomerized to the desired *trans* isomer **109** by treatment with base. $PdCl_2$-mediated introduction of a carbonyl group in

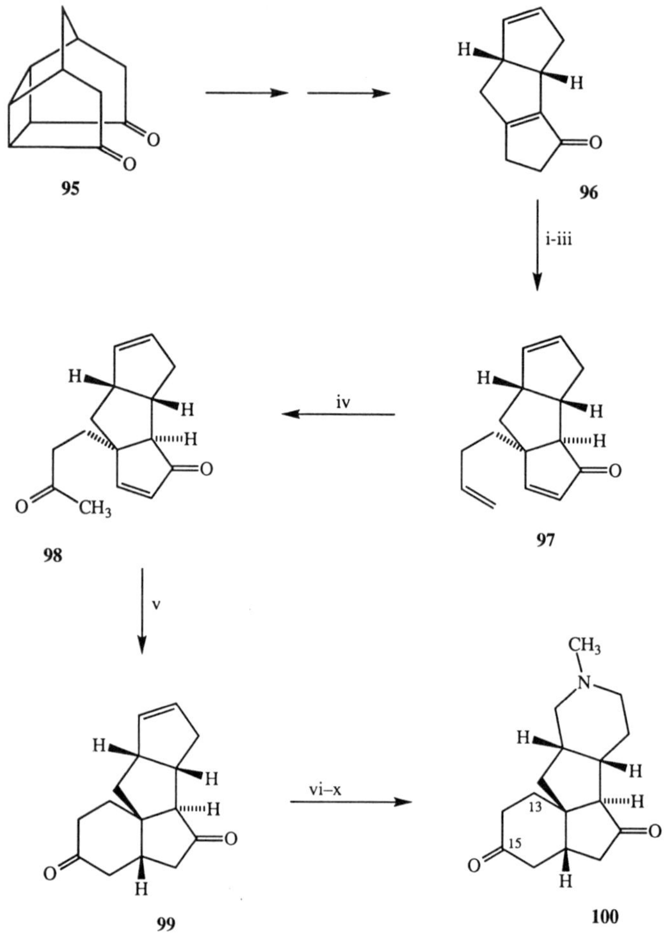

SCHEME 7. Reagents: i, 4-bromo-1-butene, Mg, Me$_2$S·CuBr, THF; ii, LHMDS, PhSeCl, THF; iii, H$_2$O$_2$, Py–CH$_2$Cl$_2$; iv, PdCl$_2$, CuCl, DMF, H$_2$O, O$_2$; v, NaH, THF; vi, HOCH$_2$CH$_2$OH, PTSA, C$_6$H$_6$; vii, O$_3$, MeOH then NaBH$_4$; viii, MsCl, Et$_3$N, CH$_2$Cl$_2$; ix, CH$_3$NH$_2$, DMSO, sealed tube; x, aq. HCl, THF.

the 3-butenyl moiety, followed by Michael addition, was used to elaborate the six-membered ring with correct stereochemistry. After carbonyl group protection, reductive ozonolysis, mesylation, insertion of tosylamide, and carbonyl deprotection, the intermediate **116** was obtained. This possesses

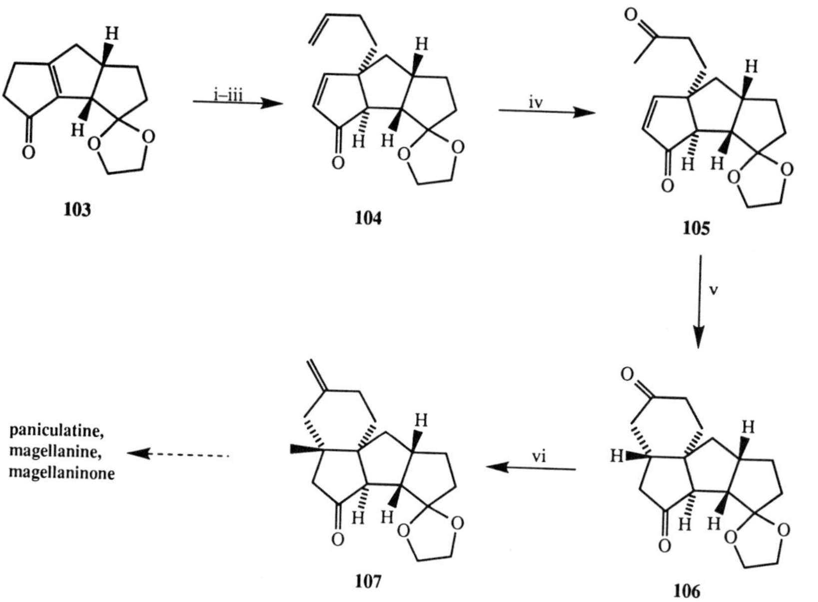

103   **104**   **105**

107   **106**

paniculatine,
magellanine,
magellaninone

SCHEME 8. Reagents: i, $CH_2 = CHCH_2CH2MgBr$, $Me_2S \cdot CuBr$, Me_2S, THF, $-78°C$; ii, $LiN(SiMe_3)_2$, PhSeCl, THF, $-78 \rightarrow -25°C$; iii, H_2O_2, THF, room temperature; iv, $PdCl_2$, CuCl, O_2, DMF, room temperature; v, NaH, THF; vi, $Ph_3P^+CH_3Br^-$, $EtC(Me)_2ONa$, $PhCH_3$, room temperature.

the desired skeleton and functionalities which may be elaborated to faw-cettimine (**4**).

The preparation of the serratinine precursor **120** utilizes the common intermediate **112** (Scheme 10). One of the carbonyl groups of **112** was

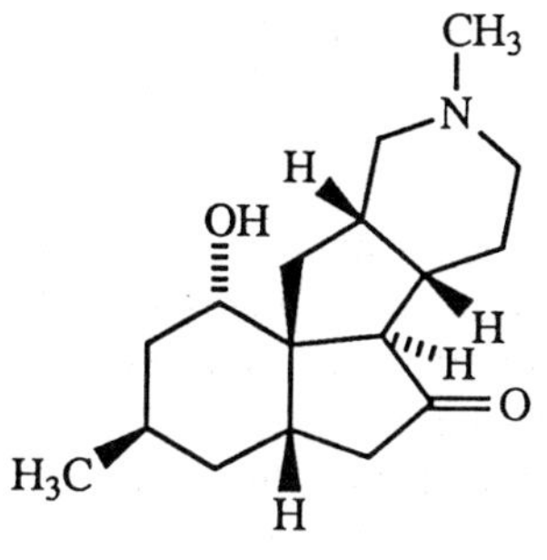

101 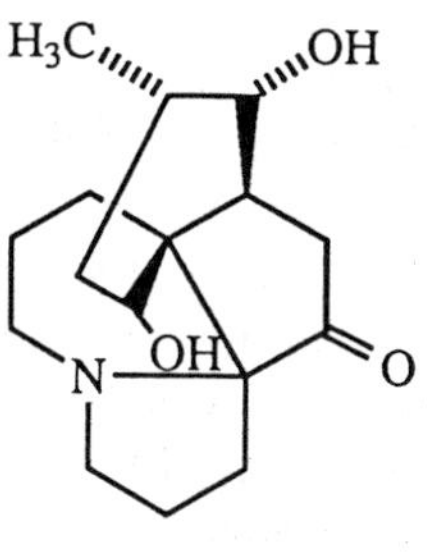   **102**

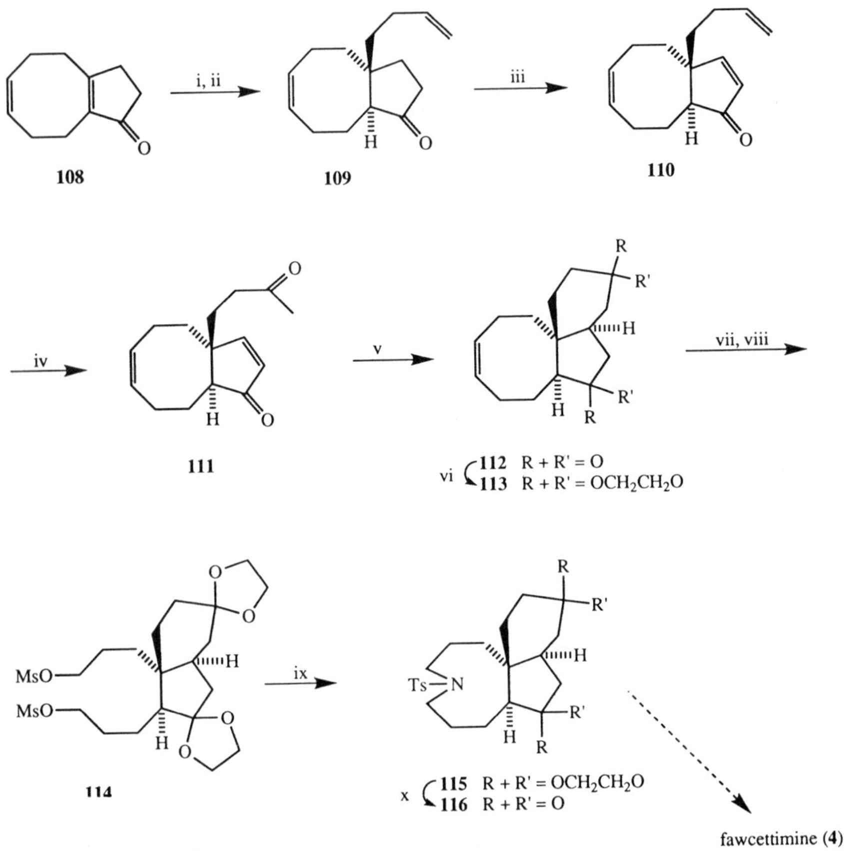

SCHEME 9. Reagents: i, $CH_2 = CHCH_2CH_2MgBr$, $CuBr \cdot Me_2S$, $BF_3 \cdot OEt_2$, Me_2S, THF, $-78 \rightarrow 0°C$; ii, 2% KOH/MeOH, room temperature; iii, $LiN(SiMe_3)_2$, PhSeCl, THF, $-78°C \rightarrow$ room temperature, 30% H_2O_2, Py-DCM, $0°C \rightarrow$ room temperature; iv, $PdCl_2-O_2$, CuCl, aq. DMF, room temperature; v, NaH, THF, 50°C; vi, $HOCH_2CH_2OH$, camphorsulfonic acid, benzene, 90°C; vii, O_3, MeOH, $-78°C$, $NaBH_4$, $-78°C \rightarrow$ room temperature; viii, $MeSO_2Cl$, Et_3N, DCM $-23°C \rightarrow$ room temperature; ix, $TsNH_2$, $^nBu_4NI$, NaOH, benzene, H_2O, 90°C; x, $Py \cdot TsOH$, acetone, 65°C.

selectively protected, and the other was reduced to the alcohol, which was silylated to give compound **117**. The nitrogen atom was introduced in the same way as in the previous scheme. Compound **118** was converted to the olefin by treatment with $BF_3 \cdot OEt_2$, the carbonyl group was protected, and the epoxide **119** was prepared. Treatment of **119** with sodium naphthalene removed the tosyl protective group. Subsequent spontaneous epoxide ring opening provided the amino alcohol **120**, which is considered to be a plausible starting material for the synthesis of serratinine (**102**).

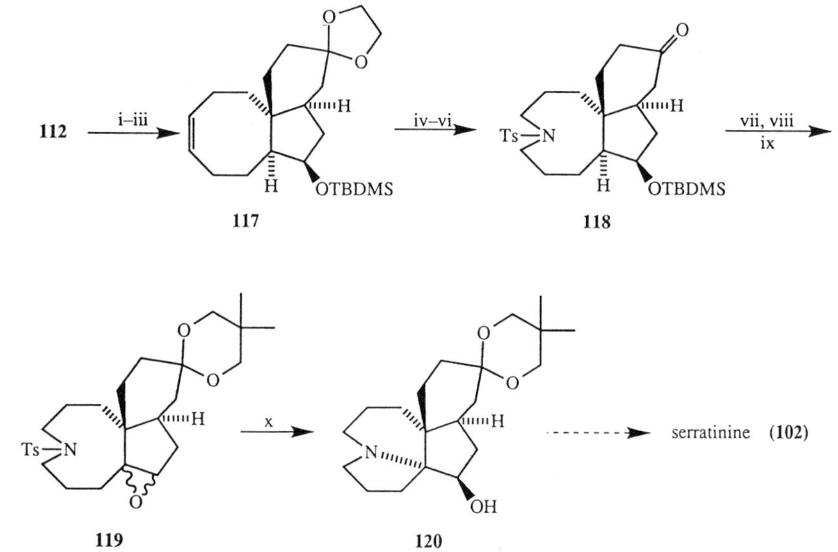

SCHEME 10. Reagents: i, HOCH$_2$CH$_2$OH, PPTS, benzene, 90°C; ii, NaBH$_4$, MeOH, room temperature; iii, TBDMSCl, DBU, DMAP, DCM, room temperature; iv, O$_3$, MeOH, −78°C, NaBH$_4$; v, MeSO$_2$Cl, Et$_3$N, DCM, −23°C → room temperature; vi, TsNH$_2$, ^{n}Bu$_4$NI, NaOH, benzene, H$_2$O; vii, BF$_3$·OEt$_2$, CHCl$_3$, room temperature; viii, HOCH$_2$C(CH$_3$)$_2$CH$_2$OH, PPTS, benzene, 90°C; ix, MCPBA, DCM, 0°C → room temperature; x, sodium naphthalene, DME, −78°C.

The preliminary results of another approach to the synthesis of magellanine (**70**), magellaninone (**71**), and paniculatine (**101**) have been reported (*46*). This approach is based on an intramolecular enone–olefin photocycloaddition. The starting chiral enones **122** and **123** were synthesized from (*R*)-(+)-methyl citronellate (**121**) in a 15-step procedure (Scheme 11). Irradiation of **122** followed by deprotection of the hydroxyl group provided the *cis*-6–4 ring fusion tricyclic compound **124**, whereas **123** gave the *trans* adduct **125**. However, neither of these possess the correct stereochemistry required for synthesis of the *Lycopodium* alkaloids of this group.

The synthesis of fawcettimine (**4**) has been described in detail (*47,48*). The starting material is the cyano–enone **126** (Scheme 12), also used as the starting material for the earlier syntheses of lycopodine (**10**), lycodine (**3**), and lycodoline (**11**) (*16*). The enone **126** was subjected to Sakurai reaction with the allylsilane **127** to give alcohol **128** as a mixture of diastereoisomers in quantitative yield. Oxidation to the aldehyde **129** and Wittig reaction with [(ethoxycarbonyl)methylene]triphenylphosphorane, followed by

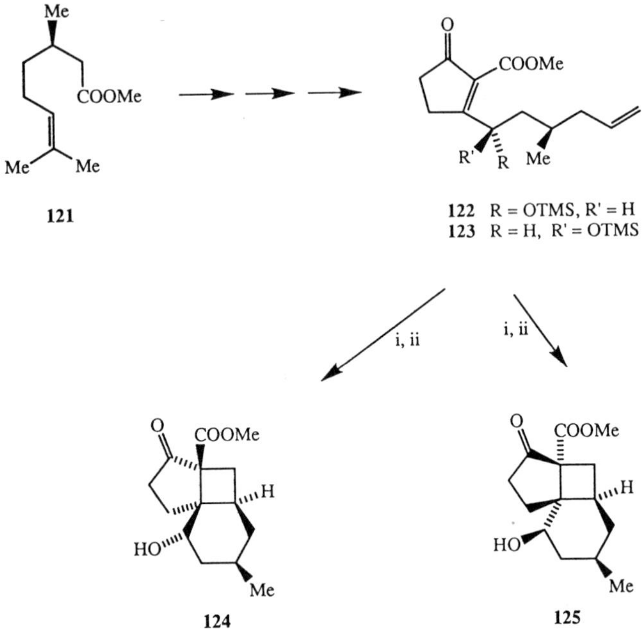

121

122 R = OTMS, R' = H
123 R = H, R' = OTMS

124

125

SCHEME 11. Reagents: i, $h\nu$ (>350 nm); ii, TsOH, THF, H_2O.

treatment with sodium ethoxide, gave hydrindanone **130** as a single diastereoisomer in close to 90% yield from **129**. Arndt–Eistert homologation of the acetic acid side chain provided **131**, containing the additional carbon required for fawcettimine. Hydride reduction and selective tosylation gave the *N,O*-bis(tosyl) derivative **132**. Treatment of **132** with tetra-*n*-butylammonium hydroxide in benzene gave the desired nine-membered heterocycle in good yield. Deprotection of the nitrogen and oxidation of the alcohol gave the aminoketone **133**. The perchlorate salt of **133** was subjected to ozonolysis at −78°C to give the perchlorate salt of the diketone. Careful neutralization of the salt with sodium bicarbonate and treatment of the resulting amine with aqueous HBr gave racemic fawcettimine hydrobromide, identical (¹H NMR and IR) with natural fawcettimine (**4 = 58**). The structure of the synthetic hydrobromide was verified by X-ray crystallography. This established that the hydrobromide is the salt of the carbinolamine form **59** of fawcettimine and proved that the stereochemistry at C-4 of fawcettimine had been assigned correctly. Because the keto olefin **133** has the opposite stereochemistry at C-4 to that of fawcettimine, this means that epimerization at C-4 is extremely facile.

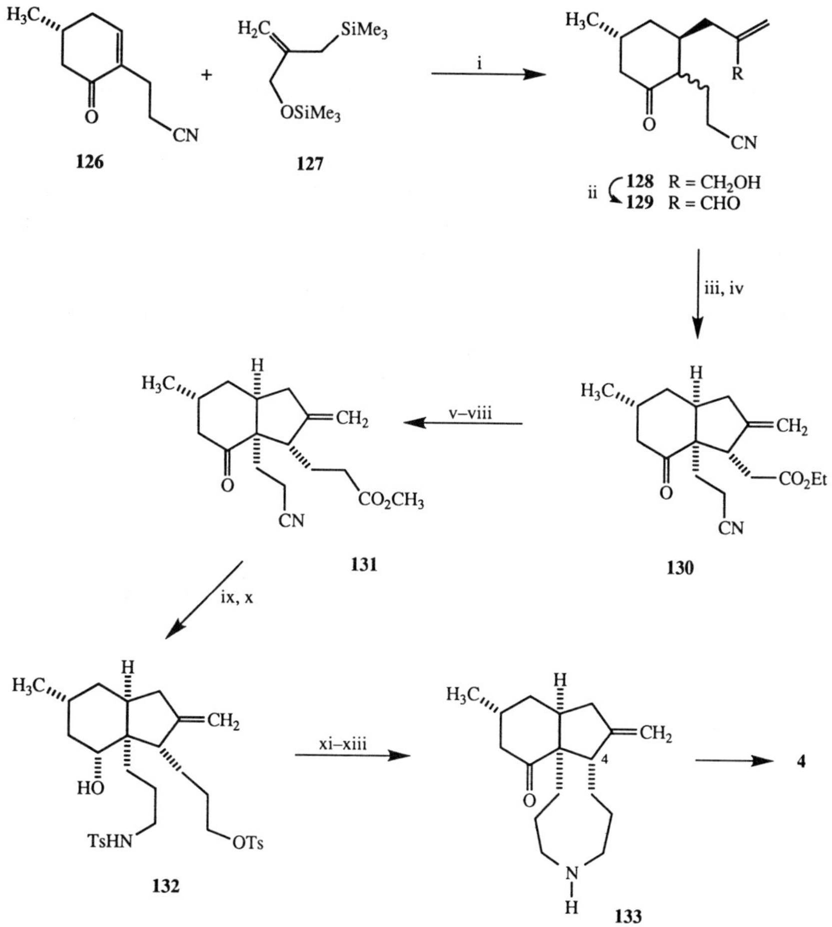

SCHEME 12. Reagents: i, TiCl$_4$, CH$_2$Cl$_2$; ii, CrO$_3$Py; iii, Ph$_3$PCHCO$_2$Et, EtOH; iv, NaOEt, EtOH; v, NaOH, EtOH, H$_2$O; vi, (COCl)$_2$, C$_6$H$_6$; vii, CH$_2$N$_2$; viii, PhCO$_2$Ag, Et$_3$N, MeOH; ix, LAH, THF; x, Ts$_2$O, DMAP; xi, Bu$_4$NOH, C$_6$H$_6$; xii, Na, Sodium, naphtholene, DME; xiii, CrO$_3$, HOAc.

Conformational analysis in the fawcettimine–4-epifawcettimine case is complicated by the possibility of concurrent keto amine–carbinolamine tautomerization. Recent work by Heathcock *et al.* (*48*) has provided valuable new information on this interesting conformational problem. Because the classic *Lycopodium* structures **58** and **59** do not lend themselves well to this discussion, fawcettimine and 4-epifawcettimine (keto amine forms) are presented in Scheme 13 as **134** and **136,** respectively. Early work (*49*) had suggested that fawcettimine exists mainly as the carbinolamine tautomer **138** [IR 3585 (OH), 1730 cm^{-1} (cyclopentanone)], but had shown that it acetylates on nitrogen to give *N*-acetylfawcettimine (**135**) [IR 1615 (amide), 1690 (cyclohexanone), 1730 cm^{-1} (cyclopentanone)]. It was reported (*50*), however, that the hydrochloride and perchlorate salts of fawcettimine showed carbonyl absorption at 1690 cm^{-1} only, suggesting that the cyclopentanone carbonyl is involved in carbinolamine formation. This is stereochemically possible only if fawcettimine has the (4*R*) configuration **136**, which allows formation of the tautomer **140**. The *N*-acetyl derivative **137** of **136** was prepared during the synthesis of fawcettimine and subjected to base-catalyzed equilibration. The *trans* epimer **135** predominates (ratio 2:1) in the equilibrium mixture. A careful study of the IR and NMR spectra of the product of ozonolysis of the hydroperchlorate of

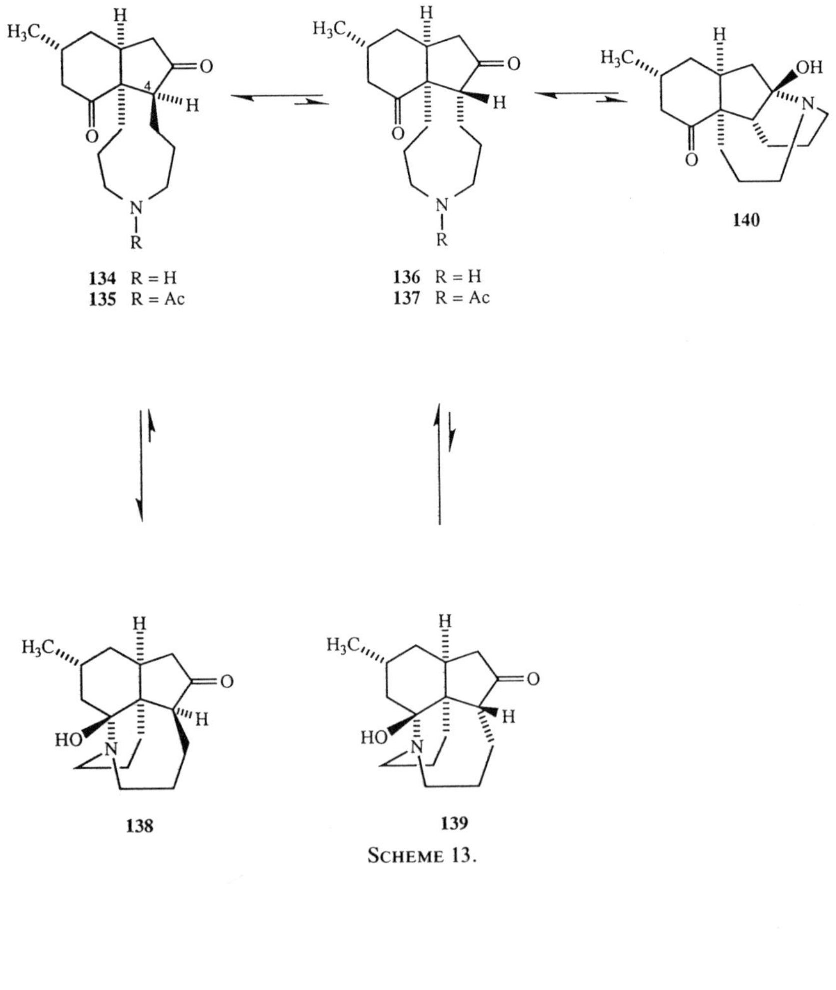

134 R = H
135 R = Ac

136 R = H
137 R = Ac

140

138

139

SCHEME 13.

synthetic **133** indicates that neutralization affords first the diketone **136** (4-epifawcettimine) in equilibrium with keto carbinolamine **140** (and perhaps some **139**). With time, **136** epimerizes to **134,** which exists in solution almost completely in the keto carbinolamine form **138** (Scheme 13). Molecular mechanics calculations (*48*) indicate that carbinolamine **138** is the lowest energy species in Scheme 13, and that, in the 4-epi series, the cyclopentanone carbinolamine **140** is preferred over the cyclohexanone carbinolamine **139.** The possibility remains that the hydrochloride and hydroperchlorate salts isolated many years ago (*50*) are salts of the cyclopentanone carbinolamine **140** of 4-epifawcettimine.

V. Miscellaneous

The miscellaneous group includes several interesting structural types such as luciduline (**141**), containing a unique 12-carbon framework, and lucidine B (**142**), a member of the uncommon $C_{27}N_3$ group (*1*) of alkaloids. Both alkaloids occur in *Lycopodium lucidulum,* which, along with *L. selago* and *L. serratum,* is placed by some in the separate genus *Huperzia* (*7*). The $C_{27}N_3$ alkaloids have been isolated to date only from *L. lucidulum* and *L. serratum,* and huperzine A occurs only in *L. serratum* and *L. selago.* The more common alkaloids such as lycopodine also occur in these species. Whether there is any taxonomical significance to the limited occurrence of luciduline and the $C_{27}N_3$ alkaloids remains to be determined.

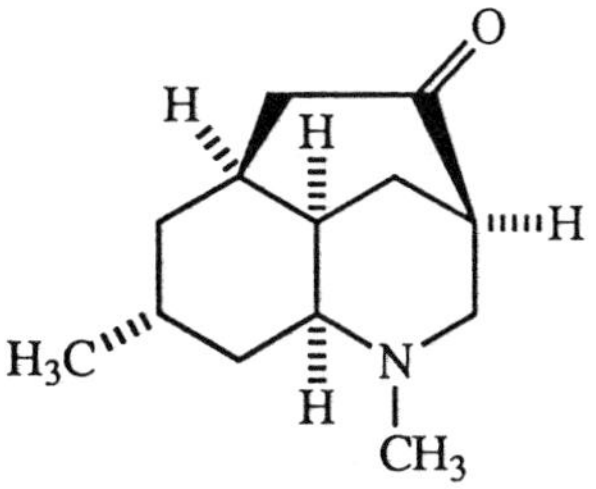

141

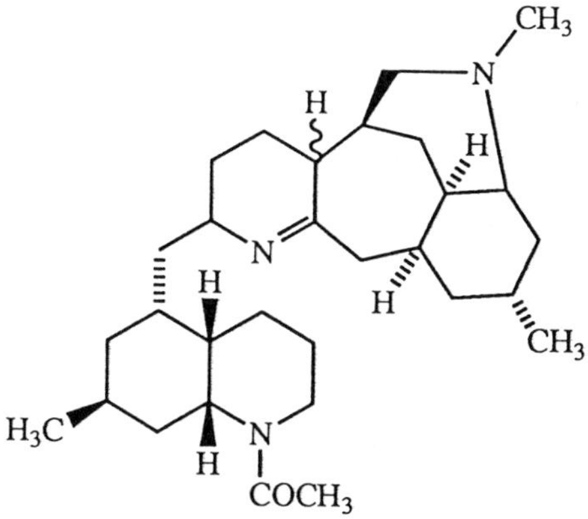

142

A new approach to the luciduline type of alkaloids containing a *cis*-decahydroquinoline fragment [e.g., luciduline (**141**) and lucidine B (**142**)] has been reported. This approach is based on a tandem intramolecular Diels–Alder, retro-Mannich reaction (*51*). In this strategy (Scheme 14) the readily available dihydropyridone **145** was converted to either the dihydropyridine **146** by reduction and dehydration or to **150** by silylation. Diels–Alder reaction of **146** occurred at reflux temperature in xylene. Catalytic

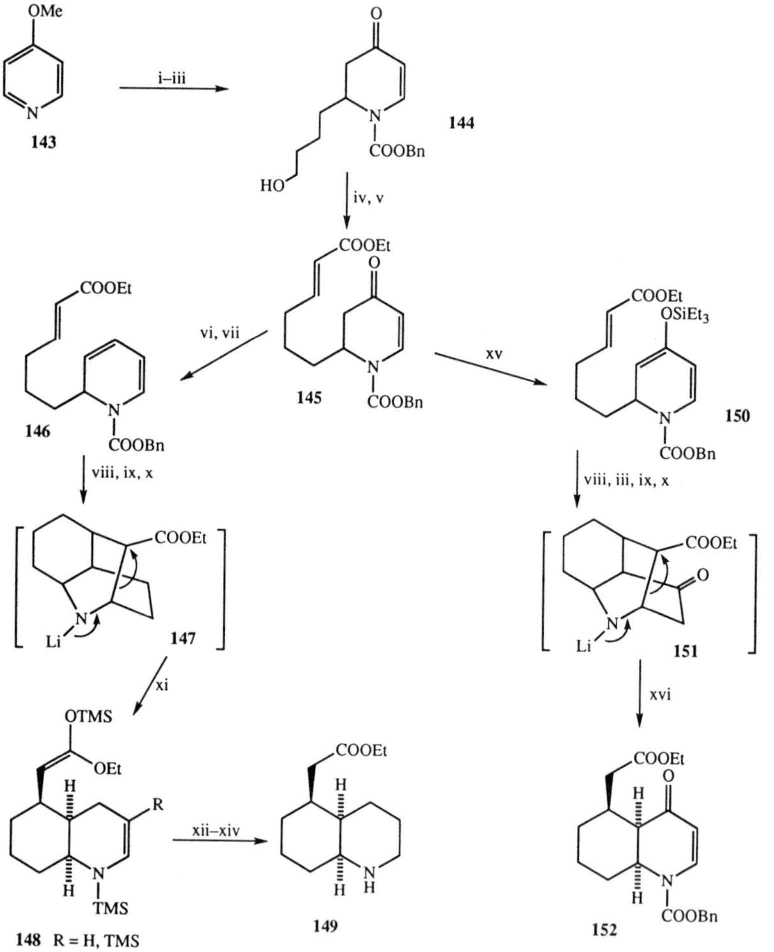

SCHEME 14. Reagents: i, EtOCH$_2$CH$_2$CH$_2$CH$_2$MgCl; ii, BnOCOCl; iii, H$_3$O$^+$; iv, Swern oxidation; v, (EtO)$_2$POCH$_2$COOEt, ^{t}BuOK, THF; vi, NaBH$_4$, CeCl$_3$; vii, MsCl, DMAP; viii, xylene, reflux; ix, H$_2$, Pd/C; x, LDA, DPA, THF, −78°C; xi, TMSCl; xii, BnOCOCl, CH$_2$Cl$_2$; xiii, H$_3$O$^+$; xiv, H$_2$, Pd/C; xv, Et$_3$SiCl, NaN(SiMe$_3$)$_2$; xvi, LDA, THF, BnOCOCl.

hydrogenation of the Diels–Alder adduct, followed by treatment with lithium diisopropylamide and diisopropylamine at low temperature, provided the lithiated intermediate **147,** which underwent spontaneous retro-Mannich ring opening to a bicyclic lithium enolate which was trapped as the silyl enol ether **148.** Nitrogen deprotection–protection, hydrolysis, and catalytic hydrogenation provided the *cis*-decahydroquinoline **149** possessing the correct stereochemistry. In a similar way, the diene **150** gave the lithiated Diels–Alder adduct **151,** which underwent ring opening to the target molecule **152.** This intramolecular Diels–Alder, retro-Mannich strategy has been successfully applied to a formal synthesis of (+)-luciduline (*51*).

153

Cernuine (**153**), representative of still another type of *Lycopodium* alkaloid ring system, has been detected recently as a constituent of the new species *L. australianum* (*9*).

VI. Biosynthesis and Biogenesis

As mentioned in the introduction, biosynthetic studies have proceeded rather slowly with the *Lycopodium* alkaloids due, at least in part, to the fact that the club mosses are very difficult to cultivate. It is clear that lysine and acetate are the main building blocks. The various biosynthetic hypotheses are well summarized in the previous review in this treatise (*1*). The only pertinent new work since that time deals with the carbon atoms of lycopodine (carbons 6, 7, 8, 14, 15, and 16) which originate from acetate. In an elegant study, Hemscheidt and Spenser (*52*) succeeded in incorporating [1,2-$^{13}C_2$]acetate and [1,2,3,4-$^{13}C_4$]acetoacetate into lycopodine (**10**) in an amount sufficient to allow the incorporation to be observed by ^{13}C-NMR spectroscopy. They also administered [1,2-$^{13}C_2$,2,2,2-$^2H_3$]acetate and

lysine
2 malonate
CO_2
[1,2-13C2]-acetate
-CO2
-CO2
-CO2
10A
10B
6
7
8
14
15
16

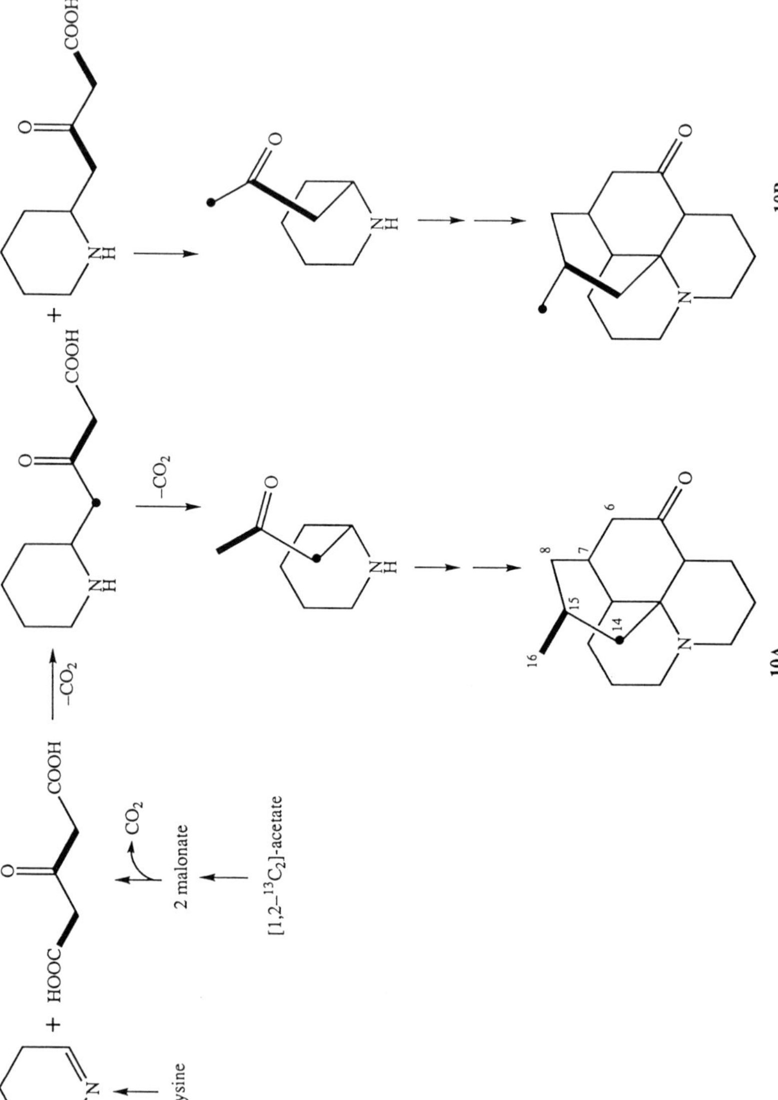

Scheme 15.

were able to show that although the ^{13}C-label was incorporated, all the deuterium was lost during the biosynthetic process. It was shown that in some molecules C-15, C-16 originated from an intact acetate unit (see **10A**) and in others C-14, C-15 were from intact acetate (**10B**). A similar dichotomy was observed for the C-6,7,8 acetate-derived portion.

To explain the 1:1 distribution of the acetate units, it is suggested that 3-oxoglutaric acid (from two malonyl-CoA units, each derived from acetate, and subsequent decarboxylation) may be the symmetrical unit which gives rise to the observed carbon atom distribution and deuterium loss. Scheme 15 illustrates how this accounts for the C-14,15,16 pattern. A similar pathway would account for the C-6,7,8 pattern. It remains to be demonstrated that 3-oxoglutaric acid is indeed on the biosynthetic path.

References

1. D. B. MacLean, *in* "The Alkaloids" (A. Brossi, ed.), Vol. 26, p. 241. Academic Press, New York, 1985.
2. W. A. Ayer, *Nat. Prod. Rep.* **8**, 455 (1991).
3. X. C. Tang, Y. F. Han, X. P. Chen, and X. D. Zhu, *Acta Pharm. Sin.* (*Yaoxue Xuebao*) **7**, 507 (1986).
4. Y. C. Cheng, C. Z. Lu, Z. L. Ying, W. Y. Ni, C. L. Zhang, and G. W. Sang, *New Drugs Clin. Remed.* **5**, 2560 (1986).
5. R. W. Zhang, X. C. Tang, Y. Y. Han, G. W. Sang, Y. D. Zhang, Y. X. Ma, C. L. Zhang, and R. M. Yang, *Acta Pharm. Sin.* (*Yaoxue Xuebao*) **12**, 250 (1991).
6. B. Ollgaard, "Index of the Lycopodiaceae," Biologiske Skrifter 34. Det Kongelige Danske Videnskabernes Selskab (The Royal Danish Academy of Sciences and Letters), Copenhagen, 1989.
7. W. J. Cody and D. M. Britton, "Ferns and Fern Allies of Canada," Chap. 1. Canadian Government Publishing Centre, Ottawa, 1989.
8. W. A. Ayer, L. M. Browne, A. W. Elgersma, and P. P. Singer, *Can. J. Chem.* **68**, 1300 (1990).
9. R. V. Gerard and D. B. MacLean, *Phytochemistry* **25**, 1143 (1986).
10. X. Zhang, H. Wang, and Y. Qi, *Zhongcaoyao* (*Chin. Tradit. Herb. Drugs*) **21**, 146 (1990); *Chem. Abstr.* **113**, 138370d (1990).
11. V. Manriquez, O. Munoz, R. Quintana, M. Castillo, H. G. von Schnering, and K. Peters, *Acta Crystallogr., Sect. C.* **44**, 165 (1988).
12. O. M. Munoz, M. Castillo, and A. San Feliciano, *J. Nat. Prod.* **53**, 200 (1990).
13. W. A. Ayer and G. C. Kasitu, *Can. J. Chem.* **67**, 1077 (1989).
14. G. A. Kraus and Y. S. Hon, *J. Am. Chem. Soc.* **105**, 4341 (1985).
15. G. A. Kraus and Y. S. Hon, *Heterocycles* **25**, 377 (1987).
16. T. A. Blumenkopf and C. H. Heathcock, *in* "Alkaloids: Chemical and Biological Perspectives" (S. W. Pelletier, ed.), Vol. 3, Chap. 5. New York, 1983.
17. D. W. Hughes, R. V. Gerard, and D. B. MacLean, *Can. J. Chem.* **67**, 1765 (1989).
18. J. S. Liu, Y. L. Zhu, C. M. Yu, Y. Z. Zhou, Y. Y. Han, F. W. Wu, and B. F. Qi, *Can. J. Chem.* **64**, 837 (1986).

19. Y. E. Wang, D. X. Yue, and X. C. Tang, *Acta Pharm. Sin.* (*Yaoxue Xuebao*) **7**, 109 (1986).
20. Anonymous, *Drugs Future* **13**, 575 (1988).
21. W. A. Ayer, L. M. Browne, H. Orszanska, Z. Valenta, and J. S. Liu, *Can. J. Chem.* **67**, 1538 (1989).
22. Z. Valenta, H. Yoshimura, E. F. Rogers, M. Ternbah, and K. Wiesner, *Tetrahedron Lett.*, 26 (1960).
23. C. H. Chen and S. S. Lee, *J. Taiwan Pharm. Assoc.* (*T'ai-wan Yao Hsueh Tsa Chih*) **36**, 1 (1984).
24. W. A. Ayer, H. Orszanska, and L. K. Ho, unpublished (1990).
25. S. Q. Yuan and T. T. Wei, *Acta Pharm. Sin.* (*Yaoxue Xuebao*) **23**, 516 (1988).
26. Z. C. Miao, Z. S. Yang, Y. A. Lu, and X. T. Liang, *Acta Chim. Sin.* (*Engl. Ed.*) **47**, 702 (1989).
27. B. M. Chu and J. Li, *Acta Pharm. Sin.* (*Yaoxue Xuebao*) **23**, 115 (1988); J. Li, Y. Y. Han, and J. S. Liu, *Chin. Tradit. Herb. Drugs* (*Zhongcaoyao*) **18**, 50 (1987).
28. Y. Xia and A. P. Kozikowski, *J. Am. Chem. Soc.* **111**, 4116 (1989).
29. L. Qian and R. Ji, *Tetrahedron Lett.* **30**, 2089 (1989).
30. A. P. Kozikowski, E. R. Reddy, and C. P. Miller, *J. Chem. Soc., Perkin Trans 1*, 195 (1990).
31. A. P. Kozikowski, F. Yamada, X. C. Tang, and I. Hanin, *Tetrahedron Lett.* **31**, 6159 (1990).
32. A. P. Kozikowski, *Heterocycles* **27**, 97 (1990).
33. Y. Xia, E. R. Reddy, and A. P. Kozikowski, *Tetrahedron Lett.* **30**, 3291 (1989).
34. I. Hanin, X. C. Tang, J. Corey, Y. Xia, E. R. Reddy, and A. P. Kozikowski, *FASEB J.* **4**, A471 (1990).
35. F. Yamada, A. P. Kozikowski, E. R. Reddy, Y. P. Pang, J. H. Miller, and M. McKinney, *J. Am. Chem. Soc.* **113**, 4695 (1991).
36. P. Hu, M. L. Gross, S. Q. Yuan, T. T. Wei, and Y. Q. Lu, *Org. Mass Spectrom.* **27**, 99 (1992).
37. R. V. Gerard and D. B. MacLean, *Phytochemistry* **25**, 1143 (1986).
38. R. V. Gerard, D. B. MacLean, R. Fagianni, and C. J. Lock, *Can. J. Chem.* **64**, 943 (1986).
39. T. Ho, R. F. Chandler, and A. W. Hanson, *Tetrahedron Lett.* **28**, 5993 (1987).
40. W. A. Ayer, Y. T. Ma, J. S. Liu, M. F. Huang, L. W. Schultz, and J. Clardy, *Can. J. Chem.* **72**, 128 (1994).
41. G. C. Hirst, T. O. Johnson, Jr., and L. E. Overman, *J. Am. Chem. Soc.* **115**, 2992 (1993).
42. L. A. Paquette, D. Friedrich, E. Pinard, J. P. Williams, D. St. Laurent, and B. A. Roden, *J. Am. Chem. Soc.* **115**, 4377 (1993).
43. G. Mehta and M. S. Reddy, *Tetrahedron Lett.* **31**, 2039 (1990).
44. G. Mehta and K. S. Rao, *J. Chem. Soc., Chem. Commun.*, 1578 (1987).
45. G. Mehta, M. S. Reddy, R. Radhakrishnan, M. V. Manjula, and M. A. Viswamitra, *Tetrahedron Lett.* **32**, 6219 (1991).
46. M. T. Grimmins and P. S. Watson, *Tetrahedron Lett.* **34**, 199 (1993).
47. C. H. Heathcock, K. M. Smith, and T. A. Blumenkopf, *J. Am. Chem. Soc.* **108**, 5022 (1986).
48. C. H. Heathcock, T. A. Blumenkopf, and K. M. Smith, *J. Org. Chem.* **54**, 1548 (1989).
49. Y. Inubushi, H. Ishii, T. Harayama, R. H. Burnell, W. A. Ayer, and B. Altenkirk, *Tetrahedron Lett.*, 861 (1969).
50. R. H. Burnell, C. G. Chin, B. S. Mootoo, and D. R. Taylor, *Can. J. Chem.* **41**, 3091 (1963).
51. D. L. Comins and R. S. Al-awar, *J. Org. Chem.* **47**, 4098 (1992).
52. T. Hemscheidt and I. D. Spenser, *J. Am. Chem. Soc.* **115**, 3020 (1993).

CUMULATIVE INDEX OF TITLES

Aconitum alkaloids, **4,** 275 (1954), **7,** 473 (1960), **34,** 95 (1988)
C_{19} diterpenes, **12,** 2 (1970)
C_{20} diterpenes, **12,** 136 (1970)
Acridine alkaloids, **2,** 353 (1952)
Acridone alkaloids, experimental antitumor activity of acronycine, **21,** 1 (1983)
N-Acyliminium ions as intermediates in alkaloid synthesis, **32,** 271 (1988)
Ajmaline-Sarpagine alkaloids, **8,** 789 (1965), **11,** 41 (1968)
Alkaloid production, plant biotechnology of **40,** 1 (1991)
Alkaloid structures
spectral methods, study, **24,** 287 (1985)
unknown structure, **5,** 301 (1955), **7,** 509 (1960), **10,** 545 (1967), **12,** 455 (1970), **13,** 397 (1971), **14,** 507 (1973), **15,** 263 (1975), **16,** 511 (1977)
X-ray diffraction, **22,** 51 (1983)
Alkaloids
forensic chemistry of, **32,** 1 (1988)
histochemistry of, **39,** 1 (1990)
in the plant, **1,** 15 (1950), **6,** 1 (1960)
Alkaloids from
Amphibians, **21,** 139 (1983), **43,** 185 (1993)
Ants and insects, **31,** 193 (1987)
Chinese Traditional Medicinal Plants, **32,** 241 (1988)
Mammals, **21,** 329 (1983), **43,** 119 (1993)
Marine organisms, **24,** 25 (1985), **41,** 41 (1992)
Mushrooms, **40,** 189 (1991)
Plants of Thailand, **41,** 1 (1992)
Allelochemical properties or the raison d'être of alkaloids, **43,** 1 (1993)
Allo congeners, and tropolonic *Colchicum* alkaloids, **41,** 125 (1992)
Alstonia alkaloids, **8,** 159 (1965), **12,** 207 (1970), **14,** 157 (1973)
Amaryllidaceae alkaloids, **2,** 331 (1952), **6,** 289 (1960), **11,** 307 (1968), **15,** 83 (1975), **30,** 251 (1987)
Amphibian alkaloids, **21,** 139 (1983), **43,** 185 (1983)
Analgesic alkaloids, **5,** 1 (1955)

INDEX

ISBN 0-12-469545-0